Digital Circuits and Systems

Digital Circuits and Systems

RICHARD L. CASTELLUCIS

RESTON PUBLISHING COMPANY, INC.
A Prentice-Hall Company
Reston, Virginia

Technical Illustrations by Robert Mosher

Library of Congress Cataloging in Publication Data

Castellucis, Richard L
Digital circuits & systems.

Includes index.
1. Digital electronics. I. Title.
TK868.D5C38 621.381 80-25389
ISBN 0-8359-1297-3

A Prentice-Hall Company
Reston Virginia 22090

1 3 5 7 9 10 8 6 4 2

Printed in the United States of America

Contents

Preface ix

SECTION I 1

1 Number Systems 3

1.1 DECIMAL NUMBERS 4
1.2 THE BINARY NUMBER SYSTEM 5
1.3 WHY USE BINARY NUMBERS? 13
1.4 OCTAL NUMBERS 13
1.5 HEXADECIMAL NUMBERS 15
1.6 THE BCD CODE 18
1.7 ALPHANUMERIC CODES 20
1.8 USE OF NUMBER SYSTEMS AND CODES 20
QUESTIONS AND PROBLEMS 21

2 Logic Gates 23

2.1 TRUE-FALSE LOGIC 23
2.2 THE AND CIRCUIT 24
2.3 THE LOGIC AND GATE 25
2.4 THE GATING FUNCTION 27
2.5 THE NOT CIRCUIT 27
2.6 THE NAND GATE 28
2.7 PROPAGATION DELAY 36
2.8 OPEN-COLLECTOR GATES 37
2.9 THE INPUT CIRCUITRY 40
2.10 THE IC AND CATE 41
2.11 INVERTERS 42
2.12 OR GATES 43
2.13 THE NOR GATE 46
2.14 THE EXCLUSIVE-OR GATE (EX-OR GATE) 46
2.15 REVIEW OF BASIC GATES 50
2.16 TRI-STATE DEVICES 50
QUESTIONS AND PROBLEMS 57

SECTION II 59

3 Flip-Flops 61

3.1 THE GENERAL FLIP-FLOP 61
3.2 TYPES OF FLIP-FLOPS 62
3.3 APPLICATIONS OF THE J-K FLOP-FLOP 68
QUESTIONS AND PROBLEMS 71

4 Shift Registers 75

4.1 SHIFT REGISTERS 75
4.2 CLOCKING AND TIMING OF A SERIAL SHIFT REGISTER 76
4.3 LEAST-SIGNIFICANT BIT 77
4.4 CIRCUIT CONSTRUCTION 77
4.5 PARALLEL SHIFT REGISTERS 77
4.6 THE IC SHIFT REGISTER 79
4.7 ADDITIONAL IC SHIFT REGISTERS 82
QUESTIONS AND PROBLEMS 87

5 Counters 88

5.1 BINARY COUNTERS 88
5.2 MODULUS COUNTERS 90
5.3 SYNCHRONOUS COUNTERS 91
5.4 OPERATION OF THE SYNCHRONOUS COUNTER 93
5.5 THE SYNCHRONOUS DOWN COUNTER 94
5.6 RING COUNTERS 94
5.7 SWITCH-TAIL COUNTERS 94
5.8 SELF-STOPPING ASYNCHRONOUS COUNTERS 96
QUESTIONS AND PROBLEMS 98

6 Integrated Circuit Counters 101

6.1 7490 DECADE BINARY COUNTERS 101
6.2 7492 DIVIDE-BY-TWELVE COUNTER 107
6.3 74190 SYNCHRONOUS UP/DOWN COUNTER 108
6.4 74192 SYNCHRONOUS UP/DOWN COUNTER 112
6.5 74176 PRESETTABLE DECADE COUNTER 115
6.6 74177 BINARY COUNTER 118
QUESTIONS AND PROBLEMS 119

7 **Encoding and Decoding** 120

7.1 ENCODERS 120
7.2 DECODERS 121
7.3 INTEGRATED CIRCUIT BCD-DECIMAL DECODER/DRIVES 123
7.4 INTEGRATED CIRCUIT BCD-SEVEN SEGMENT DECODER/DRIVERS 123
7.5 PRIORITY ENCODERS 133
7.6 PARITY 133
7.7 THE GRAY CODE 138
7.8 MULTIPLEXING (DATA SELECTIONS) 142
7.9 MAGNITUDE COMPARATORS 153
QUESTIONS AND PROBLEMS 156

8 **Counting Systems** 159

8.1 BCW COUNTER/BCD DECODER DRIVER 159
8.2 FOUR-BIT COUNTER/LATCH, SEVEN-SEGMENT DRIVER 163
QUESTIONS AND PROBLEMS 171

9 **Memory and Memory Devices** 172

9.1 TYPES OF MEMORY 172
9.2 MEMORY DEVICES 173
9.3 MAGNETIC DRUM 173
9.4 MAGNETIC TAPES 175
9.5 FERRITE-CORE MEMORY 176
9.6 BINARY WORD SELECTION 181
9.7 CORE MEMORY SYSTEM 182
9.8 SOLID-STATE MEMORIES 185
9.9 RAMS/ROMS 189
9.10 STATIC AND DYNAMIC MEMORY 199
QUESTIONS AND PROBLEMS 203

10 **Computer Arithmetic Circuits** 205

10.1 BINARY ARITHMETIC 206
10.2 THE HALF-ADDER 212
10.3 THE FULL-ADDER 213
10.4 SERIAL ADDITION 215
10.5 PARALLEL ADDITION 216
10.6 ADDER CHIPS 217
10.7 THE TYPE SN7482 TWO-BIT BINARY FULL-ADDER 217
10.8 THE TYPE 7483 FOUR-BIT BINARY FULL-ADDER 219
QUESTIONS AND PROBLEMS 228

SECTION III 231

11 The Operational Amplifier 233
11.1 THE IDEAL OPERATIONAL AMPLIFIER 234
11.2 THE OP-AMP AS A SINGLE-PACKAGE DEVICE 235
11.3 FUNCTIONS OF THE EXTERNAL CONNECTIONS 236
11.4 SOME OP-AMP TERMINOLOGY 237
11.5 OPEN-LOOP GAIN 242
11.6 CLOSED-LOOP GAIN (THE FEEDBACK AMPLIFIER) 243
11.7 SLEW RATE 246
11.8 SOME OP-AMP CIRCUITS 246
QUESTIONS AND PROBLEMS 254

12 Digital-Analog Analog-Digital Conversion Circuits 256
12.1 REVIEW OF THE SUMMING AMPLIFIER 257
12.2 DATA CONVERSION SYSTEMS: ANALOG-DIGITAL-ANALOG 261
12.3 QUALITY MEASUREMENT OF D/A CONVERTERS 263
12.4 THE ANALOG SIGNAL 264
12.5 ANALOG-TO-DIGITAL CONVERSION 265
12.6 A/D CONVERSION (UJT OSCILLATOR) 266
12.7 RAMP TYPE A/D CONVERTER 269
12.8 DUAL-SLOPE A/D CONVERTERS 272
12.9 THE SUCCESSIVE-APPROXIMATION A/D CONVERTER 274
QUESTIONS AND PROBLEMS 279

SECTION IV 281

13 Microcomputers 283
13.1 MICROCOMPUTER ARCHITECTURE 285
13.2 HOW IT WORKS 290
13.3 INPUT/OUTPUT 293
13.4 I/O INTERFACE SUMMARY 299
QUESTIONS AND PROBLEMS 302

APPENDICES

A Additional Student Exercises 303
B Useful Circuits Using NAND Gates 316
C Timing Circuits 320

Index 329

Preface

This text is intended to take the reader into the world of digital circuits and systems in a logical manner. *Digital Circuits and Systems* is intended for use at the postsecondary instruction level. The text format emphasizes two factors: (1) the student technician will learn by doing, and (2) all laboratory exercises should use realistic, up-to-date components and devices. With this in mind, many of the units are centered around a student learning activity of the "hands-on" type.

Where necessary, the reader will be led through the laboratory assignments step-by-step. At other times, the reader will be told to use his or her imagination and previous learning experiences to complete the assignment. The text is divided into four major sections and thirteen learning units. Where appropriate, end-of-unit laboratory assignments have been included. These may be detailed or merely suggestions of what the student can do to further his or her understanding of the topics.

SECTION I

Before any study of digital circuits is begun, a basic understanding of number systems is important. In Chapter 1, the foundation for much of the work in digital systems is laid. Not only is it imperative that students be familiar with the binary number system, they must also see this system as it relates to the decimal number system and to other systems used in digital circuits. Chapter 2 on logic gates is of equal importance as an introduction to the study of digital systems. This chapter goes beyond the mere use of truth tables and logic diagrams. It is a true study of the functions, capabilities, and limitations of the devices that will be used as building blocks for the later digital systems.

SECTION II

In Section II we present the major digital circuits and systems. In Chapters 3 to 10 the introduction of a number of digital IC chips is used to familiarize the student with these functions and use in the digital world. Here, the emphasis is not so much on theory, but on the actual operating characteristics of these IC devices. Pin-outs and pin functions, as well as interconnections of these devices to form a working system, are the major thrust of Section III.

SECTION III

Although it may seem strange to include analog devices in a text devoted to pulse and digital systems, it must be realized that most of the quantities measured and controlled are analog signals. Chapters 11 and 12 are used to give the student an understanding of the analog world and how it is interfaced with the digital circuits being studied.

SECTION IV

No modern text on digital systems would be complete without a chapter devoted to at least a survey of the microcomputer. This is done for the reader in Chapter 13.

The study of digital circuits is mainly the study of digital integrated circuits. Many times these devices are studied only with emphasis on their individual operation and with little regard for their function in the circuit. The intent of this text is to present not only the devices, but to give the reader an awareness of their functions, capabilities, and limitations. It should be firmly in the mind of the reader that these devices perform a unique circuit function. There is a signal that must be processed. These devices will be used to shape, store, change, display, count, time, compare, and delay the signal until it is needed at some point in the circuit. When the designer of a circuit decides what must be done with the signal, and if he or she knows which devices have that capability, then and only then can the process of digital circuit design logically begin.

As you read each chapter and study each device, keep in mind where its function might fit into the overall design. Keep all design as simple as possible. Start with the known input and desired output signals. Then, using your knowledge of digital devices, select those that will allow this change to occur.

R.L.C.

Digital Circuits and Systems

Section I

1

Number Systems

OBJECTIVE: To introduce the student to the concept of number systems. Through the study of this unit, you will become familiar with and understand the basis for several number systems used in the design and implementation of digital systems.

Introduction: Number systems are not new. The use of number systems dates to earliest times. Through the use of number systems we are able to represent a quantity of items. We all know that there is more than one number system. Probably most of us are familiar with two systems, Roman numerals and our own common decimal number system. A number system simply uses certain symbols to designate certain quantities. For example, if we wanted to write the number ten we would show it as X in the Roman system and as 10 in the decimal system. Each of these symbols represents the quantity ten. We just need to be sure that each person is thinking in the same system and that each person is capable of "decoding" the symbols of that system.

In this chapter we will explore several of the more common number systems used in the field of digital electronics and digital computers. We will also show why each is used and where they should and should not be used. This chapter will serve as the background for much of your study in this text. You, the reader, are about to undertake the study of a foreign language; one without which you will not be able to communicate successfully with others in the field or with the circuits of this new computer-oriented society.

1.1 DECIMAL NUMBERS

Before we start on our study of the foreign languages, it might be a good idea to stop and look at our own familiar language of numbers in the decimal system. Since we are most familiar with this language and use it almost constantly, we usually do not stop to think about what we are actually doing.

The decimal system consists of ten (deci) units or digits. This is called its *base*. The base of any number system is the number of unique symbols used in that system. Let's review those symbols. Table 1.1 shows the symbols used in the decimal number system. Any quantity may be represented by these ten symbols. For example, when we wish to write that we have six apples, we simply use the symbol 6. This is recognized as symbolizing the quantity six and no further explanation is necessary.

TABLE 1.1 The Decimal Number System

Written	*Spoken*	*Written*	*Spoken*
0	zero	5	five
1	one	6	six
2	two	7	seven
3	three	8	eight
4	four	9	nine

If we wish to represent a quantity larger than nine we do a very simple operation. As we count from zero to nine, we note that a quantity larger than nine simply requires starting our count over again and adding one symbol to the left of the column in which we are doing the counting. For example, the two-digit decimal number fifty-four is written as 54. What we have actually done is to arrange our digits (symbols) in columns and assign a weighted value to that column. Each column in the decimal number system is weighted as a power of ten. This makes sense since our base is ten. Therefore, the first column is given a weight of 10^0 or 1. The second column, to the left, is given a weighted value of 10^1 or 10, and so on.

So now we see that our number fifty-four was really four units in the first column weighted to a value of 10^0:

$$4 \times 10^0 = 4 \text{ units}$$

To this we had to add the weighted value of the second column

$5 \times 10^1 = 50$ units

and, column one plus column two yields

$50 + 4 = 54$ units

In a similar manner, the quantity 342 is expressed in the decimal system as follows:

	2 units at $10^0 = 2 \times 10^0 =$	2
plus	4 units at $10^1 = 4 \times 10^1 =$	40
plus	3 units at $10^2 = 3 \times 10^2 =$	300
		342

This explanation of the decimal number system is used only to refresh your understanding of the way in which a number is represented and to increase your understanding of the remainder of the number systems shown in this unit.

1.2 THE BINARY NUMBER SYSTEM

As with the decimal number system, any quantity can be represented using the symbols (or digits) of the binary system. The binary number system, as its name suggests, contains only two (bi) symbols or digits. This is a base-two system. These symbols are:

Written	*Spoken*
0	Zero
1	One

As a result, any quantities in the binary system must be written using only the two digits 0 and 1. For example, the binary number 11 (one, one; *not* eleven) is used to represent the quantity three. Using the same system of weighted columns as we did to demonstrate our decimal number, we will show how 11 (base two) is actually 3 (base 10). Since we only have two symbols in our binary number system, the weighting of the columns will be based on powers of two ($2^0 = 1$; $2^1 = 2$; $2^2 = 4$, etc.) and not on powers of ten. Therefore, for the number 11, the first column (to the left of the decimal point) has a weighted value of 2^0

or 1. The second column has a weighted value of 2^1 or 2. Thus, our number 11 in the binary number system is expressed as follows:

$$\begin{array}{ll} & 1 \text{ unit at } 2^0 = 1 \times 2^0 = 1 \\ \text{plus} & 1 \text{ unit at } 2^1 = 1 \times 2^1 = \underline{2} \\ & \qquad\qquad\qquad\qquad\quad 3 \end{array}$$

Remember, number systems are used to denote how many of an item we have, how many units of something, or how many units of a measured quantity. Ten units in the binary system are ten units in the decimal system. The written representation of these two is, however, quite different. For example:

Spoken	*Written*	
	Decimal	Binary
Ten	10.	1010.
Three	3.	11.
Eleven	11.	1011.
Twenty-one	21.	10101.

Since the binary number system consists of only two digits, it becomes quite cumbersome to express large quantities. As an example the quantity sixty-four is written in binary as 1000000.

Column Weighting of Binary Numbers

Table 1.2 shows the weighting values for both the binary and decimal number systems. Notice that in both cases a "decimal point" is shown.

TABLE 1.2 Decimal and Binary Weighting of Columns

Decimal

$10^N \ldots 10^4\ 10^3\ 10^2\ 10^1\ 10^0\ .\ 10^{-1}\ 10^{-2}\ 10^{-3}\ 10^{-4} \ldots 10^{-N}$

Binary

Radix point

$2^N \ldots 2^4\ 2^3\ 2^2\ 2^1\ 2^0\ .\ 2^{-1}\ 2^{-2}\ 2^{-3}\ 2^{-4} \ldots 2^{-N}$

This decimal point is called the *radix*. Whole numbers are shown to the left of the radix and decimal numbers are expressed to the right of the radix. In either system, to determine the quantity or number of units being represented, we simply multiply the digit in the column by its weighting factor and add each of the products.

For the binary system, the process is simplified since there are only two possible digits.

Example How many units are represented by the following numbers?

(a) 1010101

Solution Place each of the digits in its appropriate column.

$2^6 \quad 2^5 \quad 2^4 \quad 2^3 \quad 2^2 \quad 2^1 \quad 2^0 \quad .$

radix

$1 \quad 0 \quad 1 \quad 0 \quad 1 \quad 0 \quad 1 \quad .$

The following operations are performed:

$$(1 \times 2^6) + (0 \times 2^5) + (1 \times 2^4) + (0 \times 2^3) + (1 \times 2^2) + (0 \times 2^1) + (1 \times 2^0)$$

The result is

$$(1 \times 64) + 0 + (1 \times 16) + 0 + (1 \times 4) + 0 + (1 \times 1)$$

Adding these numbers yields:

$$64 + 16 + 4 + 1 = 85 \text{ units}$$

(The final answer in this example, and in those that follow, is given in the decimal form simply because this system is the one most people use most of the time.)

(b) 1101

Solution Place each of the digits in the appropriate column.

$2^3 \quad 2^2 \quad 2^1 \quad 2^0 \quad .$

radix

$1 \quad 1 \quad 0 \quad 1 \quad .$

Thus,

$$(1 \times 2^3) + (1 \times 2^2) + (0 \times 2^1) + (1 \times 2^0) \quad \text{or} \quad (1 \times 8) + (1 \times 4) + 0 + (1 \times 1)$$

Adding, $8 + 4 + 1 = 13$ units

(c) 101.101

Solution Place each of the digits in the appropriate column.

$2^2 \quad 2^1 \quad 2^0 \quad . \quad 2^{-1} \quad 2^{-2} \quad 2^{-3}$

radix point

$1 \quad 0 \quad 1 \quad . \quad 1 \quad 0 \quad 1$

Now break the problem into two. Use the decimal point (the radix) as the dividing point. To the left of the radix:

$$(1 \times 2^2) + (0 \times 2^1) + (1 \times 2^0)$$

or

$$1 \times 4 + 0 + 1 \times 1 = 4 + 1 = 5 \text{ units}$$

To the right of the radix:

$$(1 \times 2^{-1}) + (0 \times 2^{-2}) + (1 \times 2^{-3}) = 1 \times 0.5 + 0 \times 0.25 + 1 \times 0.125$$

Adding, $0.5 + 0 + 0.125 = 0.625$ units

Combine the two to obtain the complete decimal number:

$$5 + 0.625 = 5.625 \text{ units}$$

Addition of Binary Numbers

To add binary numbers, the following simple rules must be observed.

$1 + 0 =$ sum of 1 and carry of 0
$0 + 1 =$ sum of 1 and carry of 0

1 + 1 = sum of 0 and carry of 1
0 + 0 = sum of 0 and carry of 0

Example Add binary 10 to binary 11 (decimal 2 to decimal 3)

```
  2     10
 +3     11
 --     --
  5      ?
```

1. The first column to the left of the radix contains a 0 and a 1.

```
0
1
-
```

According to the binary addition rules, the sum of 0 and 1 is 1 and a carry of 0.

```
0 <---- carry
  0
  1
 --
  1
```

The carry is placed in the next column to the left.

```
0 <------ carry
1 0       original
1 1       problem
---
`  1
```

2. The second column now contains 3 digits that are added 2 at a time.

```
Add   0   second
      1   column
      1   only
      -
```

Adding the 0 and 1 as shown gives a sum of 1 with a carry of 0

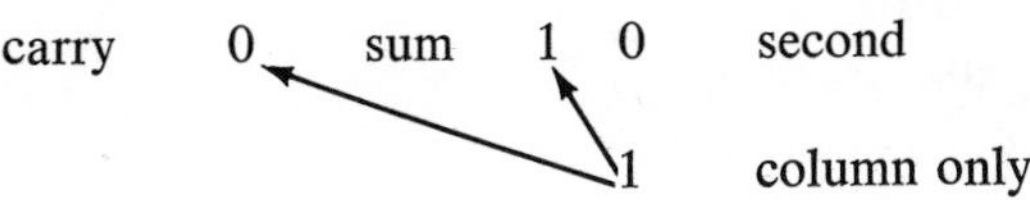

```
carry    1              1  1   add this to one,
               carry           from one and
                               zero addition
____________________________
                           0
```

This 1 is added to the third 1, giving a 0 with a carry of 1. The final 1 and 0 (the carry digits) are now added together. The final answer is as follows:

```
          0    column 2
carry                   1 0     column 1
          1             1 1
                        ---
          1             0 1
```

Add the following binary numbers:

```
010        101        111
101        110        101
---        ---        ---
```

Subtraction of Binary Numbers

To subtract binary numbers, the set of rules listed below must be followed.

$0 - 0 =$ difference of zero and borrow of zero
$1 - 0 =$ difference of one and borrow of zero
$0 - 1 =$ difference of one and borrow of one
$1 - 1 =$ difference of zero and borrow of zero

Example Subtract the binary 3 (011) from the binary 5 (101).

```
 101
-011
----
```

Start in the column of the least-significant digit (at the right) and apply the subtraction rules given.

1. 1 minus 1 gives a difference of 0 and a borrow of 0. A borrow is always made from the column to the immediate left of the column in which the arithmetic operation is occurring.

```
borrowed     101
            -011
            ----
               0
```

2. Now move to the left to the next column: 0 minus 1. The difference of this column is 1 with a borrow of 1. Since the borrow must be from the column to the immediate left, the third column is now blank. The 1 of the third column is "borrowed" to complete the arithmetic of the second column. The final answer for this subtraction problem is:

borrowed	(1)01 = 5	decimal
	0 11 = 3	equivalent
	10 = 2	

Using the rules of binary arithmetic, solve the following problems in subtraction:

111	101	110
−101	−001	−101

Subtraction by the Complement Method

Binary numbers can also be subtracted by a method called the complement method.

Example As in the problem above, subtract the binary 3 (011) from the binary 5 (101).

1. State the problem as shown:

```
 101
−011
────
```

2. The next step is to complement the subtrahend. In other words, the value of each of the digits is reversed or complemented. The complement of 1 is 0, and the complement of 0 is 1. The problem is now written in the following form.

```
101
100    complement
───
```

3. Instead of subtracting the two sets of numbers, they are now added.

```
 101
 100
────
1001
```

4. The final step is to take the most-significant digit (MSD) of the answer (digit on the left) and add it to the least-significant digit (LSD) (digit on the right).

MSD			LSD
1	0	0	1
		1	0

Thus, the final answer is 10, the binary expression for 2. The reason for the use of the subtraction by complement method will become apparent in the study of computer arithmetic operations in Chapter 11.

Conversion of Decimal Numbers to Binary Numbers

Earlier in this unit, a method was given for converting a binary number to its decimal equivalent. It is also important that the student be able to convert decimal numbers to binary numbers. The following example illustrates a method of performing this conversion.

Example Convert the number 70 from the decimal system (base 10 system) to the binary system (base 2 system).

In the base 10 system 70 is written as 70_{10}. The required conversion in the base 2 system is expressed as X_2.

The problem can then be stated as $70_{10} = X_2$. The method used to solve the problem is called the *quotient-remainder method:*

> Divide the number to be converted by the base of the number system of the answer and record the remainder. Continue to divide the new quotients and record the remainders.

Therefore, to convert 70_{10} to its binary equivalent, first divide 70 by 2 (the base of the binary system). The division yields a quotient of 35 and a remainder of 0. Record this remainder. Now divide the new quotient (35) by 2 and so on.

$2\overline{)70}$	
$2\overline{)35}$	+0
$2\overline{)17}$	+1
$2\overline{)8}$	+1
$2\overline{)4}$	+0
$2\overline{)2}$	+0
$2\overline{)1}$	+0
0	+1

Write the remainders from left to right, starting with the last remainder.

1000110

Therefore,

$70_{10} = 1000110_2$

Check the solution of Example 1.5 by converting the binary number back to its decimal equivalent.

Convert 68_{10} to its binary equivalent ($68_{10} = X_2$). Complete the following chart.

	Binary	*Decimal*		*Binary*	*Decimal*
(1)	0 0 0 1	1	(6)	0 1 1 0	___
(2)	______	2	(7)	0 1 1 1	7
(3)	0 0 1 1	___	(8)	______	8
(4)	0 1 0 0	4	(9)	______	9
(5)	______	5	(10)	______	10

1.3 WHY USE BINARY NUMBERS?

The reason for the use of the binary number system in digital electronics is obvious. Digital electronics uses primarily two levels of voltage. There is usually either a voltage (or current) present or there is not. The presence of a voltage level is said to be a binary 1 and the absence of a voltage is said to be a binary 0. With only the use of these two voltage levels, digital circuits can perform many complex functions. This will be shown in future chapters of this text. For now it will be enough to say that the binary number system is very appropriate for the digital world.

1.4 OCTAL NUMBERS

As mentioned earlier, it sometimes becomes quite cumbersome to express large quantities with the binary number system. However, the binary numbers are easy to use in our digital electronic circuits. If there were a way to show a large quantity in binary without writing all of those 1s and 0s it would be helpful. One method of doing this is via the octal number system.

The octal number system, as the name suggests, is based on eight symbols, 0, 1, 2, 3, 4, 5, 6, 7. As with the binary system, any quantity can be represented using the weighted-column method. Remember, no symbol larger than 7 may be used in the octal number system. Also, this system is a base 8 system, meaning each column is given a weight based on a power of eight; that is:

$8^0 = 1$
$8^1 = 8$
$8^2 = 64$
$8^3 = 512$

So, 31 base 8 is

$$
\begin{array}{lr}
 & 1 \times 8^0 = 1 \\
\text{plus} & 3 \times 8^1 = \underline{24} \\
 & 25
\end{array}
$$

This means that 25 units are represented as 31 (three, one) in the octal system.

Twenty-five units in the binary system would be represented as:

11001

Check this for accuracy.

There is a connection between the octal system and the binary system that makes them compatible.

Let's think back for a minute. The largest symbol used in the octal system is 7. Seven is represented in the binary system as 111. Therefore, by grouping binary numbers in groups of three (from the radix) we can represent any binary number by its octal equivalent.

This means that 25 in binary, when properly grouped, will give 31 in octal. To illustrate,

011	001	binary for 25
3	1	octal for 25

Example Convert 32 octal to binary and decimal equivalents.

Solution We simply work backwards to go from octal to binary; thus:

3	2	octal
11	010	binary

To go to decimal, we can either convert the binary or octal to decimal. We shall do the conversion using the octal first.

32 octal

means

$$
\begin{array}{lrl}
 & 2 \times 8^0 = 2 \times 1 = & 2 \\
\text{plus} & 3 \times 8^1 = 3 \times 8 = & \underline{24} \\
 & & 26 \quad \text{decimal}
\end{array}
$$

Converting from the binary we have:

11010 binary

means

$$
\begin{array}{lrl}
 & 0 \times 2^0 = 0 \times 1 = & 0 \\
\text{plus} & 1 \times 2^1 = 1 \times 2 = & 2 \\
\text{plus} & 0 \times 2^2 = 0 \times 4 = & 0 \\
\text{plus} & 1 \times 2^3 = 1 \times 8 = & 8 \\
\text{plus} & 1 \times 2^4 = 1 \times 16 = & \underline{16} \\
 & & 26 \quad \text{decimal}
\end{array}
$$

So we have: twenty-six = 26 decimal, 32 octal, and 11010 binary.

To avoid any future confusion we shall subscript all numbers to denote the proper base. Since you will from now on in your studies be dealing with more than one number system, it will be a good idea to follow this practice anytime there might be a chance a number could be misinterpreted. Thus

twenty-six = 26_{10}; 32_8; and 11010_2

1.5 HEXADECIMAL NUMBERS

Any quantity may be expressed as a number in a system with any base. However, certain number systems lend themselves more readily to certain applications than others. One of these is the hexadecimal number system. The hexadecimal system (Hex) as the name might suggest, is to the base 16. It is also, as the name might not suggest, a hybrid system. It uses letters as well as digits. The decimal gives us the familiar 10 digits, 0 through 9, while the hexa gives us the first six letters of the alphabet. Therefore, the Hex number system is represented by:

zero	0	eight	8
one	1	nine	9
two	2	ten	A
three	3	eleven	B
four	4	twelve	C
five	5	thirteen	D
six	6	fourteen	E
seven	7	fifteen	F

Why A Hex System?

To answer this question, we need to jump ahead a bit. Most microcomputer systems use an eight-bit binary word to handle data transfer. An eight-bit binary word is simply eight bits of binary information such as 10110011_2. Again notice that this is a cumbersome quantity to handle. Also think back to our octal number and notice that if we were to break this eight-bit word into two halves, each half would have a maximum total of 15 and that is our largest hex symbol. Therefore, any eight-bit binary word can be represented by two hex digits. As you can see, it would be easier to give our computer two hex digits than eight binary bits. So our binary 10110011_2 can be represented in hex as:

```
1011  0011
  B     3
```

Therefore, one-hundred seventy-nine units may be represented as:

179_{10}	decimal
$0011\ 0011_2$	binary
263_8	octal
$B3_{16}$	hex

Hex Conversion

For all practical purposes, conversion to or from the Hex number system is something that is best done through familiarization with the Hex code. Table 1.3 shows the Hex number system and the equivalent decimal and binary representation. Since the Hex system is merely (as far as computers are concerned) just an extension of a set of four binary digits, it is best to become familiar with the number system by means of this table.

TABLE 1.3 Three Common Number Systems

Decimal	*Binary*	*Hexadecimal*	*Decimal*	*Binary*	*Hexadecimal*
0	0	0	8	1000	8
1	1	1	9	1001	9
2	10	2	10	1010	A
3	11	3	11	1011	B
4	100	4	12	1100	C
5	101	5	13	1101	D
6	110	6	14	1110	E
7	111	7	15	1111	F

Example Convert $04C5_{16}$ to a binary number.

Solution From the table, we simply assign a 4-digit binary equivalent to each Hex digit. Therefore:

hex	0	4	C	5
binary	0000	0100	1100	0101

Example Add $2A_{16}$ to 33_{16}.

Solution

$2A_{16}$	00101010
33_{16}	00110011
5D	01011101

$$A_{16} + 3_{16} = A + 1 + 1 + 1$$
$$A + 1 = B$$
$$B + 1 = C$$
$$C + 1 = D$$

The use of Hex numbers will be a time-saver when dealing with large binary numbers. As we will see in a later chapter, through the use of a hexadecimal-to-binary converter, we will be able to enter data to a computer in Hex and have the computer receive the data in binary form.

1.6 THE BCD CODE

The BCD code is a special form of binary code. In this case, each digit of a decimal number is represented by a four digit binary number. For this reason, the BCD code is called *binary-coded-decimal*. Since the largest decimal digit is 9, we see we need a binary 1001 to represent this digit. To represent the decimal 637 we simply change it as follows:

6	3	7
0110	0011	0111

and to represent 945_{10} we write:

9	4	5
1001	0100	0101

One feature of this code is the fact that it gives us some error detecting ability. Since no binary larger than 9 (1001) will ever be shown for any set of four digits, we can assume that if one were ever to show up, that bit of code is in error.

Straight Binary Code and BCD

You must realize that binary and binary-coded-decimal will not yield the same result for a given quantity. For example, the decimal number 142 will not yield the same string of binary digits.

$142_{10} = 10001110_2$ straight binary
$142_{10} = 000101000010$ BCD

Note also that the number of digits used by the straight binary was only 8, whereas the BCD required 12. So while we may gain a bit of error checking via the BCD, we lose a bit of economy. This is something you will become familiar with throughout your study of electronics. There will always be trade-offs and you will be required to weigh and judge what is best for the given situation.

BCD Addition

For the most part, BCD addition is the same as binary addition. That is, the rules apply. For example, add

5_{10}	0101
4_{10}	0100
9_{10}	1001

However, if we try to add 6_{10} to 7_{10}, we notice something else happens.

6_{10}	0110	
7_{10}	0111	
13_{10}	1101	BCD

Notice that the BCD equivalent is now an invalid group. That is, 13_{10} cannot be represented in BCD by four digits. The solution to this problem is as follows. We correct by the use of an *add 6;* so we now have:

6_{10}		0110	
7_{10}		0111	
13_{10}		1101	Invalid
		0110	Add 6
	0001	0011	New BCD
	1	3	

The use of the add 6 is a correction factor and is added to any BCD group larger than 1001.

Example

Add	67_{10}	0110	0111	
	55_{10}	0101	0101	
	122_{10}	1011	1100	Invalid
		1	0110	Add 6
		1100	0010	Invalid
		0110		Add 6
		1 0010	0010	
		1 2	2	Correct

Notice that each time we had to add the 6 correction factor it created a carry into the next BCD grouping. This carry must be added to the grouping before a 6 correction factor is added to that grouping. In other words, for the example shown, do not add the correction factor to both groupings, even though both were invalid from first addition. Until the first 6 is added, its carry cannot be added to next group to the left.

1.7 ALPHANUMERIC CODES

The next code or number system we shall look at is a very important system. To this point, we have looked at systems whose primary function was to represent numerical data. To use a computer, however, it is usually necessary to allow operators to enter data not only in numerical fashion but as alphabetic data.

One such code that allows us to do this is the ASCII (American Standard Code for Information Interchange). The ASCII code uses seven bits of binary information, therefore it can be used to represent 2^7 or 128 different characters or symbols. For example, to send the word *HELLO* we would need to send five groups of data as:

100 1000	100 0100	100 1100	100 1100	100 1111
H	E	L	L	O

This code is shown in Table 1.4.

1.8 USE OF NUMBER SYSTEMS AND CODES

This unit is intended as an introduction to number systems. It is hoped you will study these systems and become familiar with their differences and similarities. The use of these systems will be shown in the following chapters of this text, when we talk about such things as circuits, which will enable us to change from one system to another, and when we talk about performing arithmetic functions using one or more of the systems studied in this chapter. While all of the uses and implications of these systems may not be clear at this time, it is hoped that you will have an understanding of the systems themselves.

SUMMARY

Chapter one was an introduction to the concept and use of number systems. Through the understanding and proper use of number systems, the material presented in this text will have meaning, and the user will have a clearer understanding of the world of logic and digital circuits and systems.

TABLE 1.4 Alphanumeric Code

Character	*7-Bit ASCII Code*	*Character*	*7-Bit ASCII Code*
A	100 0001	Y	101 1001
B	100 0010	Z	101 1010
C	100 0011	0	011 0000
D	100 0100	1	011 0001
E	100 0101	2	011 0010
F	100 0110	3	011 0011
G	100 0111	4	011 0100
H	100 1000	5	011 0101
I	100 1001	6	011 0110
J	100 1010	7	011 0111
K	100 1011	8	011 1000
L	100 1100	9	011 1001
M	100 1101		
N	100 1110	blank	010 0000
O	100 1111	.	010 1110
P	101 0000	(	010 1000
Q	101 0001	+	010 1011
R	101 0010	$	010 0100
S	101 0011	*	010 1010
T	101 0100	)	010 1001
U	101 0101	—	010 1101
V	101 0110	/	010 1111
W	101 0111	,	010 1100
X	101 1000	=	011 1101

QUESTIONS AND PROBLEMS

1.1 Convert the hexadecimal number B6 to its binary equivalent.

1.2 What is the similarity between octal and binary numbers?

1.3 What is the difference between a binary number and a binary-coded-decimal number?

1.4 Convert binary 001101 to decimal.

1.5 Convert binary 0011001 to octal.

1.6 Add the binary 101101 to the binary 110110 and give answer in

a. binary

b. octal

c. hexadecimal

d. decimal

1.7 Subtract the binary 1011101 from the binary 1111111.
1.8 Name one use of the hexadecimal number system.
1.9 What is the difference between hexadecimal and alphanumeric?
1.10 Subtract 33_8 from 46_8 and give results in octal.
1.11 Subtract 13_8 from 32_8 and give results in octal.
1.12 Convert 0.32 decimal to binary.
1.13 Convert 32 decimal to binary.
1.14 Convert 0.101 binary to decimal.
1.15 Convert 101.0 binary to decimal.
1.16 Convert 32 decimal to BCD.
1.17 Convert 0.32 decimal to BCD.
1.18 Add 0.101 binary to 0.011 binary and give results in binary.
1.19 Add 0.101 BCD to 0.011 BCD and give results in BCD.
1.20 Add 32 decimal to 41 decimal and give results in BCD.
1.21 What is meant by a radix?
1.22 What is the base for the hexadecimal number system?
1.23 What is meant by a weighted value?
1.24 "Write" the word *number* using the ASCII code.
1.25 Can we "write" the word *hello* using hexadecimal coding? Think! Try and write it.

Logic Gates

2

OBJECTIVE: To become familiar with the types and uses of logic gates. To become familiar with the characteristics of integrated circuit devices used as logic gates.

Introduction: The study of digital circuits and systems is essentially the study of logic gates. No matter how complex the final circuit or system, it can be reduced to a logic-gate configuration. Before you can hope to explore the entire digital system, it is important that you fully understand the uses and the characteristics of logic gates. This unit will study circuit analysis of the integrated circuit logic gates as well as the function of various types of gates that are the basis for all of our digital systems.

2.1 TRUE-FALSE LOGIC

Just as the binary number system that we saw in Chapter 1 was suited to the digital field, so is a type of logic often called true-false or yes-no logic.

If we were to establish a table and call it a true-false table, we might want to use other terms to designate the true-false logic. This is shown in Table 2.1. Let's carry our analogy a bit further. Look at the circuit of figure 2.1. Here we see a simple series DC circuit with a battery, a switch, and a lamp. When the switch is closed, the lamp will light. When the switch is open, the lamp is off.

TABLE 2.1 True-False Logic Designations

True	*False*
Yes	No
High	Low
Switch closed	Switch open
On	Off
Binary 1	Binary 0

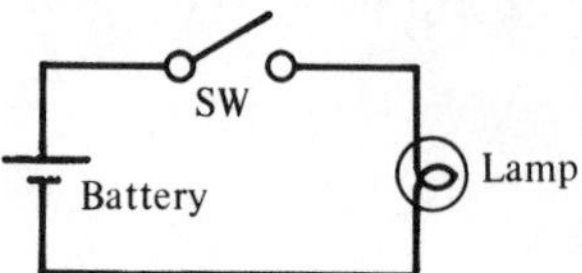

Figure 2.1 Simple True-False Circuit

2.2 THE AND CIRCUIT

Figure 2.2 carries our logic a bit further. Here we have added a second switch in series with the first. Now it takes a closure of both switches to light the lamp. In other words, it takes SW_A and SW_B closed to light the lamp. The truth table shown here is a method of denoting the characteristics of the circuit.

Truth Table

SW_A	SW_B	*Lamp*
Open	Open	Off
Open	Closed	Off
Closed	Open	Off
Closed	Closed	On

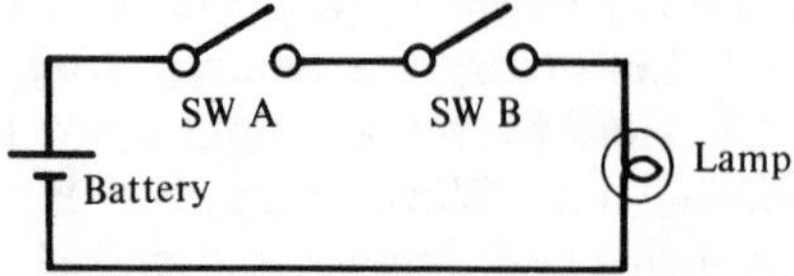

Figure 2.2 AND Circuit

By means of the truth table we have listed all possible combinations of switch conditions and the appropriate lamp condition for each combination.

Let us now redo the truth table and substitute the values of binary 0 and binary 1 for the conditions of the switches and the condition of the lamp. We shall let a binary 0 denote an open switch and the off lamp condition and a binary 1 a closed switch and on lamp condition. We now have:

SW_A	SW_B	*Lamp*
0	0	0
0	1	0
1	0	0
1	1	1

Remember, this truth table shows that SW_A and SW_B must be closed to light the lamp. If we just called SW_A, "A" and SW_B, "B" we could write an equation that would say the same thing.

$A \cdot B = 1$ where the dot (·) denotes AND

2.3 THE LOGIC AND GATE

The circuit symbol for the AND function is:

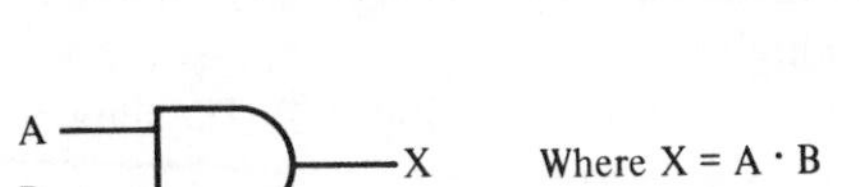

Input

Output (X)

If we denote one of the input leads as A and the other as B we can show it as:

A

B

X

Where $X = A \cdot B$

That is to say, that the output X is a logic 1 (high level) when the inputs to both A and B are logic 1. This is called positive logic. The truth table for this positive-logic AND gate is given as:

A	B	X
0	0	0
0	1	0
1	0	0
1	1	1

Positive AND logic

The use of such a gate should be quite evident. Suppose we wish to enable a circuit only when two conditions are met. At a plant, for example, it is necessary to monitor the oil pressure and the oil temperature at a machine. Only when each of these is at a safe level will we allow power to be connected to the machine. The A and B inputs to the gate then may come from an oil-pressure switch and an oil-temperature switch. When each is within the normal operating limits, the switches close, producing a high to the inputs of the gate. The output then produces a high that activates a relay allowing power to be connected to the machine. AND Gate Truth Tables:

A	B	X
0	0	0
0	1	0
1	0	0
1	1	1

A	B	X
Low	Low	Low
Low	High	Low
High	Low	Low
High	High	High

A	B	X
False	False	False
False	True	False
True	False	False
True	True	True

A	B	X
No	No	No
No	Yes	No
Yes	No	No
Yes	Yes	Yes

The Three-Input AND Gate

Just as was the case with the two-input AND gate, with the three-input gate, all input levels must be logic 1 (or high) in order for the output level to be a logic 1 (high).

The logic symbol for the three-input AND gate is similar to the two-input:

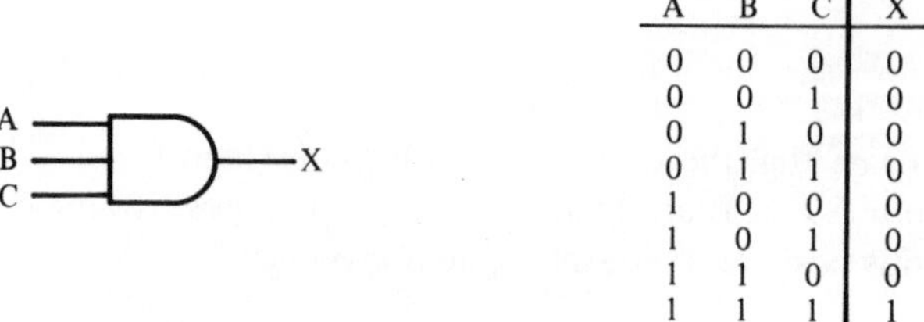

A	B	C	X
0	0	0	0
0	0	1	0
0	1	0	0
0	1	1	0
1	0	0	0
1	0	1	0
1	1	0	0
1	1	1	1

Figure 2.3 Three-Input Gate and Truth Table

The AND gate may have several inputs. Again, in order to generate a logic 1 output, all inputs must be at a logic 1.

2.4 THE GATING FUNCTION

A gate will be a circuit that will pass an input signal to the output for certain conditions. An AND gate will pass a signal through if all other signals to its inputs are at logic 1 when our signal comes along. As shown in figure 2.4 we have a signal that is actually a pulse. In order to allow this signal to reach the output, we must allow the other input lead of the gate to be at a logic 1. Notice that the first four pulses of the signal at input A do not pass through to the output. When the input at B goes positive (t_1), the input pulses at A are passed through the gate.

So, by allowing input A to be the signal input and input B to be the "gating" input, we can decide when and if to allow the signal to pass through the gate. The gate then is just as its name applies, and we, the user, can decide when to open and when to close the gate, allowing (or not allowing) a signal to pass through.

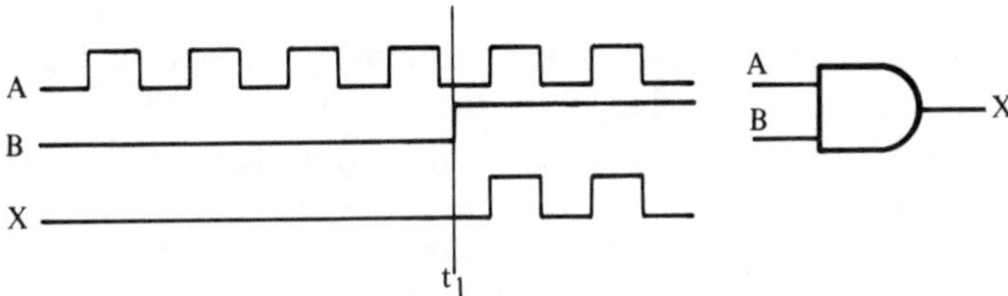

Figure 2.4 Gating Action

2.5 THE NOT CIRCUIT

Before we continue with our study of gates and gate circuits, we need to look at a very useful and simple circuit. This circuit is called the NOT circuit. The NOT circuit is sometimes called an *inverter circuit.* The symbol for the NOT circuit is shown in figure 2.5. The truth table for the NOT circuit shows its function. When A (the input) is 0, X (the output) is 1 and when A is 1, X is 0. This circuit allows the user to change the logic level of the signal anywhere in the system. A feature that will come in handy, as we shall see later. Note the equation to the right of the truth table, $X = \bar{A}$. The bar over the A denotes the NOT function. This equation is read as "X equals not A." In some texts the bar may be replaced by a prime sign ($'$), and the equation will look like this; $X = A'$. Both of these equations mean the same thing. The signal at A has been inverted when it appeared at the output X.

A	X
0	1
1	0

$\therefore X = \overline{A}$

Figure 2.5 NOT Circuit

2.6 THE NAND GATE

Probably no single gate has proved as versatile as the NAND gate. To understand the NAND gate, we need just look at figure 2.6, which shows the basic two-input AND gate with a NOT circuit attached to its output lead. This will invert all output signals from the AND gate. As the truth table shows, the output of the circuit is exactly opposite that for the AND gate. Is it any wonder then that this gate is called the NOT-AND (NAND) gate? Notice also, that the output expression for this gate is given as $y = \overline{A \cdot B}$ or y equals not A and B. This of course means (in our positive-logic world) that in order to have a logic 1 (high) output from the gate, the two inputs must not be logic 1. If either is a logic 1, that is fine, since the equation says that y = not A and B. Figure 2.6(b) shows the circuit symbol for the NAND gate. The symbol looks exactly like the AND gate with the exception of the small ○, which means invert. One good rule of thumb to keep in mind when using these NAND gates is that a 0 on any input line will produce a logic 1 at the output line.

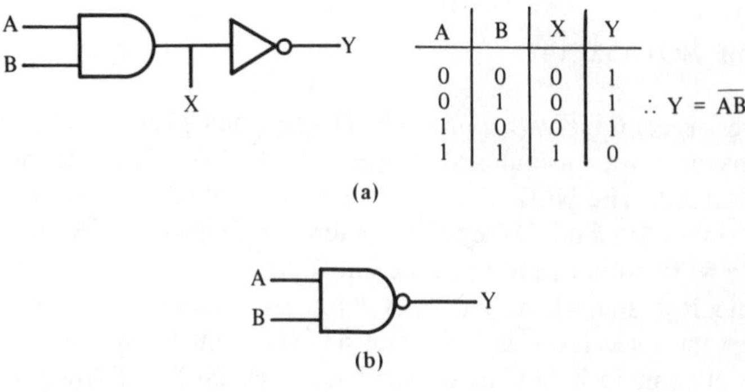

A	B	X	Y
0	0	0	1
0	1	0	1
1	0	0	1
1	1	1	0

$\therefore Y = \overline{AB}$

(a)

(b)

Figure 2.6 NAND Gate

The Integrated Circuit NAND Gate

This text will deal primarily with one "family" of integrated circuit devices. Probably the most widely used family of discrete logic circuits (outside of the microcomputer field) is TTL (transistor-transistor-logic). Figure 2.7 shows the circuit for the basic TTL gate. It is around this gate that most of the other gates and devices in the family are based. Notice that if both inputs (A and B) are held high (a logic 1), Q_A will be off. Q_A's collector will also be high, turning Q_B on. This will force Q_D into saturation, causing Y to go low (0.4 volts in this case). On the other hand, should either input A or B go low, Q_A will be on. This will turn Q_B off and consequently Q_D will go off. With Q_D off and Q_C now on, point Y will go high to approximately 3.6 volts. (Assuming 0.7 volt at both the diode and base-emitter of Q_C.) If we were to write a truth table for the circuit of figure 2.7 it would be the same as that of the NAND gate of figure 2.6. This circuit is a TTL NAND gate and is fabricated on a single chip to produce the integrated circuit gate. The TTL designation is arrived at since both the input circuit (Q_A) and the output circuit (Q_C and Q_D) are transistors. There have been other logic families called RTL (resistor-transistor-logic) and DTL (diode-transistor-logic), in which the input circuitry was primarily a resistor or a diode. These circuits are no longer much in use, and the reader is referred to the library for an explanation of these circuits.

Study the circuit of figure 2.7. Keep in mind two things when using these gates, first, the input is achieved at the emitter of a transistor, which means that a low at the input turns the circuit on. Second, the output is the collector of a transistor.

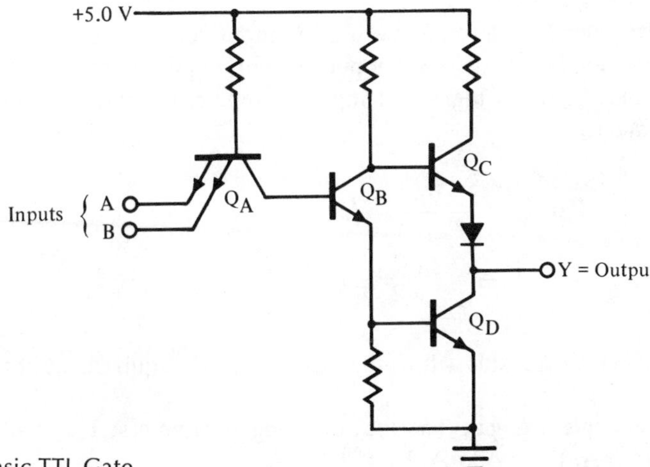

Figure 2.7 Basic TTL Gate

Figure 2.8 shows two connected gates. When Q_D gate 1 is on, it allows a path for current from Q_A gate 2. As can be seen, this arrangement limits the number of gates that can be tied to the output of one gate.

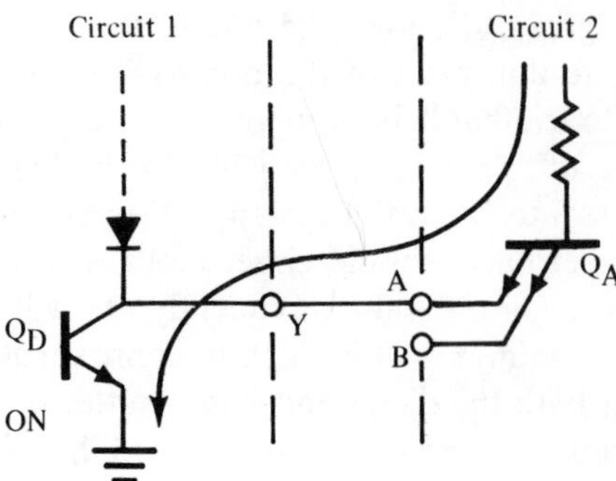

Figure 2.8 Current Sinking

Unit Loads

A unit load (UL) has been established to let the user calculate the drive capability and input requirements of the gating circuits. These variables are sometimes referred to as *fan-in* and *fan-out, fan-in* being the input current requirement, and fan-out the output drive capability of the gate.

One unit load (UL) has been defined as:

$$1 \text{ UL} = \begin{cases} 40\ \mu\text{A output high} \\ 1.6\text{ mA output low} \end{cases}$$

In order to determine the fan-in (input current requirement) for a gate, we divide its low-level input current requirement by 1.6 mA (1 unit load), and its high-level input current requirement by 40 μA (1 unit load).

$$\text{Fan-In} = \frac{I_{IH}}{40\ \mu\text{A}} = \text{UL}$$

$$= \frac{I_{IL}}{1.6\text{ mA}} = \text{UL}$$

The lowest result will be the drive (fan-in) requirement of the circuit.

Example A gate has the following parameters: $I_{OH} = 400\ \mu\text{A}$, $I_{OL} = 16\text{ mA}$, $I_{IH} = 40\ \mu\text{A}$, $I_{IL} = 1.6\text{ mA}$

a. Determine the fan-in and fan-out for this gate.
b. How many similar gates could one gate drive?

Solution

$$\text{Fan-in} = \frac{40\ \mu\text{A}}{40\ \mu\text{A}} = 1.0 \text{ unit load}$$

$$= \frac{1.6\text{ mA}}{1.6\text{ mA}} = 1.0 \text{ unit load}$$

$$= 1.0 \text{ unit load}$$

$$\text{Fan-out} = \frac{400\ \mu\text{A}}{40\ \mu\text{A}} = 10.0 \text{ unit loads}$$

$$= \frac{16\text{ mA}}{1.6\text{ mA}} = 10.0 \text{ unit loads}$$

$$\text{Fan-out} = 10 \text{ unit loads}$$

a. Fan-in = 1.0 unit load; fan-out = 10.0 unit loads
b. Drive capability equals

$$\frac{\text{Fan-out}}{\text{Fan-in}} = \frac{10}{1} = 10 \text{ similar gates}$$

The 7400 Logic Family

The Texas Instruments Company (TI) in 1964 introduced a standard line of TTL gates and other integrated circuits. They called this logic family the 5400/7400 series. The 5400 designators are used mainly by the military. We shall focus much of our attention in this text on the 7400 series. This series is one of the most widely used series of devices and circuits ever manufactured for the electronics industry. Although other manufacturers now produce the same series of circuits, some use their own numbering sequence. For ease of use, most data books provide a cross-reference index. For reasons of clarity, we will not use any other numbers other than the standard 7400 number unless there is no equivalent 7400 number for a particular device. The reader is urged to use a cross-reference file when using devices of other manufacturers.

The 7400 NAND Gate

While the 7400 designator has been given to identify an entire series of devices, the one device in the series that best typifies it also has as its identification the 7400 designation. Figure 2.9 shows the symbolic representation and pin configuration of this device. The 7400 is called a quadruple two-input positive-NAND gate. This means that there are four identical gates (quad), each with two inputs, manufactured on a single chip. Each of these gates operates according to the truth table given in figure 2.6. In addition, each gate symbol represents one complete circuit as shown in figure 2.7. In order to use any, or all, of the gates housed in this package, the user must supply V_{CC} and ground (pins 14 and 7) to the chip. All four circuits operate from the V_{CC} supply (+5.0 volts) and ground bus. The user of this integrated circuit device need not worry about such things as transistor bias levels, saturation and cut-off parameters, and so on. As long as the fan-in and fan-out requirements are not exceeded, and as long as the +5.0 volt and ground is supplied, the circuits will function properly. All that the user need be concerned with is the proper connections to perform the required logic functions of the circuit.

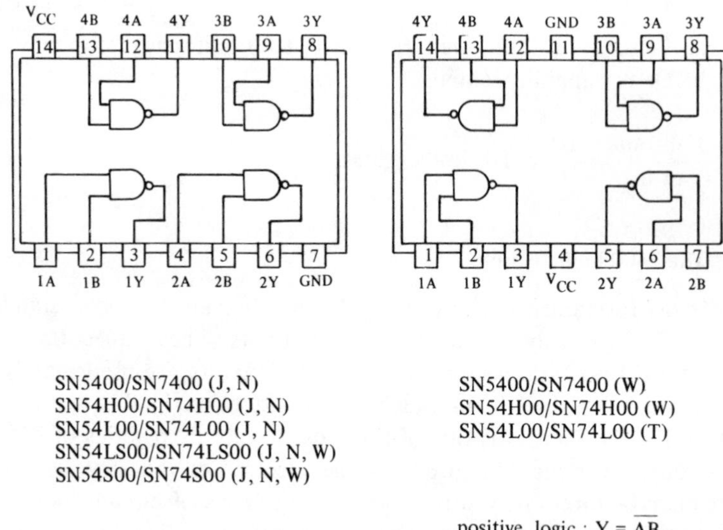

Figure 2.9 Quadruple Two-Input Positive-NAND Gates

Figure 2.10 is a chart that gives the recommended operating conditions of the 7400. Notice that this chart also mentions a series of 7400, 74L00, 74H00, and 74S00 devices. These three other series have been developed to give the user a wider choice of speed and power-dissipation characteristics.

74L00 Series — This is a low-power version of the 7400 series. All resistor values are increased, thereby decreasing power dissipation but increasing propagation delay. A good use of the 74L00 series would be in battery-operated circuits.

74H00 Series — This is a high-speed device where all resistors are decreased, thereby increasing power dissipation but decreasing propagation delay. Where fast switching times are essential and increased power consumption is no problem, this series may be selected.

74S00 Series — Through the use of a Schottky barrier diode (SBD), which is connected from base to collector of each transistor, high-speed operation is achieved. Where high-speed switching is critical, these devices may be used.

As a means of comparison, note the various switching times (given as propagation delay) for the various devices. A comparison of the parameters given in figure 2.10 will enable the user to select the device best suited to the design requirements. One further consideration that should be given to any design project is the *cost* factor. A recent catalog puts the price of these devices as:

Device	Cost per chip (Cost per gate = CPC/4)
7400	16¢
74H00	19¢
74S00	30¢
74L00	30¢
*5400	50¢

*Military version

As you see, there is better than an 87 percent increase in price from the 7400 to the 74S00 or 74L00. This increase in price may or may not be warranted, depending on the project being designed. If I were going to use 100,000 of these devices in the manufacture of 10,000 units my cost would rise from $1600 to $3000 and my cost per unit would rise by 87 percent or $1.40 per unit built.

The cost per gate for a 7400 is only 4 cents. At this price the designer can use an extra gate here and there to solve a difficult logic problem rather than spend the labor hours required to design for a minimum number of gates.

However, as is always the case, a number of factors must eventually be weighed before the final design, including the series of devices to be used, is set. Certainly cost, speed, power consumption, and chip count will be among the factors used in the final decision.

Electrical characteristics over recommended operating free-air temperature range (unless otherwise noted)

Parameter	Test Figure	Test Conditions†		Series 54 Series 74 '00, '04, '10, '20, '30,			Series 54H Series 74H 'H00, 'H04, 'H10, 'H20, 'H30,			Series 54L Series 74L 'L00, 'L04, 'L10, 'L20, 'L30,			Series 54LS Series 74LS 'LS00, 'LS04, 'LS10, 'LS20, 'LS30,			Series 54S Series 74S 'S00, 'S04, 'S10, 'S20, 'S30, 'S133,			Unit
				Min	Typ‡	Max	Min	Typ‡	Max	Min	Typ‡	Max	Min	Typ‡	Max	Min	Typ‡	Max	
V_{IH} High-level input voltage	1,2			2			2			2			2			2			V
V_{IL} Low-level input voltage	1,2		54 Family			0.8			0.8			0.7			0.7			0.8	V
			74 Family			0.8			0.8			0.7			0.8			0.8	
V_I Input clamp voltage	3	V_{CC} = Min, I_I = §				*-1.5			*-1.5						-1.5			-1.2	V
V_{OH} High-level output voltage	1	V_{CC} = Min, V_{IL} = V_{IL} max, I_{OH} = Max	54 Family	2.4	3.4		2.4	3.5		2.4	3.3		2.5	3.4		2.5	3.4		V
			74 Family	2.4	3.4		2.4	3.5		2.4	3.2		2.7	3.4		2.7	3.4		
V_{OL} Low-level output voltage	2	V_{CC} = Min, V = 2 V, I_{OL} = Max	54 Family		0.2	0.4		0.2	0.4		0.15	0.3		0.25	0.4			0.5	V
			74 Family		0.2	0.4		0.2	0.4		0.2	0.4		0.35	0.5			0.5	
I_I Input current at maximum input voltage	4	V_{CC} = Max, V_I = 5.5 V				1			1			0.1			0.1			1	mA
I_{IH} High-level input current	4	V_{CC} = Max	V_{IH} = 2.4 V			40			50			10							µA
			V_{IH} = 2.7 V												20			50	
I_{IL} Low-level input current	5	V_{CC} = Max	V_{IL} = 0.3 V									-0.18							mA
			V_{IL} = 0.4 V 'LS30												-0.4				
			V_{IL} = 0.4 V Others			-1.6			-2						-3.6				
			V_{IL} = 0.5 V																
I_{OS} Short-circuit output current◆	6	V_{CC} = Max	54 Family	-20		-55	-40		-100	-3		-15	-6		-40	-40		-100	mA
			74 Family	-18		-55	-40		-100	-3		-15	-5		-42	-40		-100	
I_{CC} Supply current	7	V_{CC} = Max		See table on next page															mA

Recommended operating conditions

		Series 54 / Series 74			Series 54H / Series 74H			Series 54L / Series 74L			Series 54LS / Series 74LS			Series 54S / Series 74S			Unit
	54 Family / 74 Family	'00, '04, '10, '20, '30,			'H00, 'H04, 'H10, 'H20, 'H30,			'L00, 'L04, 'L10, 'L20, 'L30,			'LS00, 'LS04, 'LS10, 'LS20, 'LS30,			'S00, 'S04, 'S10, 'S20, 'S30, 'S133,			
		Min	Nom	Max	Min	Nom	Max	Min	Nom	Max	Min	Nom	Max	Min	Nom	Max	
Supply voltage, V_{CC}	54 Family	4.5	5	5.5	4.5	5	5.5	4.5	5	5.5	4.5	5	5.5	4.5	5	5.5	V
	74 Family	4.75	5	5.25	4.75	5	5.25	4.75	5	5.25	4.75	5	4.75	4.75	5	5.25	
High-level output current, I_{OH}	54 Family						- 500			-100			- 400			- 1000	μA
	74 Family			- 400			- 500			- 200			- 400			- 1000	
Low-level output current, I_{OL}	54 Family			16			20			2			4			20	mA
	74 Family			16			20			3.6			8			20	
Operating free-air temperature, T_A	54 Family	-55		125	-55		125	-55		125	-55		125	-55		125	°C
	74 Family	0		70	0		70	0		70	0		70	0		70	

† For conditions shown as Min or Max, use the appropriate value under recommended operating conditions

‡ All typical values are at $V_{CC} = 5V$, $T_A = 25\,^{\circ}C$.

§ $I_I = -12mA$ for SN54'/SN74', -8mA for SN54H'/SN74H', and -18mA for SN54LS'/SN74LS' and SN54S'/SN74S'.

◆ Not more than one output should be shorted at a time, and for SN54S'/SN74S', duration of short-circuit should not exceed 1 second

* The input clamp voltage specification is effective for Series 54/74 and 54H/74H parts date-coded 7332 or higher.

Figure 2.10 Positive-NAND Gates and Inverters with Totem-Pole Outputs

2.7 PROPAGATION DELAY

One other factor that must come into play when using the TTL devices is propagation delay. To put it simply, propagation delay is the time it takes a signal to pass through the gate or device. There are two propagation delay times defined in the operating characteristics of the gates. These are called:

1. t_{plh}: This is the low (logic 0) to high (logic 1) delay time.
2. t_{phl}: This is the high (logic 1) to low (logic 0) delay time.

Figure 2.11 is an illustration of propagation delay times for a typical 7400 gate. Figure 2.11(a) shows the symbolic representation of the gate with appropriate input and output signal levels. Figure 2.11(b) shows a time diagram representation of the signal levels. Input level B is held at a logic 1 for the illustration. As input A goes from a low to a high, we would expect the output to go from a high to a low. However, there is a lagging of the change in output with reference to the change in the input level. This lag is due to the time required for the various transistors within the gate to change states.

Propagation delay is a fact of life with circuits and should be understood by anyone who wishes to design logic circuits. While it is true that the delay through one gate may be no more than 10 or 11 nanoseconds, this could produce a delay of 50 or 60 nanoseconds by the

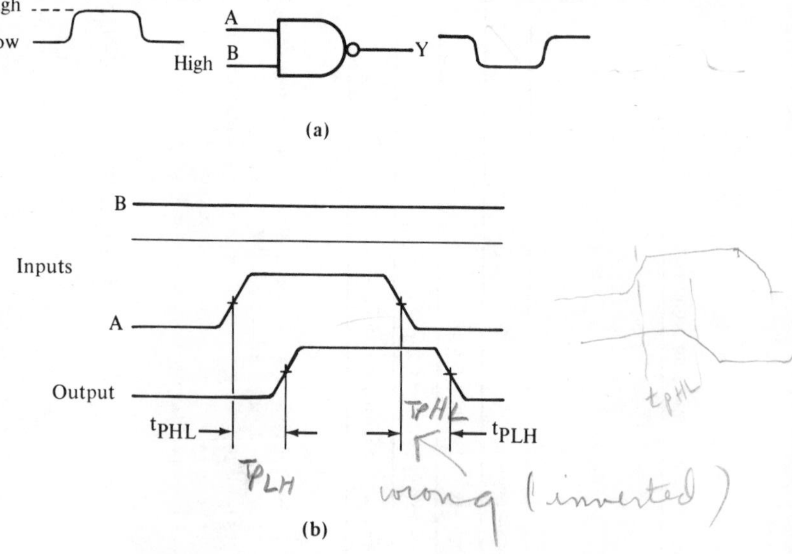

Figure 2.11 Propagation Delay

time we have processed this signal through only five gates. This could seriously affect the operation of a circuit where the timing of two pulses was critical. We will look at propagation delay and its problems as we develop more complex circuits in later chapters of this text.

2.8 OPEN-COLLECTOR GATES

Before we finish our discussion of the NAND gate we need to look at another important characteristic of the 7400 family. Figure 2.12 is a symbolic representation and pin diagram of a 7401 device. This device is also a quad 2-input NAND gate. Its logic equation is the same as that of the 7400; $Y = \overline{AB}$. It would appear that the 7401 and the 7400 are identical. They are in logic function, but they are not in circuit operation. Notice the label of this figure says "Quadruple 2-input positive-NAND gates with open-collector outputs." It is this open-collector out-

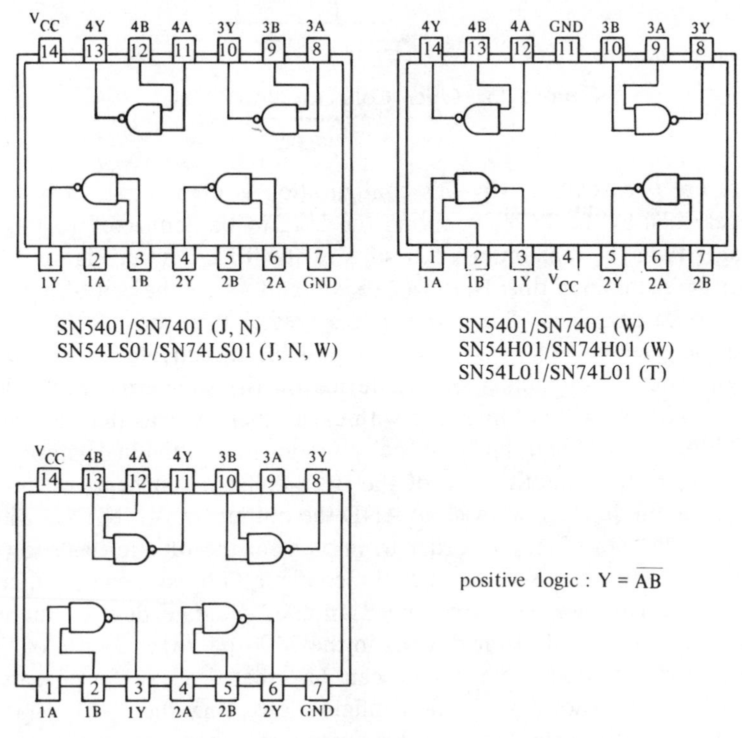

Figure 2.12 Quadruple Two-Input Positive-NAND Gates with Open-Collector Outputs

put that is the unique feature of this gate. As we shall see later, several other devices also have this feature. Just what is the open-collector feature and what can it do? Refer to figure 2.13. Here we see a circuit that looks identical to figure 2.7, the exception being that Q_C is missing. The output collector is therefore open and must be connected through a lead back to V_{CC} in order to develop a logic level at the collector.

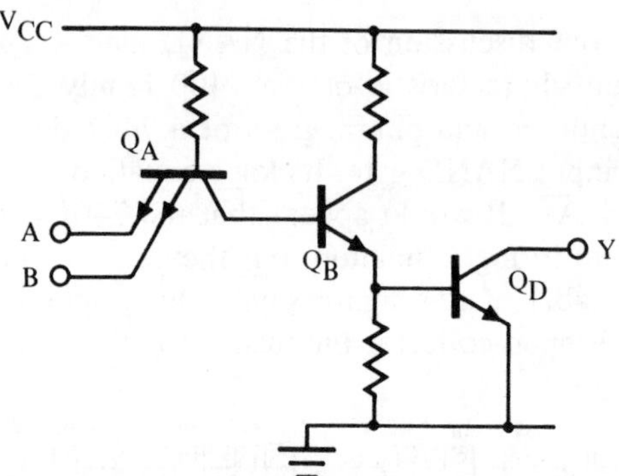

Figure 2.13 Open-Collector NAND Gate

There are two reasons why this configuration is used. First, in certain devices such as the 7426, it allows the load to be connected to larger voltages than the maximum +5.0 volts of the TTL device. Figure 2.14 shows an example of this. Here we see a circuit where the coil of a relay needs to be energized. The coil requires greater than the 5.0 volts. By using the open-collector concept, it is possible to activate the coil and still operate the remainder of the device on the standard +5.0 volts. The two circuits will not interface with each other. As was the case with the 7400, the 7401 and the 7426 each contains four individual gates.

To illustrate another use of the open-collector concept, we must back up a bit. Suppose we wish to AND the outputs of two NAND gates as shown in figure 2.15. In order to implement the function as shown, three gates must be used. Even if the configuration as shown in figure 2.15(b) is used, we would not need another separate device, but we would need to use all of the devices in the 7400 package. By use of the 7401, however, this same circuit can be created using fewer gates. Figure 2.16(a) shows a circuit configuration using the 7401 (open-collector) NAND gate. Since we have open collectors, we can connect the collectors together and by use of a pull-up resistor establish the same circuit function as was given in figure 2.15. Figure 2.16(a) symbolizes the wired-AND connection. There is in reality no physical gate

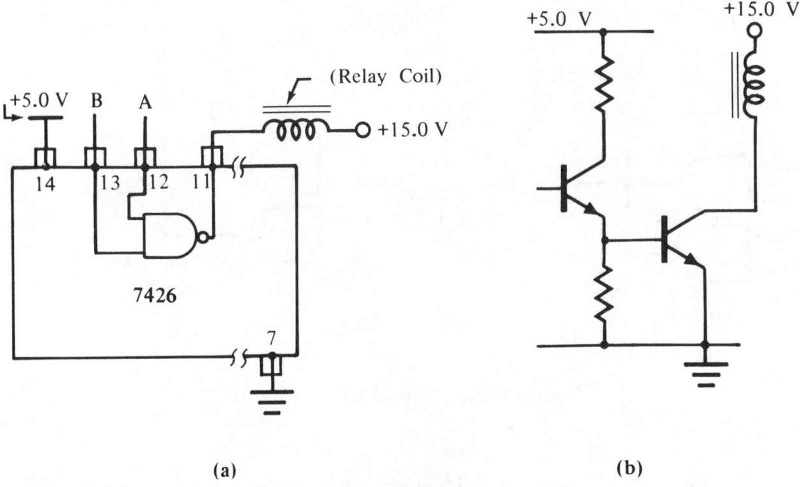

Figure 2.14 Open-Collector Uses

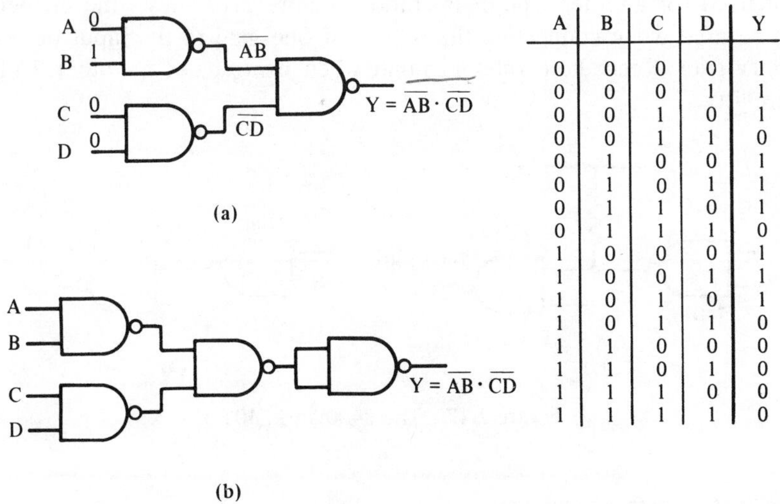

A	B	C	D	Y
0	0	0	0	1
0	0	0	1	1
0	0	1	0	1
0	0	1	1	0
0	1	0	0	1
0	1	0	1	1
0	1	1	0	1
0	1	1	1	0
1	0	0	0	1
1	0	0	1	1
1	0	1	0	1
1	0	1	1	0
1	1	0	0	0
1	1	0	1	0
1	1	1	0	0
1	1	1	1	0

Figure 2.15 Conventional "ANDIND" of Gate Outputs

at this junction. As an exercise it is suggested that the reader verify the output expression of figure 2.16(a) and prove it is the same as that of the truth table of figure 2.15.

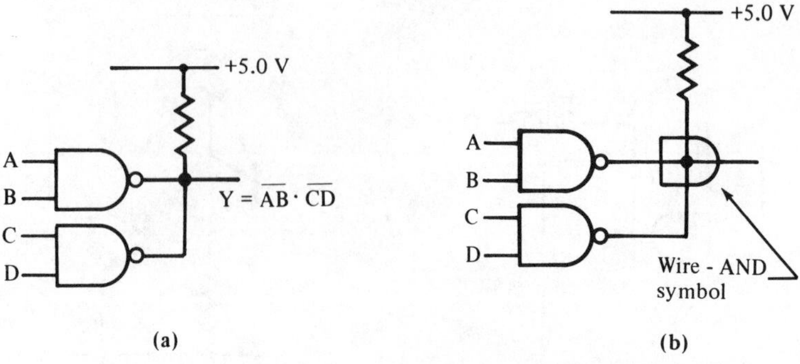

Figure 2.16 Wire-AND Concept

If the circuit of figure 2.16 were to be connected using the 7400, the output transistor (Q_D) could be damaged due to excessive current when Q_C of one gate and Q_D of the other gate were both in the on state. *Never* connect 7400 gate outputs together.

The 7400 using conventional output circuitry, called *Totem-pole* is intended for one type of operation, while the 7401, open-collector, is intended for another type of operation. Figure 2.17 shows the method to be used when connecting the output of one gate to the input of another gate. Keep these rules in mind when using gates on other TTL circuitry.

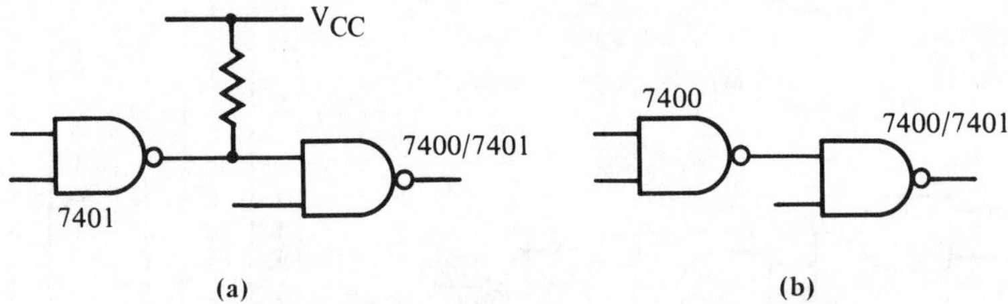

Figure 2.17 The 7400 and 7401

2.9 THE INPUT CIRCUITRY

As a reminder of the input circuitry, figure 2.18(a) shows that the input of a TTL gate is the emitter lead of a transistor. When this lead is left open (unconnected), it is at a logic 1. The normal method of activating this type of input is to pull the input low. There are times when it is necessary to use an active high to activate the gate. That is, keep the input low until we wish to trigger it. To do this, refer to figure 2.18(c). Here we see the emitter to ground through a resistor. If the resistor is sufficiently small, the drop on the emitter will be close to ground and the IC will believe it is at ground level, a logic 0. For TTL circuits, this level (logic 0) is said to be less than 0.4 volt. The manufacturer will guarantee the device to be low if the voltage is below 0.4 volt. If I_{IB} is 1.6 milliamperes, the value of R_{max} will be 250 ohms, to insure a voltage of 0.4 or less. Then, when the switch is closed, the voltage will increase to V_{CC} (at least 3.6 volts for a TTL logic high). If the value of R is too small, however, there will be excessive current drawn from the V_{CC} supply. As a general rule, you should always connect the input leads to something, never leave them floating. Remember an unconnected TTL input is at a logic high (logic 1).

Refer back to figure 2.15. Notice here that a 7400 was shown with its two input leads tied together. This circuit functions the same as an inverter or NOT circuit, so by feeding a NAND gate into a NOT circuit, an AND gate was created.

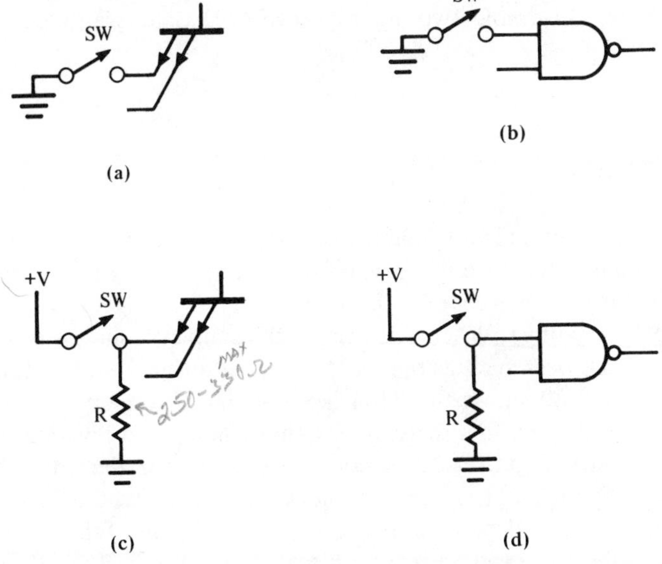

Figure 2.18 Activating with a High

2.10 THE IC AND GATE

Figure 2.19 shows the symbolic configuration and pin assignment for the 7409. This is a quad 2-input AND gate with open-collector output.

All of the considerations given to AND gates and to open collector TTL circuitry apply to this circuit.

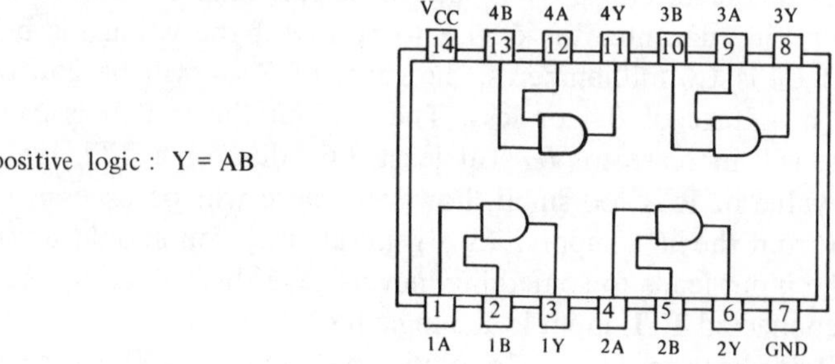

Figure 2.19 Quadruple Two-Input Positive AND Gates with Open-Collector Outputs

2.11 INVERTERS

The use of inverters is a valuable and integral part of the study of digital logic circuits. There are times when the wrong logic signal is present at a particular gate (wrong only in the sense that it necessitates a different approach to gating). When this occurs, the simplest solution is to invert the signal. Devices such as the 7404 hex inverter are used for this function. Figure 2.20 shows the 7404 hex inverter. The name hex inverter comes from the fact that there are six (hex) individual inverters in each 7404 package. Figure 2.21 demonstrates a use for the 7404 hex inverter. In figure 2.21(a), when point A is a logic low, the LED (light-emitting diode) will be forward-biased and will glow. When point A is a logic 1, the LED will be reverse-biased and will not glow. If we want to use the LED as an indicator of the logic level of point A, we need to reason in reverse; light off = 1, light on = 0. Somehow this is contrary to what we normally expect. By adding the inverter, we now see that when point A is high, the cathode of the diode is low and the diode will light. This is the way it should be, a high at point A corresponds to a lit LED.

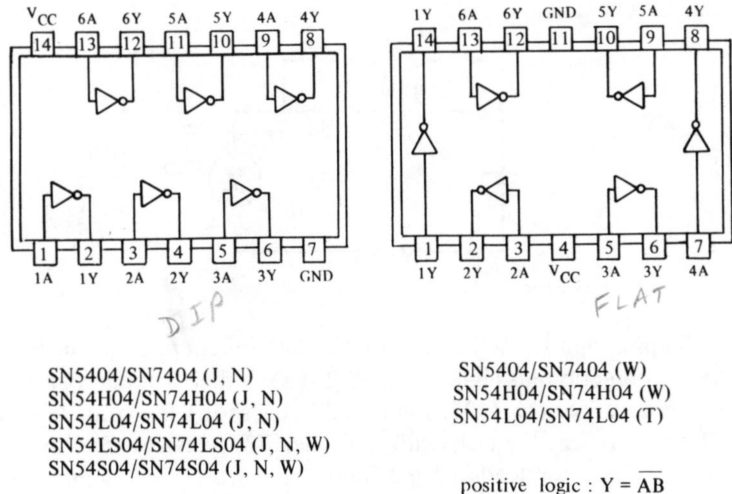

Figure 2.20 Hex Inverters

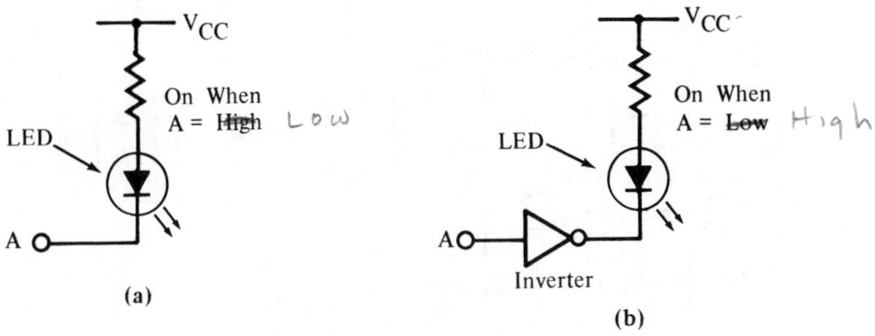

Figure 2.21 Use of Inverters

2.12 OR GATES

If an output might occur if either path A or path B is chosen, the circuit that accomplishes this is called an OR circuit. Referring back to our truth tables and logic definitions, this means an output will occur if either statement is true, if either switch is closed, etc. Figure 2.22 symbolizes the OR circuit using switches. Here we see that if either switch is closed, the LED will light. The equation for the OR circuit is given as:

$$Y = A + B$$

Figure 2.22 OR Circuit

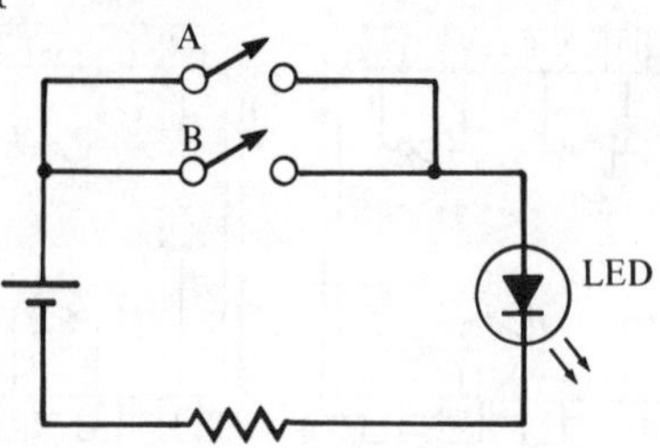

where the plus sign is used to signify the OR function. Remember, this is not an arithmetic plus sign. Figure 2.23(a) is the symbolic configuration and pin out for the 7432. Figure 2.23(b) shows the truth table for the OR gate. As was the case with the other TTL devices, the inputs to the 7432 are emitter leads of the input transistor. This means that a logic 1 (or an input left disconnected) on either input will cause the output to go to a logic 1. Only if both inputs are at logic 0 will the output be at a logic 0.

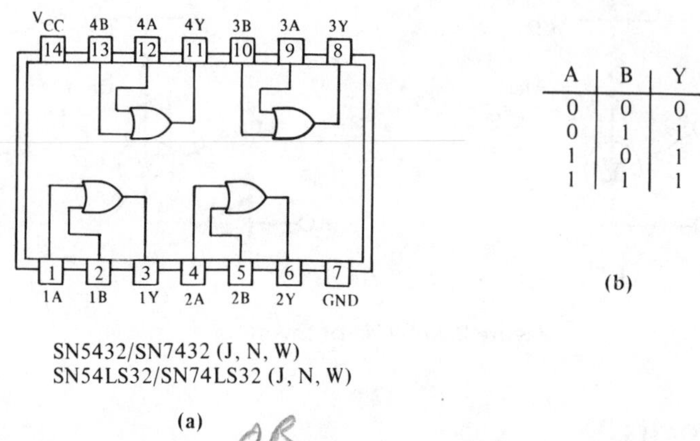

A	B	Y
0	0	0
0	1	1
1	0	1
1	1	1

(b)

Figure 2.23 **(a)** The 7432 Package Configuration. **(b)** Truth Table for OR Gate

Using the OR-Gate

Suppose we wish to activate a circuit (sound an alarm, light an LED, etc.) when either pair A or pair B of switches is activated. A simple circuit to do this could be the one shown in figure 2.24. This circuit will light the LED if either inputs CD or inputs EF are 1.

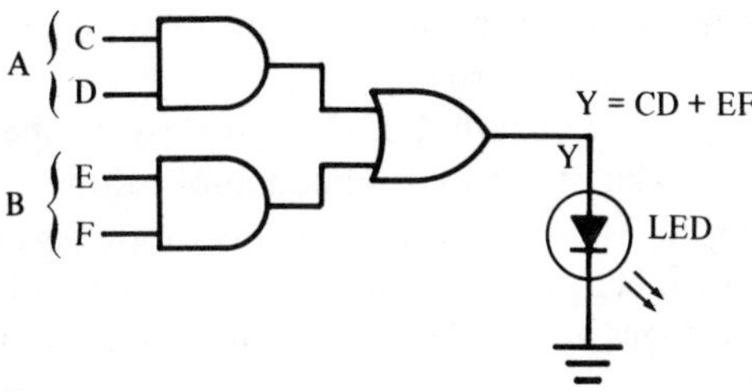

Figure 2.24 Circuit Using OR Gate

Here is a more complex problem. We saw earlier that we could perform addition of binary numbers. Why not devise a circuit that will do this addition for us? By using three of the gates we have seen thus far, this addition can be accomplished. Figure 2.25 shows the circuit configuration. This circuit is called a *half-adder*. It will be covered in detail in a later chapter. For now, let's just say it has the function of adding two binary digits and producing a sum and a carry. The truth table of figure 2.25(c) will help us to analyze the circuits. If we wish to add a logic 1 to a logic 1, we know this will give us:

1_2	one	1_{10}
1_2	one	1_{10}
10_2	two	2_{10}

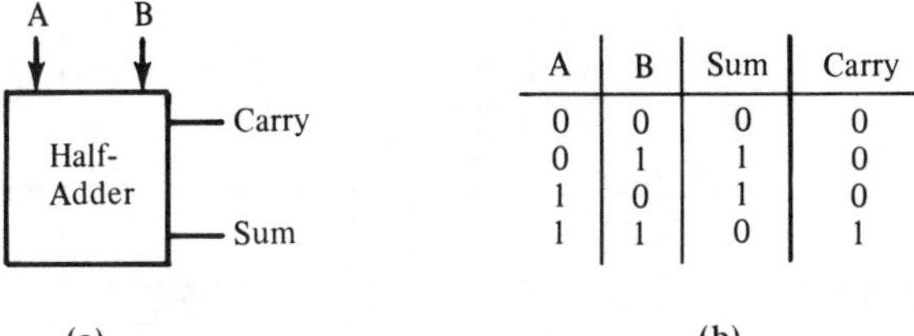

(a)

A	B	Sum	Carry
0	0	0	0
0	1	1	0
1	0	1	0
1	1	0	1

(b)

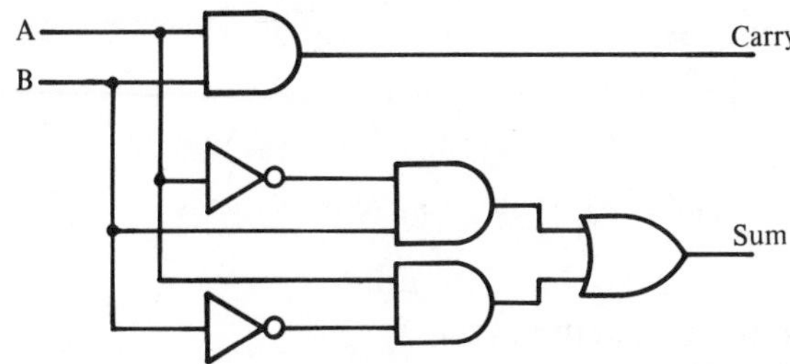

(c)

Figure 2.25 Binary Half-Adder

That is, it will produce a sum of 0 and a carry of 1. Therefore, 1 and 1 = 10 (one, zero).

Using figure 2.25(b), go through the logic and prove for yourself that each of the four input combinations will produce the correct output result. Notice the use of the inverters. As an exercise, it is suggested that you label each individual gate as part of an IC package. How many packages are needed to implement the circuit of figure 2.25 as shown?

2.13 THE NOR GATE

Just as the AND gate had a counterpart with its output inverted, so does the OR gate. Figure 2.26 is the IC package for the 7402, NOR gate. The logic expression for each of these gates is of course:

$$Y = \overline{A + B}$$

It is interesting to note that in the numbering scheme, the NOR gate is numbered 7402 while the AND gate was numbered 7408. Of course, the truth table for the NOR gate is the negative of that of the OR gate. As a point of interest, notice that the truth table for the NOR gate is that of the AND gate inverted (not the negative of the AND gate).

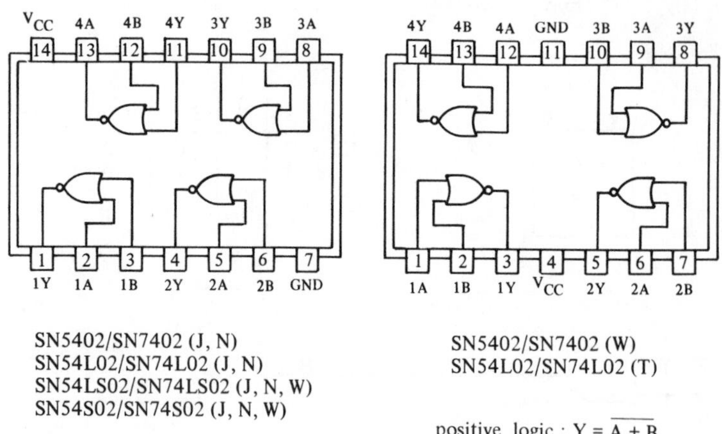

Figure 2.26 Quadruple Two-Input Positive-NOR Gates

2.14 THE EXCLUSIVE-OR GATE (EX-OR GATE)

A gate that has a rather unique function is the exclusive-OR gate. This gate will produce a logic 1 output if either one of, but not both of, the

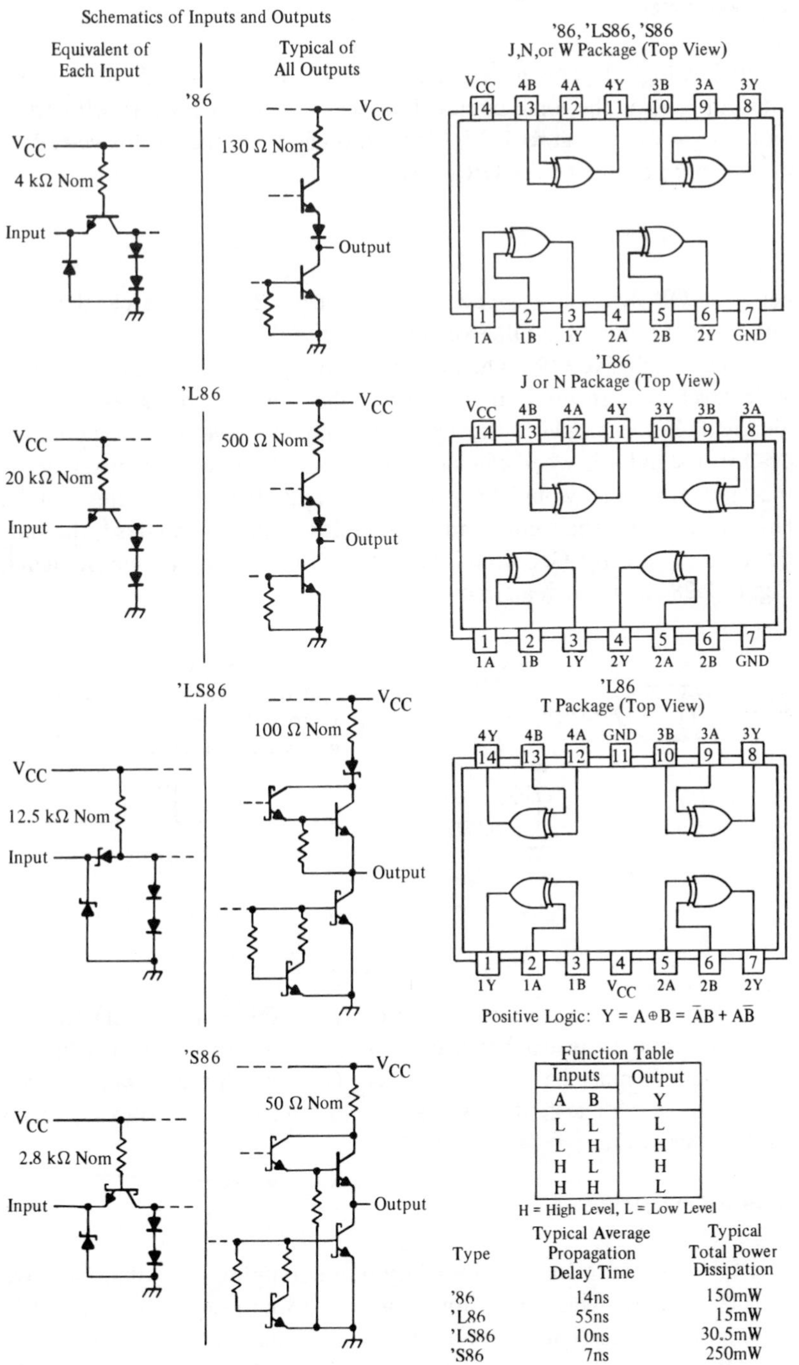

Function Table

Inputs		Output
A	B	Y
L	L	L
L	H	H
H	L	H
H	H	L

H = High Level, L = Low Level

Type	Typical Average Propagation Delay Time	Typical Total Power Dissipation
'86	14ns	150mW
'L86	55ns	15mW
'LS86	10ns	30.5mW
'S86	7ns	250mW

Figure 2.27 SN7486, SN74L86, SN74LS86, SN74S86 Quadruple Two-Input Exclusive-OR Gates

inputs is a logic 1. This gate has a number of functions and will become a useful tool of the logic circuit designer. The 7486 Ex-OR with truth table is shown in figure 2.27. The expression for the exclusive-OR is quite similar to that of the OR gate:

$$Y = A \oplus B$$

here, the OR symbol + is enclosed by a circle. It is this circle that distinguishes the exclusive-OR from the straight OR gate.

The exclusive-OR gate may be used if we wish to examine two logic levels and receive an indication when these two signals are not identical. For example, if we were checking the level of a logic signal against a standard, when the signal we were checking was the same as the standard there would be a logic 0 output from the exclusive-OR. However, should the signal under test differ from the standard, the output of the exclusive-OR would go high, possibly giving an alarm signal. Figure 2.28 shows such a circuit.

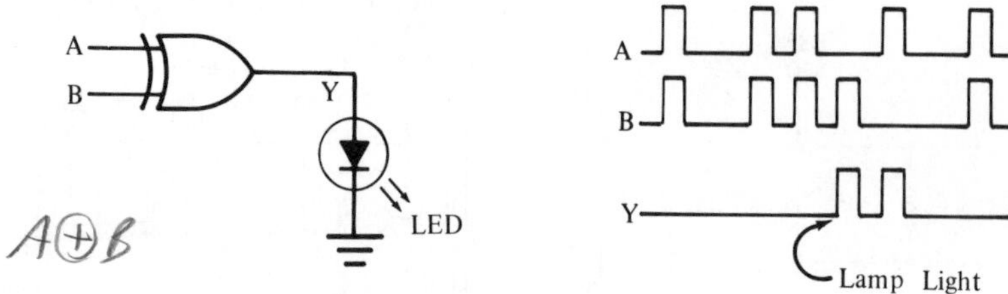

Figure 2.28 Exclusive-OR as a Test Circuit

You are encouraged to use an exclusive-OR (or several) to construct a circuit to test a batch of 7400 NAND gates against a known good 7400. It must be a simple device. One switch for power and not more than one for test. It must be easy to read (i.e., the status of each 7400 under test).

Exclusive-OR/NOR Gate

It is time to take a look at a slightly more complex gate than we have seen up to now. This gate allows the user to choose its function. The gate, as the name signifies, is a dual gate. It may be used as an exclusive-OR gate, or it may be used as an exclusive-NOR gate. Figure 2.29 shows the pin-out and symbolic representation of the 74135. Here we see by the truth table that pin C is the user control pin. For example, if we wish to use pins 1, 2, and 3 as an exclusive-OR gate, then pin 4 will be held low. If, on the other hand, we wish to use pins 1, 2, and 3 as an exclusive-NOR, pin 4 will be held at a logic 1 (high). Of what value is a device such as the 74135?

- Fully Compatible with Most TTL and TTL ,MSI Circuits
- Fully Schottky Clamping Reduces Delay Times .... 8 ns Typical
- Can Operate as Exclusive-OR Gate (C Input Low) or as Exclusive-NOR Gate (C Input High)

J or N Dual-In-Line or W Flat Package (Top View)

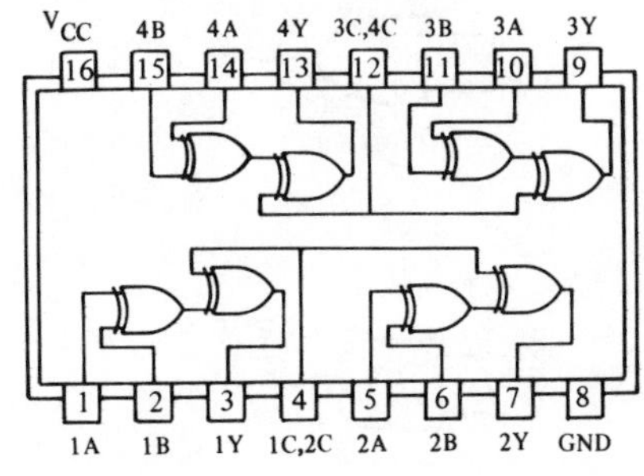

Function Table

Inputs			Output
A	B	C	Y
L	L	L	L
L	H	L	H
H	L	L	H
H	H	L	L
L	L	H	H
L	H	H	L
H	L	H	L
H	H	H	H

H = high level, L = low level

Positive Logic : $Y = (A \oplus B) \oplus C = \overline{A}\,\overline{B}C + \overline{A}B\overline{C} + A\overline{B}\,\overline{C} + ABC$

Schematics of inputs and outputs

Equivalent of each Input

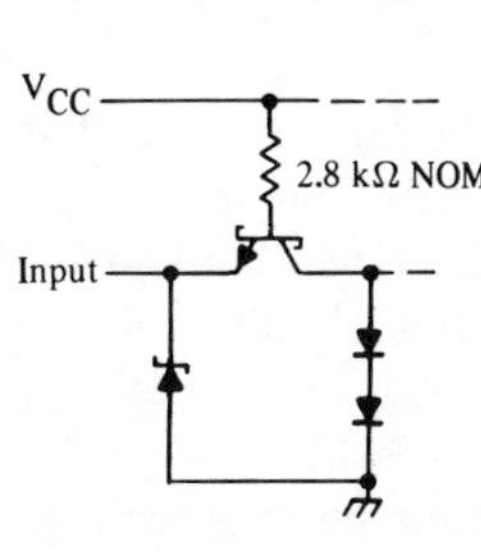

Typical of all Outputs

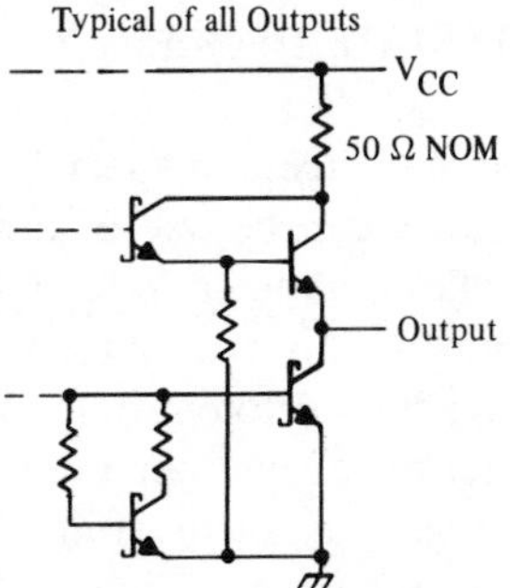

Absolute maximum ratings over operating free-air temperature range (unless otherwise noted)

Supply voltage, V_{CC} (see Note 1)	. . . .	7 V
Input voltage	. . . .	5.5 V
Operating free-air temperature range:	SN54S135	−55° C to 125° C
	SN74S135	0° C to 70° C
Storage temperature range	. . . .	−65° C to 150° C

Note 1 : Voltage values are with respect to network ground terminal.

Figure 2.29 SN74S135 Quadruple Exclusive-OR/NOR Gates

2.15 REVIEW OF BASIC GATES

Figure 2.30 is a review of the various gate configurations with logic expressions and device numbers.

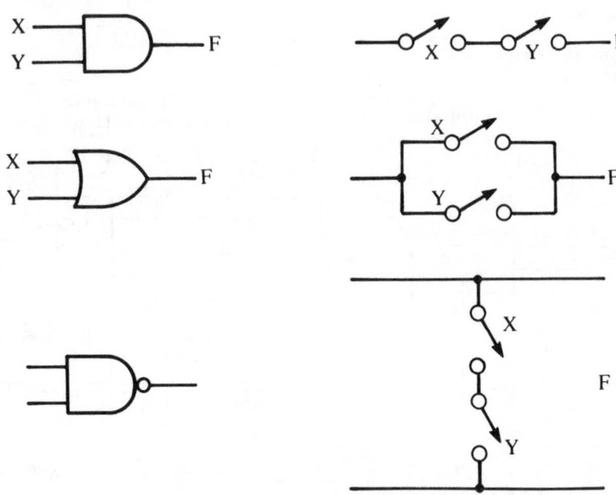

Figure 2.30 Switch and Gate Circuits

2.16 TRI-STATE DEVICES

A tri-state device is a special device which has three states. That is, the output level can be one of three states. At first, this may seem strange when talking about a logic device. Figure 2.31 will explain the unique operation of this type circuit. Notice that this circuit looks similar to the conventional totem-pole inverter. One major difference, however, is the inclusion of another input line (G). If G is high (logic 1) the circuit operates as an inverter. That is, a low at point A will produce a low at Y. When the G is held high, this forces the second emitter of Q_1 low as well as the base of Q_4. Q_4 is then off. Since the collector of Q_2 is also low, Q_2 is off and so is Q_3. With both Q_3 and Q_4 in the off state, point Y is a high-impedance (Hi-Z) state.

One circuit configuration where this type circuit is used is shown in figure 2.32. Here, we see two tri-state gates whose outputs are tied to a common line. By selecting via the G control line, we can switch a logic signal from A or from B to the common line. This permits us to do two things. First, the selection or routing of a signal is achieved. Second, we have successfully isolated the unwanted gate from the sys-

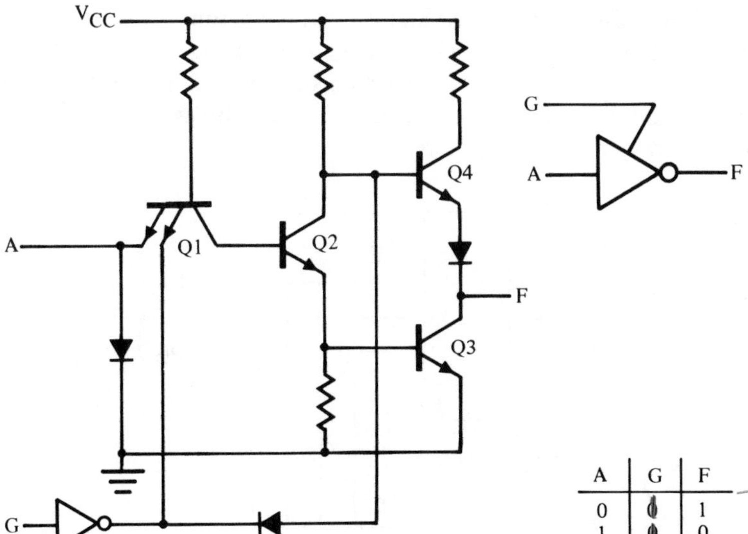

(a)

A	G	F
0	0	1
1	0	0
X	1	*

*High - Impedance

(b)

Figure 2.31 Tristate Gate

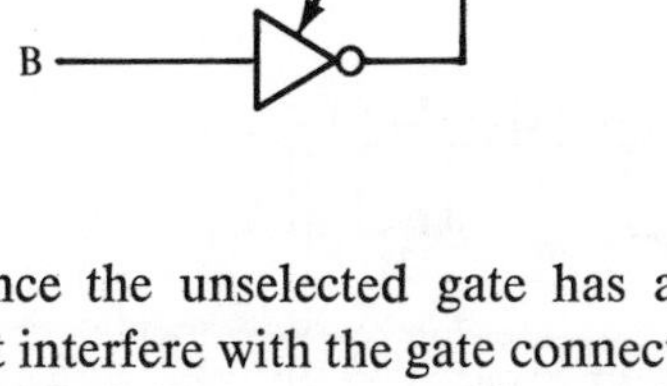

Figure 2.32 Tristate on Bus

tem. Since the unselected gate has a high-impedance output level, it does not interfere with the gate connected to the system.

Figure 2.33 is an expansion of the idea shown in figure 2.32. Suppose we have two sets of data, or we are receiving data from two separate sources. If we only wish to transmit one set of this data on a data bus to the rest of the system, the tri-state device can again be used. By using a single logic level at G, either tri-state device A or tri-state device B can be put in the Hi-Z state. If for example, G is a logic 1, A will be in the high-impedance state. This means that device B is allowed access to the bus. (Each device, A and B, contains four tri-state inverter gates.) So the data on the bus are then the data D_4, D_5, D_6, D_7. The term *bus* is used here to denote a group of signals that are usually examined together. Examples of busses are such things as multibit data words, address lines to computers, etc.

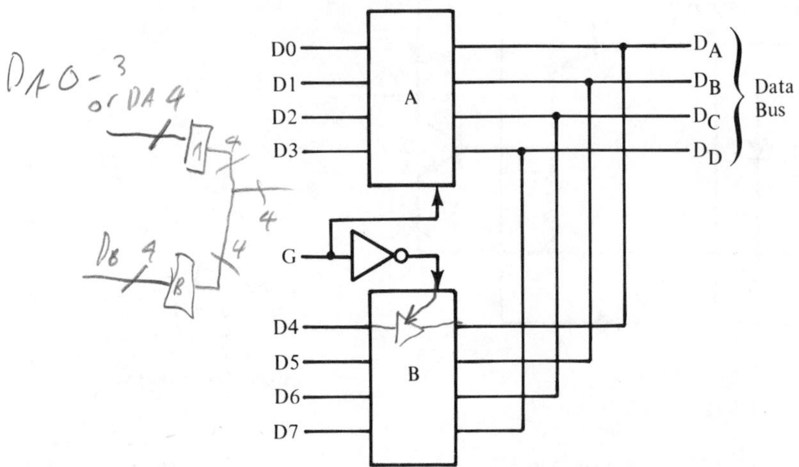

Figure 2.33 Tristate Data System

Some Real Tri-state Devices

Figures 2.34 and 2.35 show two quad gates that are tri-state. The difference in these two devices is that one is disabled by a high and the other is disabled by a logic low.

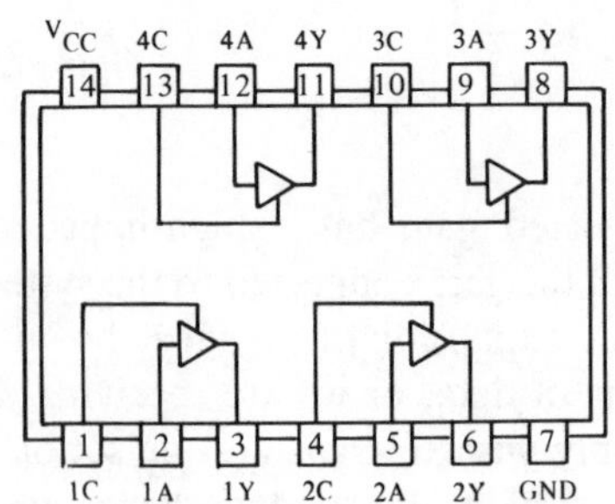

Figure 2.34 The 74426 Package Configuration

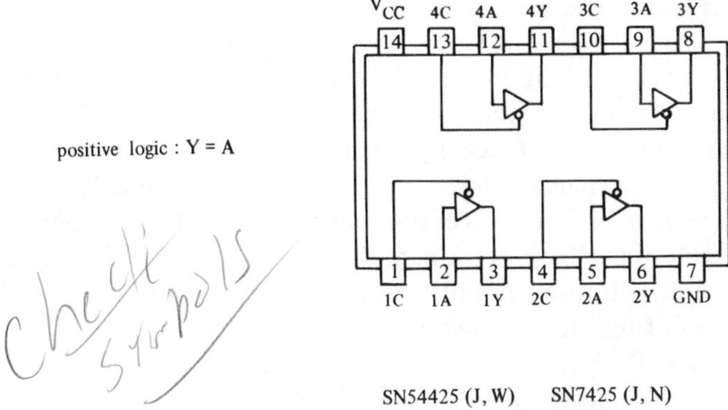

Figure 2.35 Three-State Outputs Active-High Enabling Quad Gates

Figure 2.36 shows two other types of hex-bus drivers. In figure 2.36(a) we see a device which has six separate bus drivers. Four are enabled by one gate and two by a separate gate. Figure 2.36(b) shows an arrangement where all six bus drivers are enabled at one time. Here, however, it takes a logic low at both pins 1 and 15.

As you can see, the tri-state bus driver is a very important tool for the logic circuit designer. It is good practice to become as familiar with the operation of as many of these devices as you can.

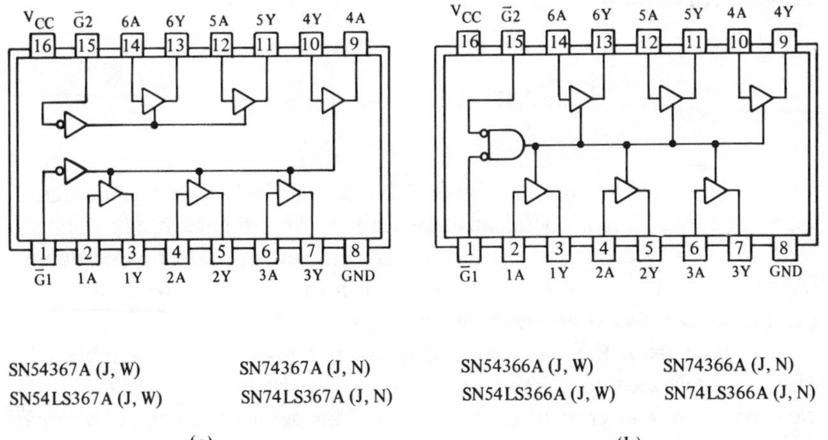

Figure 2.36 Hex Bus Drivers. **(a)** Noninverted Data Outputs, Four-Line and Two-Line Enable Inputs, Three-State Outputs. **(b)** Inverted Data Outputs, Gated Enable Inputs, Three-State Outputs

Tri-state Bus Transceivers

The next device we will look at is also similar to the tri-state bus drivers. One important feature of these new devices is their ability not only to select a bus, but to decide which direction data are traveling on the bus. Thus, the name *transceiver.* Figure 2.37 represents one such gate arrangement. This gate has two buffers per unit. Each buffer has its own enable line. If we say a high on the enable line will enable the gate and a low will disable (put output in a Hi-Z state), we can show the operation of the circuit. Figure 2.37(a) has a logic 1 (H_1) at enable E1 and a logic 0 (L_0) at enable E2. This will put gate 2 in the Hi-Z state. The circuit will be enabled in the X to Y direction. Any signal at X will proceed through gate 1 to Y. The signal cannot travel backward through gate 2. Likewise, any signal on Y cannot travel backward through gate 1, nor can it arrive at X since gate 2 is disconnected (by a H1-Z) from the point X. To move data from point Y to point X simply entails changing the logic levels at E1 and E2. It should be obvious that E1 and E2 cannot be held at the same logic level.

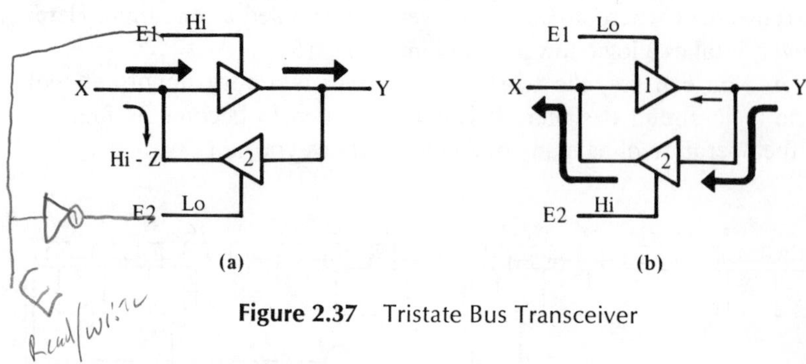

Figure 2.37 Tristate Bus Transceiver

A practical version of figure 2.37 is the 74245 shown in figure 2.38. The 74245 is an octal bus transceiver. This means it can connect eight lines at once. It is very useful for transferring data in microcomputers since those data are usually in the form of an eight-bit word. So the 74245 can access an eight-bit bus.

Notice from the truth table that we not only have an enable line ($\bar{E}$), we have a direction line (Dir). Through the use of these two lines we have complete control of the device. The device is enabled when E is a low. Data move from A to B or B to A depending on the level of the signal at pin 1 (Dir). Figure 2.39 illustrates a bus system of a microcomputer using the 74245. As we know, a computer needs to be able to either send data to memory or retrieve data from memory. This

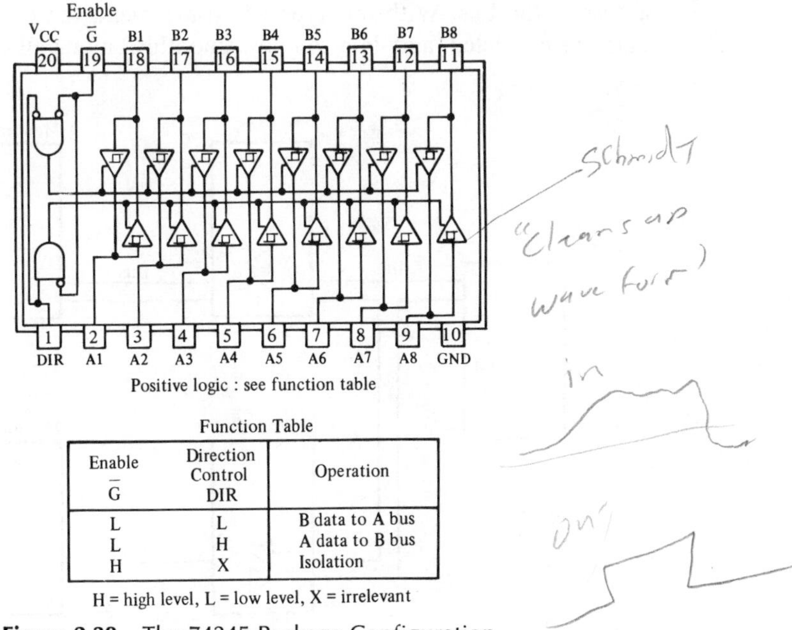

Function Table

Enable $\overline{G}$	Direction Control DIR	Operation
L	L	B data to A bus
L	H	A data to B bus
H	X	Isolation

H = high level, L = low level, X = irrelevant

Figure 2.38 The 74245 Package Configuration

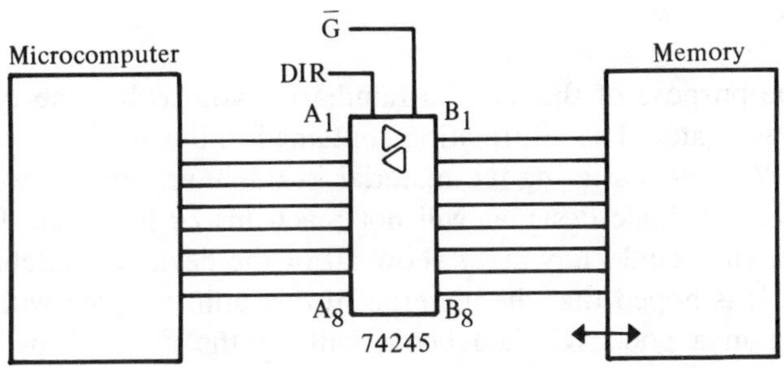

Figure 2.39 Bus System Using 74245

type of system will allow us to have data in the bus only when requested and to have the data going in the correct direction. Another system using the transceiver concept is shown in figure 2.40. Here we see a single microcomputer accessing two separate memory devices. By proper enabling and disabling of the two 74245's, not only can the direction of data flow be controlled, but we can control which memory device has access to the bus. With this type of arrangement, the computer has its data bus protected, and data can flow smoothly among all devices.

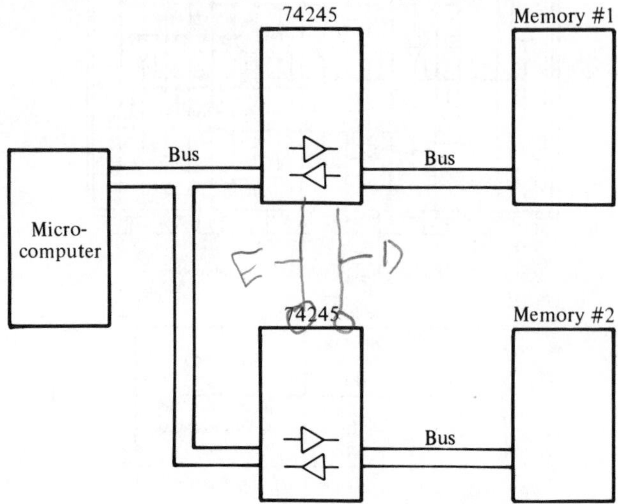

Figure 2.40 Bus System for Two Memory Devices

SUMMARY

It was the purpose of this unit to familiarize you with some types and uses of logic gates. The information contained in this unit referred to IC devices. Without mastering the material in this unit, anyone wishing to be a successful logic designer will not reach his or her goal. This unit could not (nor could any text) show all of the devices available to the designer. It is hoped that the material in this unit, coupled with the information in a good IC data book, will be the cornerstone of your learning.

QUESTIONS AND PROBLEMS

2.1 What is meant by true-false logic?
2.2 What is meant by positive logic?
2.3 What is meant by negative logic?
2.4 What are the logic functions of a gate?
2.5 What is the maximum fan-out of a 7405 device?
2.6 How many 7402 inputs can be safely driven by one 7405? How is this proved?
2.7 What is the definition of a unit load?
2.8 Exactly, what does the term *fan-in* mean?
2.9 In the circuit below, what is the difference in drive requirement (if any) of the device #1?

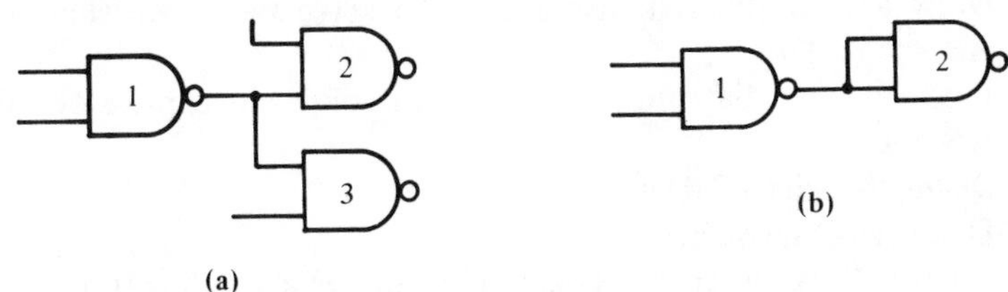

2.10 Define propagation delay.
2.11 In the circuit below, draw the output pulse and give its pulse width. The input pulse is 50 nanoseconds wide.

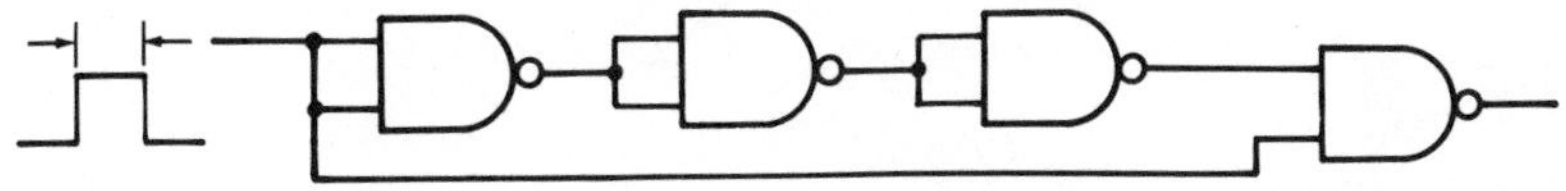

2.12 Name two advantages of open-collector gates.
2.13 What is a possible disadvantage of the open-collector gate?
2.14 What is the function of a pull-up resistor?
2.15 What constitutes the input circuitry of TTL gates?
2.16 Draw the circuitry (using exclusive-OR gates) to test a batch of 7400 NAND gates. Show the test conditions via an LED output arrangement.
2.17 For a 7400 (single gate) give the minimum and maximum values for V_{IH} and V_{IL} and typical high and low output-voltage levels.
2.18 Draw a switch representation for a NAND gate.
2.19 Draw a switch representation for a NOR gate.

2.20 Draw a circuit using gates to simulate an automobile ignition system. It must show oil-pressure, ammeter (light only), and seat belts and doors. Note all lights must come on if engine is not running.

2.21 Prove that the circuit shown below will serve as a latch or bounceless switch.

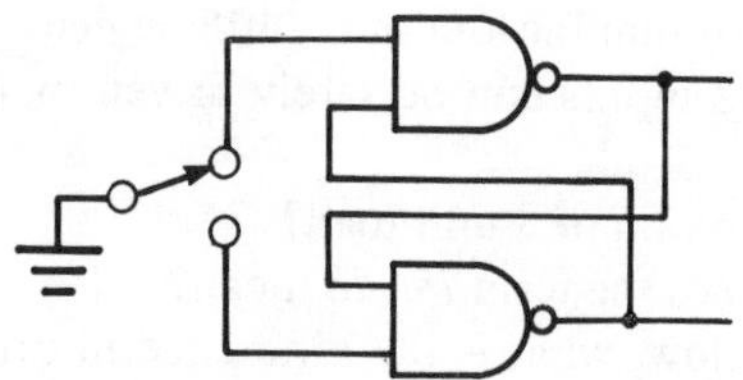

2.22 Draw a logic diagram using gates to solve the expression: $F = AB + C + DE$.

2.23 Draw a logic diagram using gates to solve the expression: $F = \overline{AB} + \overline{C} + DE$.

2.24 Define the wire-AND concept.

2.25 Define tri-state devices.

2.26 What is the difference between a bus driver and a bus transceiver?

Section II

Flip-Flops

3

OBJECTIVE: To familiarize the student with the characteristics and the operation of various types of flip-flops.

Introduction: The flip-flop (FF), or bistable, is a circuit with two stable states. It will remain in one of its stable states until a signal from outside the circuit causes the FF to change its state.

3.1 THE GENERAL FLIP-FLOP

Figure 3.1A shows the general symbol for a flip-flop. The outputs of the flip-flop circuit are labeled Q and $\bar{Q}$ (not Q). Since the bistable is to be used with binary or digital arithmetic, these outputs are sometimes called 1 and 0. Figure 3.1B is a truth table depicting the operating states of the general flip-flop. For figure 3.1A the connections to the flip-flop are as follows:

C: this is the trigger or clock input to the FF. This is the signal that will cause the device to change states.

S: this is the *set* input. A proper signal here will cause the Q output to become a logic level of 1. This input is sometimes called the preset.

R: this is the *reset* input. A proper signal here will cause the Q output to become a logic level of 0. This input is sometimes called the clear input.

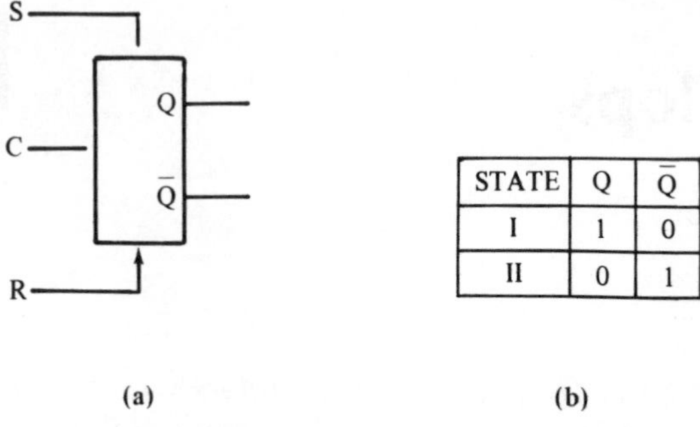

STATE	Q	$\overline{Q}$
I	1	0
II	0	1

(a) (b)

Figure 3.1 General Flip-Flop

3.2 TYPES OF FLIP-FLOPS

There are four basic types of flip-flops; they are the D, T, R-S, and J-K. Flip-flops may have synchronous inputs, asynchronous inputs, or both. Asynchronous inputs are those inputs that can affect the output state of the flip-flop independent of a clock or timing pulse. Synchronous inputs do not have direct control of the output. They can only affect the output in conjunction with a clock pulse.

D-Type Flip-Flops

The D-type flip-flop has a single data input (D) and a clock input. It may also have preset and clear inputs, which are asynchronous. A 0 state at the preset input will cause Q to go to logic state 1, a 0 at the clear input will cause Q to go to 0. The D-type flip-flop is shown in Figure 3.2A and the truth table appears in Figure 3.2B. The basic D-type flip-flop is often used as a latch, transferring the logic state at the data input (D) to the output (Q) when the clock pulse goes high, to a 1.

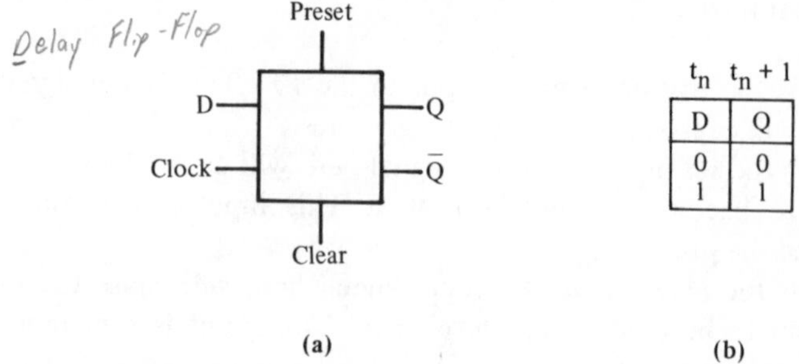

t_n	$t_n + 1$
D	Q
0	0
1	1

(a) (b)

Figure 3.2 **(a)** D-Type Flip-Flop. **(b)** Truth Table for D-Type Flip-Flop

The Texas Instruments SN7474 D-type flip-flop is a positive-edge triggered device. As the clock pulse goes from logic 0 to logic 1, the data input at D is transferred to Q. The pin connections for the SN7474 are given in Figure 3.3. Using Figure 3.3, verify the truth table for the D-type flip-flop.

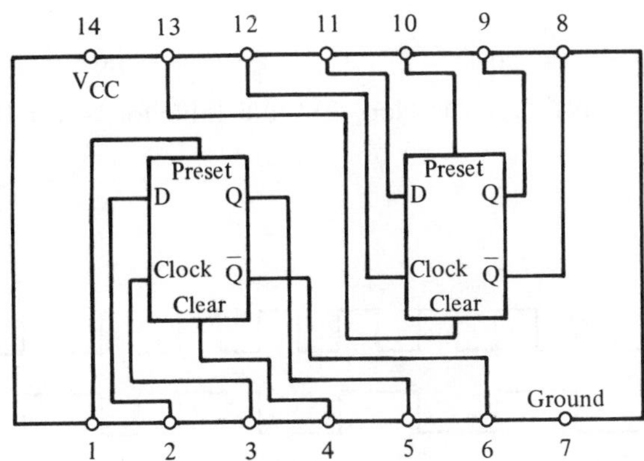

Figure 3.3 Pin Connections for SN7474

T-Type Flip-Flop

The T-type or toggle flip-flop also has one data input and a clock input. It also may have asynchronous inputs such as preset and clear. As shown in the truth table in Figure 3.4B, the T flip-flop performs a simple operation, toggling or changing states. When the T input is logic 0 prior to a clock pulse, no change will occur at Q after clocking. However, if T is logic 1 and the device is clocked, the output will change state regardless of Q's state before clocking. The T flip-flop differs from the D flip-flop in that when its data input, the T input, is a logic level 1, the Q output will change states with each input clock pulse.

The T-type flip-flop is often used in binary-coded counters because of its inherent divide-by-two capability. When the clock pulse is applied, the output changes state every cycle (if T is logic 1), thus completing one cycle of output change for every two input cycles. This is illustrated in Figure 3.5.

The T-type flip-flop is not generally available as such, but rather is obtained by modifying one of the other types of flip-flops as is shown in Figure 3.6.

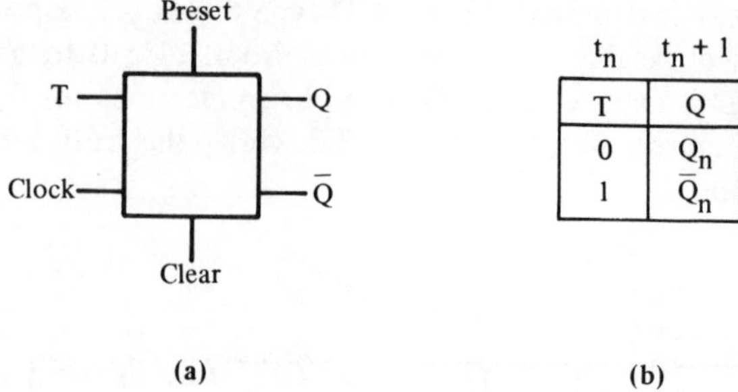

t_n	$t_n + 1$
T	Q
0	Q_n
1	$\overline{Q}_n$

Figure 3.4 (**a**) T-Type Flip-Flop. (**b**) Truth Table for T-Type Flip-Flop

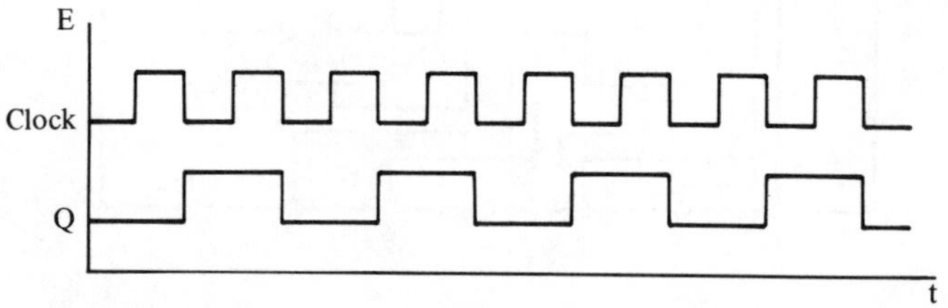

Figure 3.5 Waveshapes for T Flip-Flop

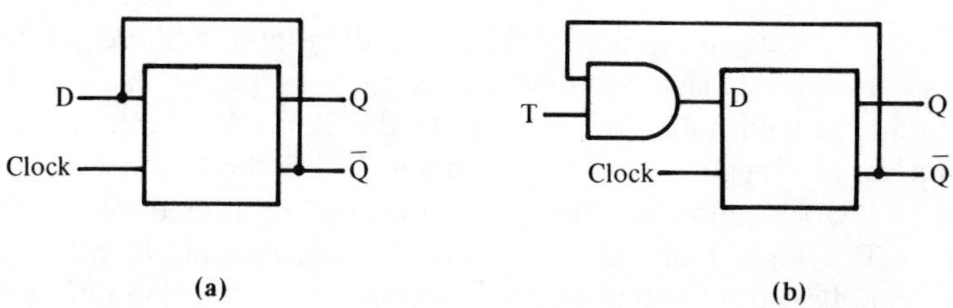

Figure 3.6 Converting D to T Flip-Flop. (**a**) Toggle Only. (**b**) With a T Data Input

R-S Flip-Flop

The R-S flip-flop has two synchronous inputs, R and S, and a clock input; preset and clear may also be provided. The R-S flip-flop and its truth table are shown in Figure 3.7. Note that if both inputs (R and S)

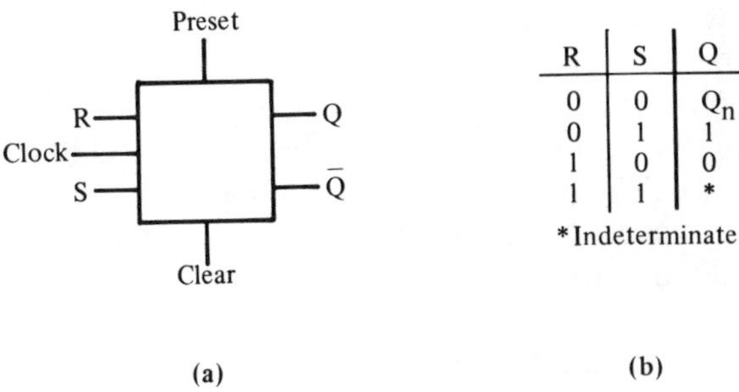

R	S	Q
0	0	Q_n
0	1	1
1	0	0
1	1	*

Figure 3.7 The R-S Flip-Flop

are at logic 0, the output will not change after clocking. However, if both inputs are at logic 1 the output is indeterminate, which means this state must be avoided. If Q is logic 0 and S is logic 0, Q will stay at logic 0 regardless of the state of R. Also, if Q is logic 1 and R is logic 0, regardless of the state of S, Q will remain at logic 1 after clocking. The R-S flip-flop can be constructed from NAND gates to perform its function with the use of a clock pulse. Figure 3.8 shows two NAND gates connected as an R-S flip-flop. If initially Q is at a logic 1 and $\bar{Q}$ is at a logic 0, then pin 3 is at logic 1 and pin 2 is at logic 0. Since any zero to the input of a NAND gate creates a 1 at its output this will hold Q at logic 1 regardless of the level of input S. If both inputs S and R are open (or at logic 1 for 7400 TTL logic) then the flip-flop will have a 1 at Q and a 0 at $\bar{Q}$. Grounding input R (logic 0) will cause the $\bar{Q}$ output to go to a logic 1. This will put a logic 1 at both pins 1 and 2. A 1 and 1 will create a 0 out of the NAND gate causing Q to be 0 and $\bar{Q}$ to be 1. By swapping the input ground (logic 0) the R-S flip-flop can be made to change states, setting Q to logic 1 when the S input is 0, and resetting Q to logic 0 when the R input is 0. Grounding both the R and S inputs will cause both Q and $\bar{Q}$ to go to logic 1. This state of course is not allowed for true flip-flop operation.

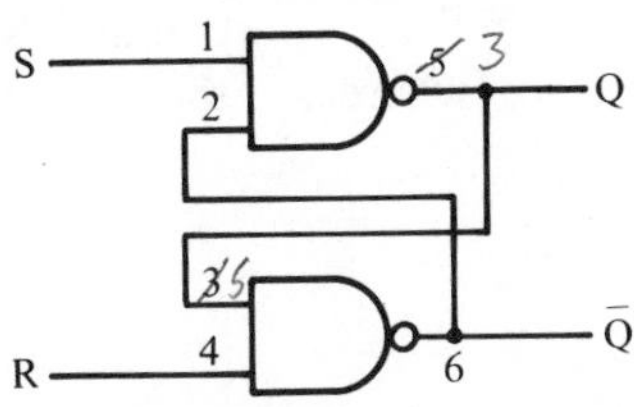

Figure 3.8 NAND Gate R-S Flip-Flop

NAND-Gate-Clocked R-S Flip-Flop

Figure 3.9 shows a further refinement of the NAND gate R-S flip-flop. Here the circuit of figure 3.9 has been coupled to an additional pair of gates to form a NAND-gate-clocked R-S flip-flop.

How does the NAND gate R-S FF differ from the general R-S FF as previously discussed.

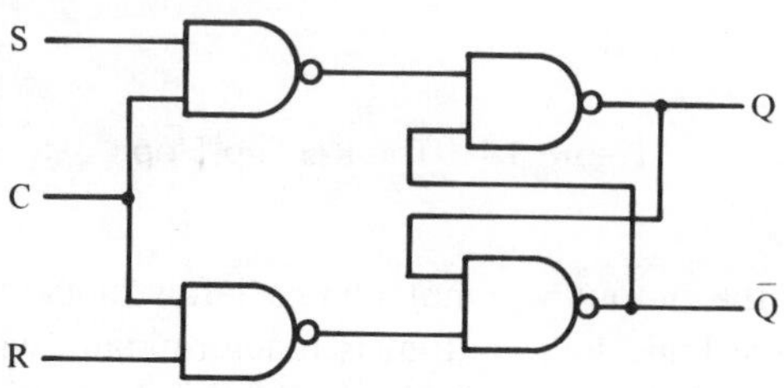

Figure 3.9 NAND Gate R-S Flip-Flop

J-K Flip-Flop

The J-K flip-flop has two data inputs and a clocking input; it also may have preset and clear capabilities. Many J-K flip-flops also have internal gates on the inputs to provide multiple J-K inputs. The J-K and its truth table are shown in Figure 3.10. Note that all four possible input states are defined. If both J and K are 1 the flip-flop operates similar to a T-type flip-flop, toggling on each clock pulse. The J-K inputs may be thought of as coding inputs. These logic levels will determine the operation of the flip-flop when it is clocked. It should be understood that the J-K inputs will not in themselves cause any change in the output states of Q and $\overline{Q}$. The set (or preset) and clear inputs to the J-K will override any coding at the J-K inputs. What this means is that to set the Q output to 0, a pulse to the clear input is all that is needed. For figure 3.10B, the notation t_n means prior to the clock pulse and t_{n+1} means immediately after the clock pulse. A particular J-K may operate from a negative-going clock pulse, whereas others require a positive-going clock pulse.

Master-Slave J-K

Many J-K flip-flops are of the master-slave type. The master-slave type of flip-flop is basically two latches connected serially. The first latch is

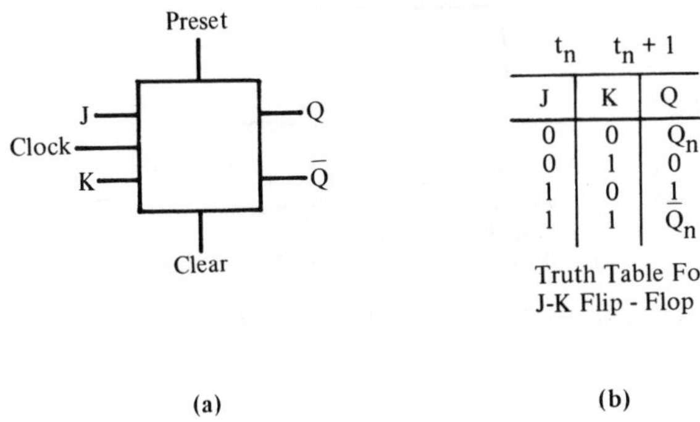

t_n J	K	t_{n+1} Q
0	0	Q_n
0	1	0
1	0	1
1	1	$\bar{Q}_n$

Truth Table For J-K Flip - Flop

(b)

Figure 3.10 **(a)** J-K Flip-Flop. **(b)** Truth Table for J-K Flip-Flop

called the master and the second is the slave. The normal action during clocking is shown in figure 3.11.

Figure 3.12 shows a conventional connection for a master-slave J-K flip-flop. As you can see, the outputs of this device are the outputs of triple-input NAND gates. Any 0 to the input of a NAND gate will put its output to a logic 1. Therefore, placing a 0 or grounding the preset, for example, will cause the Q output to go to a logic 1. If the J and K inputs are at logic 1 (the toggle made for the flip-flop) and Q is not logic 1, the sequence of events is as follows:

1. The left-hand, four-input AND gate will have 1s on three of its inputs. The preset is open, or high, also.
2. A clock pulse will put the final input to a logic 1, causing the output to become a logic 1.
3. This will cause the left-hand NOR gate to have a logic 0 at its output, putting the left-hand clock transistor to the off condition.
4. The left-hand NAND gate now has two 1s at its inputs.
5. At this time, the right-hand NAND gate has two logic 1s (clear and Q).
6. The clock pulse is now back to 0, turning the right-hand transistor on, and causing its collector to go to 0.
7. This 0 is applied to the right-hand NAND gate and causes $\bar{Q}$ to become a 1.
8. This 1 couples to the left-hand NAND gate, causing Q to become a 0.

This action coupled with the explanation of figure 3.11 should serve to aid in the understanding of the master-slave J-K flip-flop.

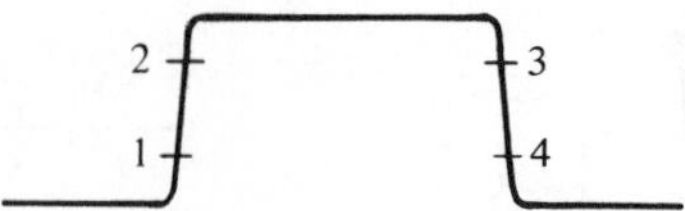

1. Isolate slave from master
2. Enable data inputs to master
3. Disable data inputs
4. Transfer data from master to slave

Figure 3.11 Master-Slave Action as Affected by Clock Pulse

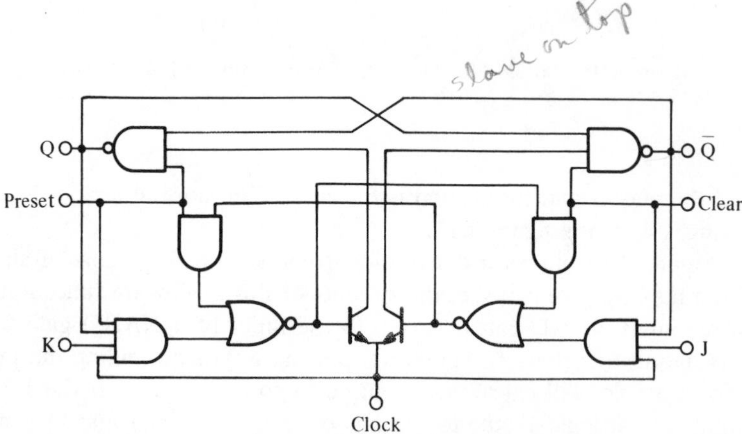

Figure 3.12 Master-Slave J-K Flip-Flop

An important feature of this type of clocking is that the data inputs are never directly connected to the outputs at any time during clocking. This provides total isolation of outputs from data inputs. An important fact to remember is that the *data inputs should never be changed while the clock pulse is high,* as this will cause erratic outputs. Also, master-slave flip-flops are not susceptible to racing. Racing occurs when two or more inputs to a gate are applied almost instantaneously, as to the NAND gate in Figure 3.13.

At times t_1 and t_3 the output condition is defined, but at time t_2, A and B are neither high nor low so the output condition at C is indeterminate. The dip at C might be large enough to cause false triggering.

3.3 APPLICATIONS OF THE J-K FLIP-FLOP

The J-K flip-flop is the most commonly used and most versatile flip-flop. It can easily be adapted to perform the operations of either a D- or

T-type as shown in figure 3.14. In future units on counters and shift registers, we will see the use of the J-K and a J-K modified as both the D- and T-type flip-flop.

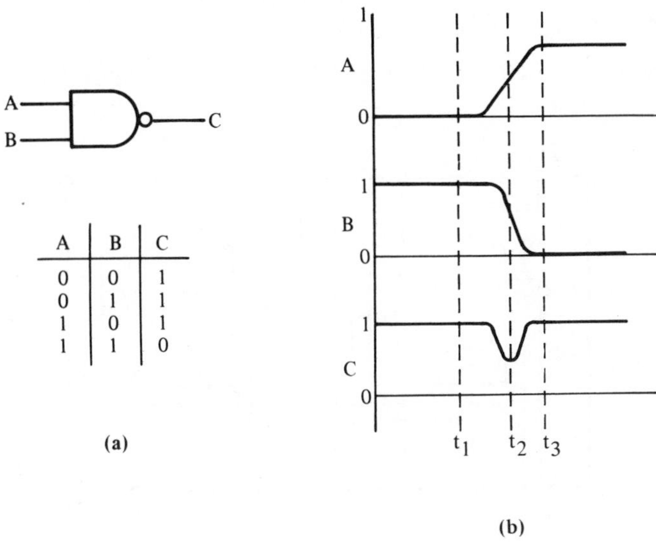

A	B	C
0	0	1
0	1	1
1	0	1
1	1	0

Figure 3.13 (**a**) NAND Gate and Truth Table. (**b**) Waveforms Showing Race Problems

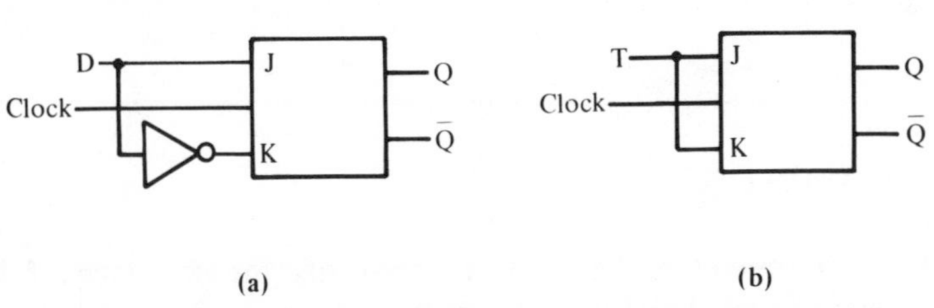

Figure 3.14 Converting J-K Flip-Flop to D- or T-Type. (**a**) J-K to D. (**b**) J-K to T.

3.4 SUMMARY

This unit has taken a look at one of the most versatile digital circuit building blocks. The flip-flop or bistable multivibrator will play an important part in future circuits. It can be used for counting, storing, and timing. The devices we have studied were in the general form.

EXERCISES

1. Using the Texas Instruments, SN7476 or equivalent dual J-K flip-flop, verify the truth table for the J-K type. Also after adapting the J-K to perform the functions of a T-type as shown in Figure 3.14B, verify the truth table for the T-type.

The pin connections for the SN7476 are given in figure 3.15.

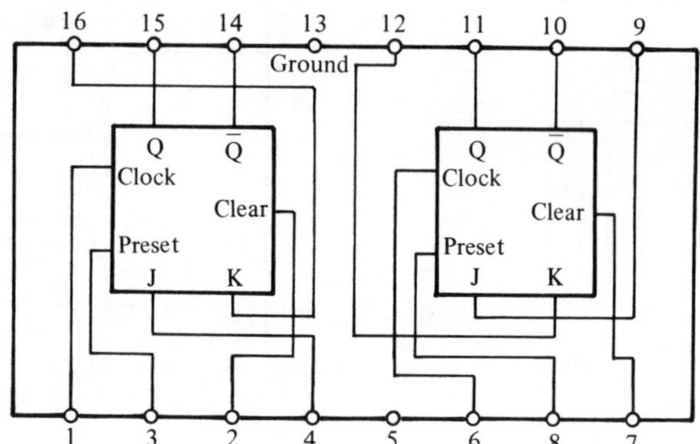

Figure 3.15 Pin Connections for SN7476

2. Place a logic 1 on the clock and change the levels of the J-K inputs. Does this verify the material that was presented in the unit?

STUDY TOPIC

Construct the circuit of figure 3.12 and verify the sequence of logic events for the operation of the clock pulse as shown in figure 3.11.

QUESTIONS AND PROBLEMS

3.1 Match the following:

(a) D Type Flip-Flop ________

(b) R-S Flip-Flop NOR ________

(c) Clocked R-S Flip-Flop ____

(d) R-S Flip-Flop NAND ______

I

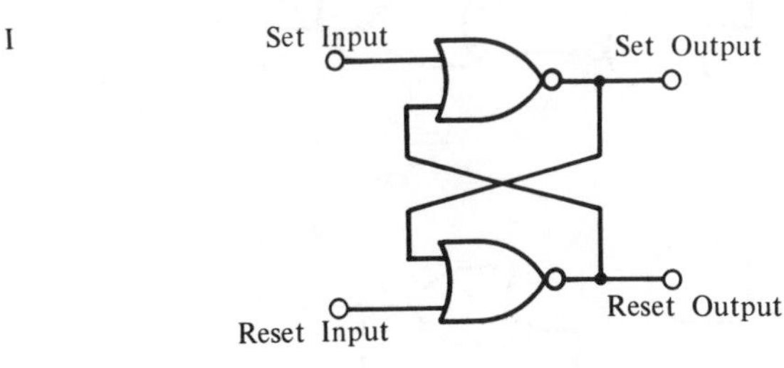

II

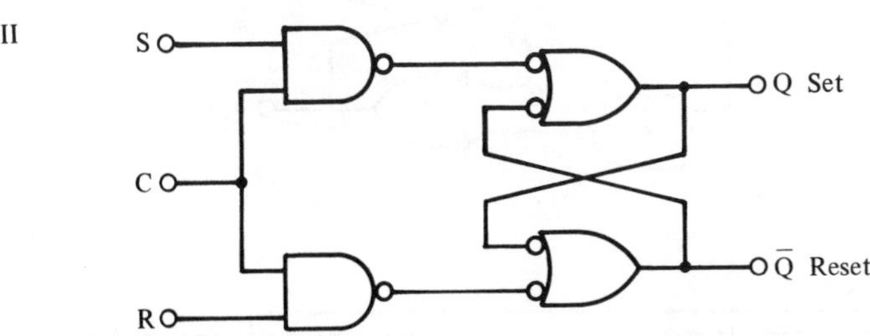

III

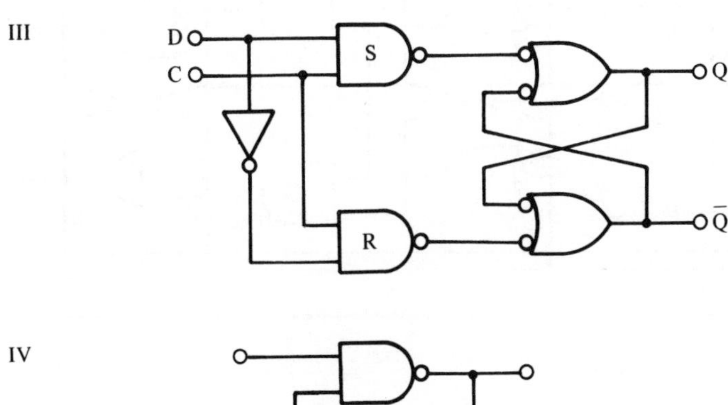

IV

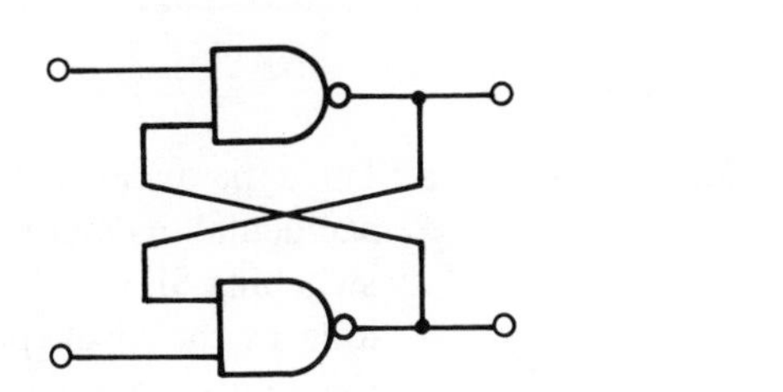

3.2 Which of the following is the correct circuit diagram for a J-K flip-flop?

(a)

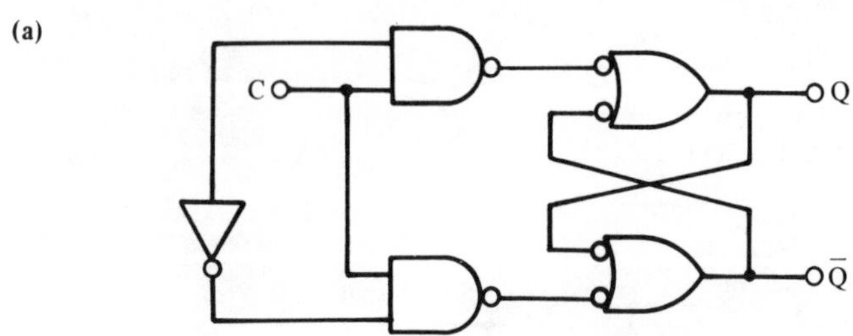

(b)

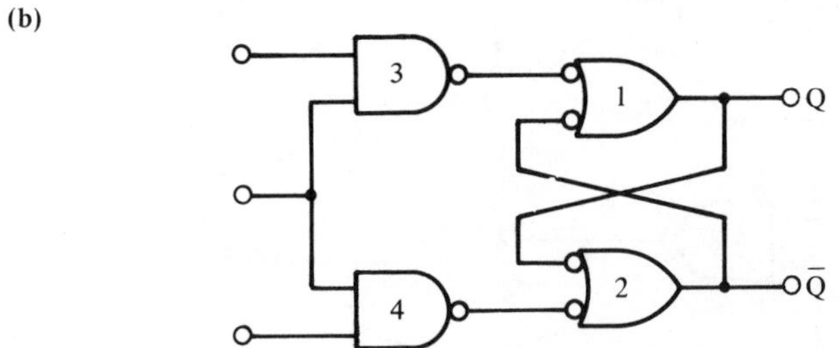

(c)

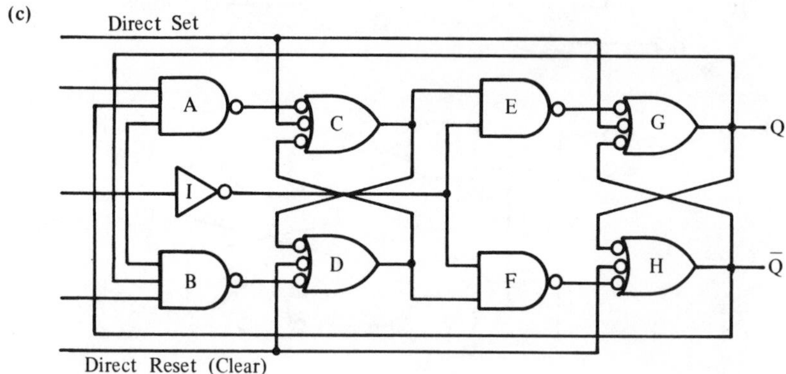

3.3 Match the following:

1. D flip-flop

a. This type avoids the possibility of an accidental indeterminate state when switching SET and RESET inputs as long as the clock pulse occurs after the inputs are properly enabled.

2. R-S flip-flop — b. This type eliminates the possibility of an indeterminate state occurring by making sure the R and S inputs are always complements of each other.

3. Clock R-S flip-flop — c. This type is limited in its use because of problems that arise due to the timing of input pulses or levels.

3.4 Whatever state is present at the set output is also the state of the flip-flop.
True ______
False ______

3.5 Which of the following truth tables is the truth table for a J-K flip-flop?

(a)

CONDITION BEFORE CLOCK PULSE				CONDITION AFTER CLOCK PULSE	
Q	$\overline{Q}$	J	K	Q	$\overline{Q}$
0	1	0	0	0	1
0	1	1	0	1	1
0	1	0	1	0	1
0	1	1	1	0	0
1	0	0	0	1	1
1	0	1	0	0	0
1	0	0	1	1	1
1	0	1	1	0	0

(b)

INITIAL OUTPUT		INPUT		NEW OUTPUT	
S	R	S	R	S	R
0	1	0	0	INDETERMINATE	
0	1	0	1	1	0
0	1	1	0	0	1
0	1	1	1	0	1
1	0	0	0	INDETERMINATE	
1	0	0	1	1	0
1	0	1	0	0	1
1	0	1	1	1	0

(c)

CONDITION BEFORE CLOCK PULSE				CONDITION AFTER CLOCK PULSE	
Q	$\overline{Q}$	J	K	Q	$\overline{Q}$
0	1	0	0	0	1
0	1	1	0	0	1
0	1	0	1	0	1
0	1	1	1	0	1
1	0	0	0	1	0
1	0	1	0	1	0
1	0	0	1	0	1
1	0	1	1	0	1

3.6 What is meant by positive-edge triggering?

3.7 What is the primary function of a D-type flip-flop?

3.8 Explain the difference between synchronous and asynchronous inputs.

3.9 Of the various types of flip-flops discussed, one is said to be a divide-by-two flip-flop. Which is it?

3.10 Name two advantages of the J-K master-slave flip-flop.

3.11 Define racing and give an example of how this might affect circuit operation.

3.12 A trouble indicator circuit has both buzzer and warning light to indicate a malfunction. The technician may disable the buzzer by a switch, but the light indication must stay on until the malfunction is corrected. Draw such a system using gates and flip-flops.

3.13 Using the TTL data book, determine the number of 7400 NAND Gates, the Q output of a 74LS175 D-Type flip-flop could drive.

3.14 Explain the operation of a master-slave J-K flip-flop.

Shift Registers

4

OBJECTIVE: To understand the functions and uses of the series of circuits known as shift registers. To show the use of shift registers as storage devices, and to become familiar with a family of single-chip shift-register units.

Introduction: Shift registers are circuits used to transfer data from one location to another. They may be categorized in several ways, such as: serial-input/serial-output; serial-input/parallel-output; parallel-input/serial-output; and parallel-input/parallel-output. Serial shift registers may shift their data to the left or to the right. Shift registers, being storage devices, are constructed of bistable flip-flops.

4.1 SHIFT REGISTERS

The shift register is a series of flip-flops connected so that each bistable stores one bit of data of a data word. The length (size) of the register is given as the number of bits it can store at one time. Thus, a four-bit register is comprised of four flip-flops. These may be J-K flip-flops or they may be D-type flip-flops. Each bit of the data is stored at a location with a binary weight. The bits may be shifted one location at a time to a location within the register. In this manner, a binary word may be shifted into the register one bit at a time, starting with the least-significant bit. This will be shown in a later example.

4.2 CLOCKING AND TIMING OF A SERIAL SHIFT REGISTER

To insure proper operation of the serial shift register, each flip-flop must be clocked at the right time. This is accomplished by having the clock input tied to a common line. This means each FF will clock at the same time. The definition of a serial shift register is as follows: Information enters and exits the register one binary bit at a time. When constructed, the register shown in figure 4.1 will also have its entire contents accessible at one time. The output can be taken from each of the Q outputs of the flip-flops giving an indication of the contents of the register. The register can be clocked and its contents removed one bit at a time from the Q output of the last flip-flop.

Operation of a Serial Shift Register

Using figure 4.1 we shall study the operation of a serial shift register. At the initial or starting conditions, each Q output will be at logic 0.

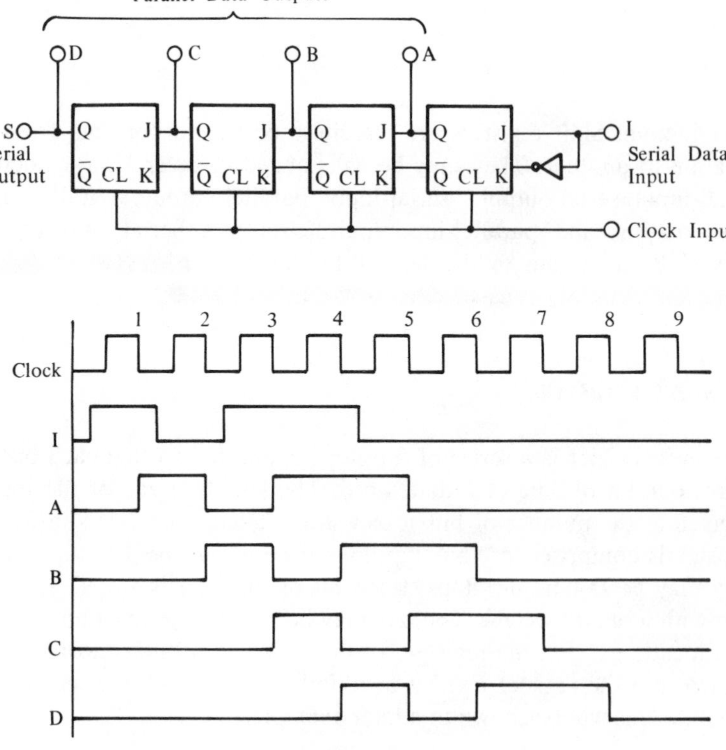

Figure 4.1 Shift-Left Register and Waveforms

This will mean that FF B, FF C, and FF D, have a 0 at their J input and a 1 at their K inputs. A clock pulse then will cause the Q outputs to become 0 (no change in this case). FF A, being a D flip-flop, will simply transfer its input to its Q output on each clock pulse. We will have our flip-flops operate on a negative-going clock pulse. Prior to clock pulse 1, the input will go high (to a logic 1). This will cause FF A to have its Q output go to a logic 1 on clock pulse 1. This in turn causes the J-K input to FF B to have a 1-0. The input pulse returns to 0. Clock pulse 2 now causes FF A to return to 0 and FF B to have its Q output go to 1 (J-K properly coded). The waveforms of figure 4.1 show the appropriate logic levels on each flip-flop for nine clock pulses and several changes of the input signal. Note that the serial output is also the Q output of FF D. Also note that the data input (I) must be changed during the time that the clock pulse is at logic level 0.

4.3 LEAST-SIGNIFICANT BIT

Serial shifting involves shifting one bit at a time into the register. Remember, the first data bit to leave a shift register (serially) is the least-significant bit. For example, the decimal number 13 is written in binary code as 1101 with the most-significant bit to the left and least-significant bit to the right. Since the first bit to leave a register must be the first bit to have entered the register, enter the least-significant bit first.

4.4 CIRCUIT CONSTRUCTION

To understand this particular circuit properly, the student should construct the circuit of figure 4.1. Use two SN7476 dual J-K flip-flops.

Connect the proper inputs and complete Table 4.1. Use LEDs to monitor parallel data outputs A, B, C, and D. Enter the binary number 1101 serially. Remember, D is the least-significant bit.

4.5 PARALLEL SHIFT REGISTERS

While we saw that figure 4.1 was a serial shift register, it was noted that the entire contents of the register could be observed at one time (in parallel). In reality, then, this circuit could have been a serial-in, parallel-out shift register. The major difference between a serial and a parallel shift register is in the manner in which information is entered and retrieved. Figure 4.2 is a circuit diagram of a parallel-in–parallel-out

TABLE 4.1 Truth Table

Clock Pulse	*I*	*A*	*B*	*C*	*D*
0	0	0	0	0	0
1	1				
2	0				
3	1				
4	1				
5	0				
6	0				
7	0				
8	0				
9	0				

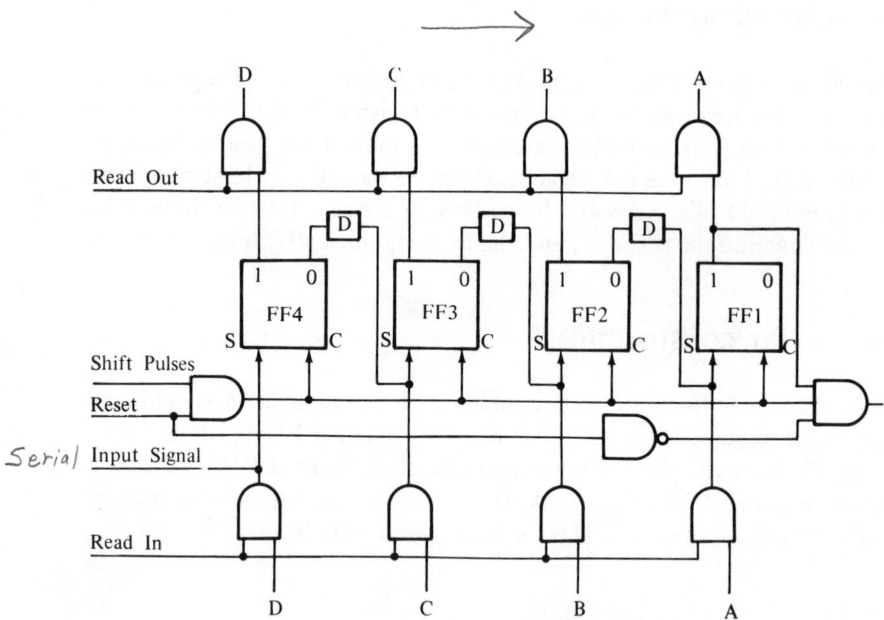

Figure 4.2 Shift Register with Parallel Load or Unload

shift register. To understand the operation of this type of circuit, figure 4.3 shows a portion of the shift register. The individual data bits are entered to the D inputs. When each input has the proper logic level, the enable circuit is given a pulse. The enable pulse along with the AND gate sends the data bit to the J-K inputs of the flip-flop. At the proper time, the flip-flop will be clocked, transferring the data to the Q output. For a parallel-in shift register, the flip-flops will be triggered simultaneously. This will cause each flip-flop to receive the input data at one time. This is in contrast to the serial mode of operation, where a clock pulse is required to enter one bit of data. The parallel-input shift register then enters data at a faster rate. A sixteen-bit serial-in shift register would require sixteen clock pulses to enter data to each flip-flop. The parallel-in register requires only one enable pulse plus one clock pulse to achieve the same results.

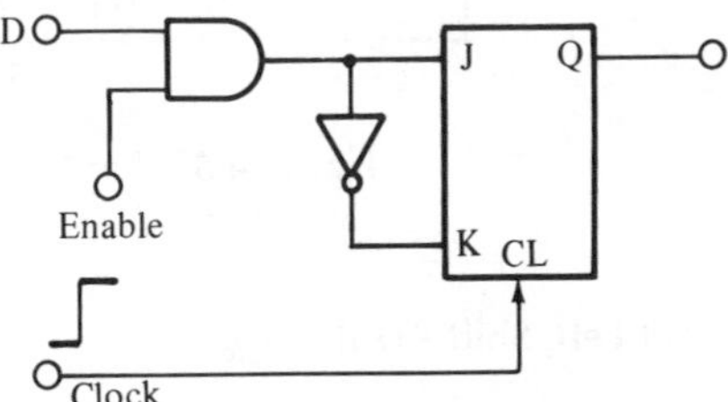

Figure 4.3 Input-Enable Circuitry

A further refinement of the parallel-in–parallel-out shift register is shown in figure 4.4. Here the data out from each flip-flop is inhibited from the output by an AND gate. A signal (output enable) is needed to activate the gate and produce the data out only when required.

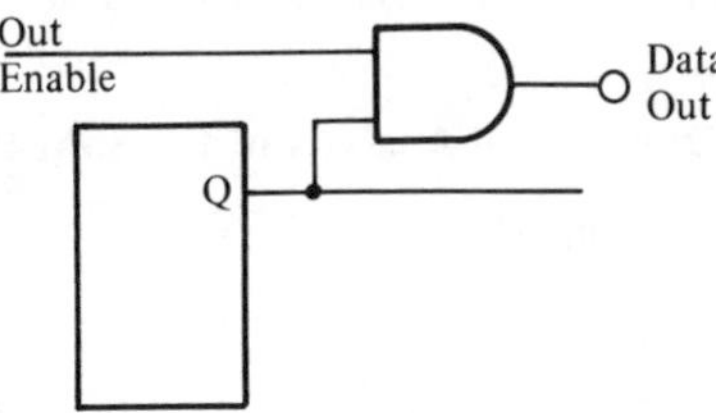

Figure 4.4 Output-Enable Circuitry

4.6 THE IC SHIFT REGISTER

Integrated circuit manufacturers have simplified the task of assembling shift registers for the user. There are several complete shift registers contained in one IC package. We shall examine one of these packages, the 7495 a serial-in, serial-out, parallel-in, parallel-out shift register. Figure 4.5 is the pin connection diagram of a 7495. The device is self-contained in a fourteen-pin package. Remember, to build the shift register of figure 4.1 required two 7476 devices. Here, on one chip, is not only a four-bit shift register, but one capable of both serial and parallel operation.

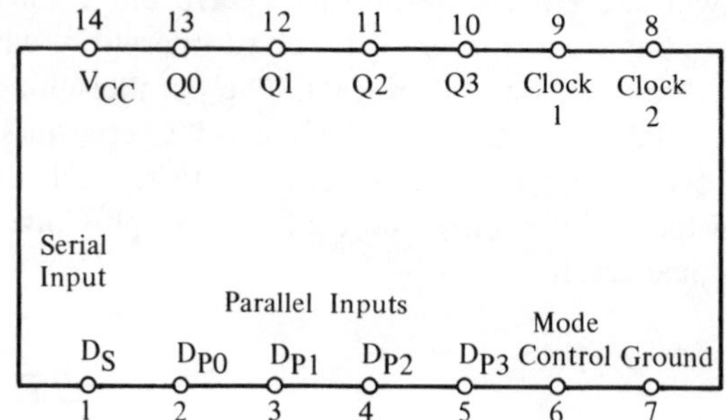

Figure 4.5 Pin Connections for SN7495

Shift Left, Shift Right

The 7495 shift register has an additional feature that needs to be mentioned before we analyze its operation. The 7495 has the capability of serially shifting its data to the left or to the right. This ability is useful for certain arithmetic operations where it is the function of the shift register not only to store the data, but to shift it in one or more bits in either direction or command. A shift register that has this full capability is sometimes referred to as a *universal shift register.*

Operational Analysis of the Shift Register

Parallel

The functional logic diagram for a universal shift register is shown in figure 4.6. We will first explore the parallel-in–parallel-out mode of operation. The data to be entered in parallel are presented at the parallel inputs (pins 2, 3, 4, and 5). The mode control (pin 6) is then switched from a low (0) to a high (1). This will enable all of the B AND gates and place the proper coding on the J-K inputs of each flip-flop. Clocking pin 8 (clock 2) will cause the clock pulse to appear at each clock input and enable the flip-flops.

Example Placing a logic 1 at pin 2 will cause AND gate B_1 to have a 1 at one of its inputs. Holding the mode control (pin 6) high will place a logic 1 at one input of B_5 and place the second one at the other input

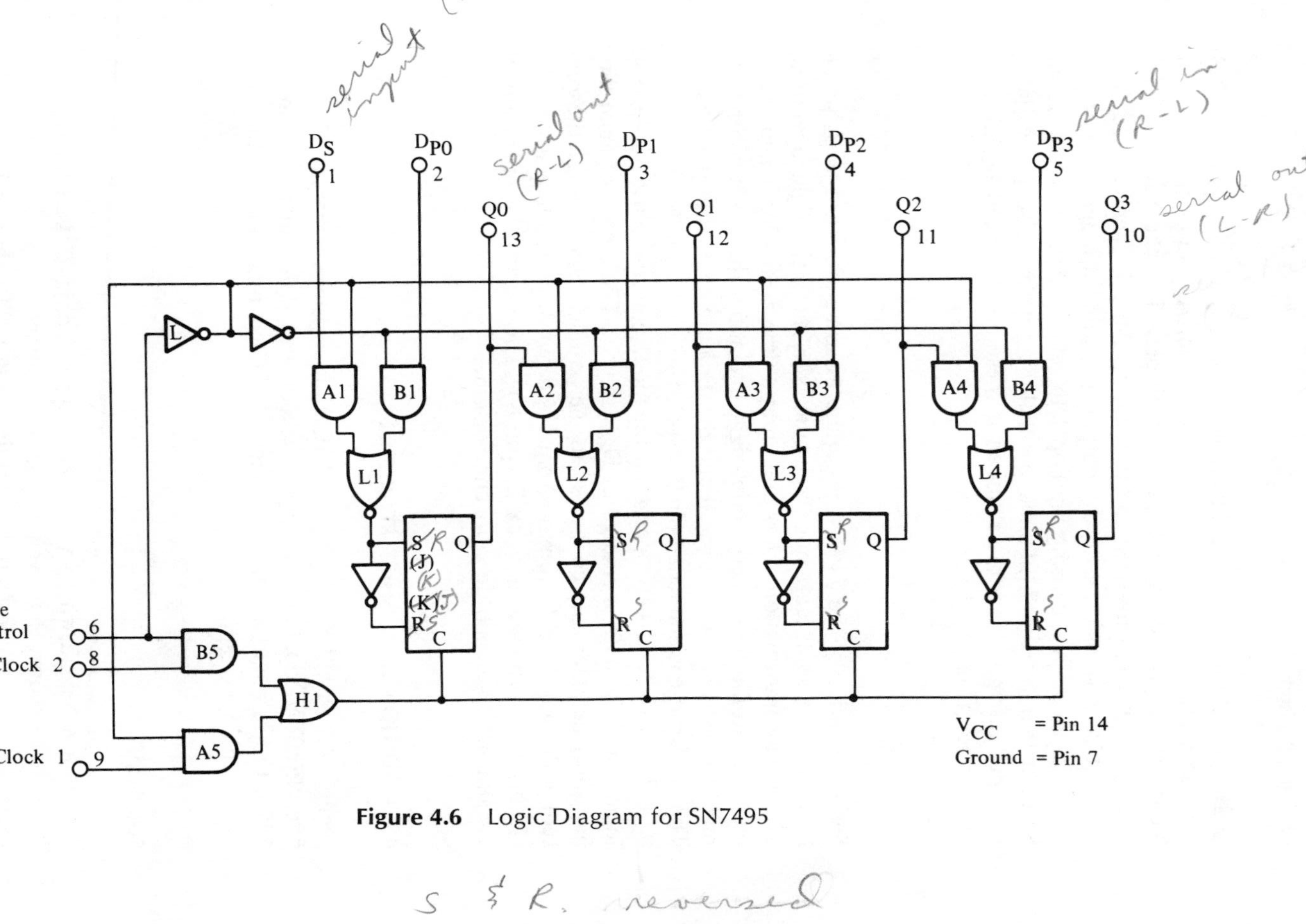

Figure 4.6 Logic Diagram for SN7495

of B_1 (a 1 through the two inverters remains a 1). The output of B_1 then goes high. This high, with the low from A_1 causes the output of NOR gate L_1 to become a 1. A 1 is then placed on the J input of flip-flop 1 and this 1 is inverted to the K input. Pin 8 is then clocked. This clock pulse is transferred to the output of B_5 and through the OR gate H_1 to the clock line, causing flip-flop 1 to set a 1 at its Q output. It can be seen that for the 7495, the parallel outputs are always available. External gating is therefore required if these outputs need be inhibited prior to actual use.

Serial

Serial operation of the 7495 is not quite as straightforward as its parallel mode. We shall start our discussion with the serial-right operation. This means information enters one bit at a time at flip-flop 1 and moves bit by bit to flip-flop 4. For serial-right operation, the mode control input (pin 6) is held on a logic 0 and the data are entered at pin 1 (D_S). The clock pulse is entered at pin 9 (clock 1).

For serial shift-left operation, the mode control (pin 6) is held at logic 1 and the data is entered at D_{P3} (pin 5). The output of each flip-flop must be externally connected to the parallel data input of the preceding flip-flop (pin 10 to pin 4, pin 11 to pin 3, pin 12 to pin 2). The clock pulse is entered at clock 2 (pin 8). This makes the operation of the shift-left mode a little more cumbersome.

4.7 ADDITIONAL IC SHIFT REGISTERS

Figures 4.7 to 4.9 show additional IC shift registers. Whereas pin configuration and operation of these devices may differ slightly from the 7495, they are all basically shift registers. Their functions are the same in a circuit or system.

SUMMARY

This unit has dealt with a very special type of circuit, the shift register. It has been shown that even on the most complex of these circuits, the basis was the bistable (flip-flop) multivibrator. The shift register is thus an extension of the bistable used as a storage device. Further, it is the D flip-flop used to transfer data on command. Keeping these simple concepts in mind, it should cause no problem to visualize the function or operation of even the most complex of these useful circuits.

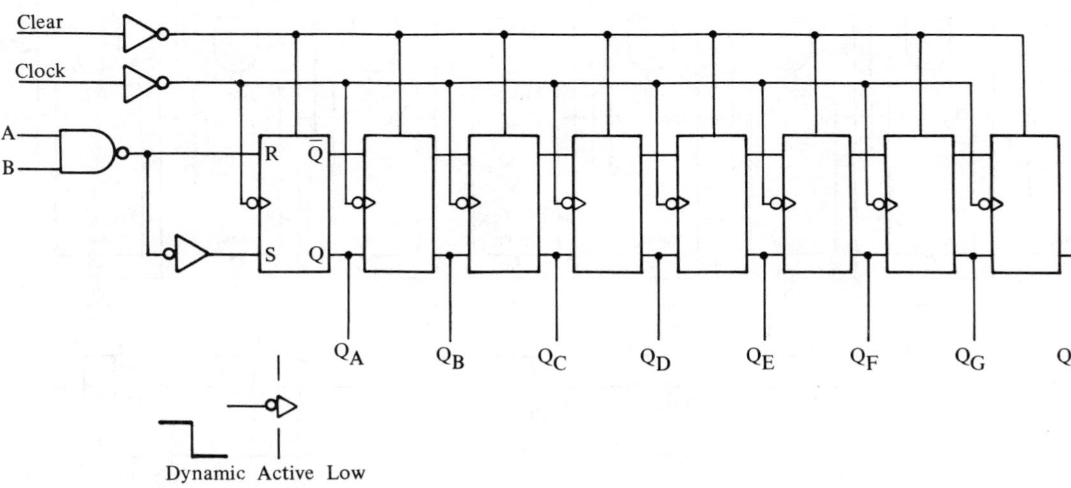

Figure 4.7 SN74164 8-Bit Parallel-Out, Serial Shift Register

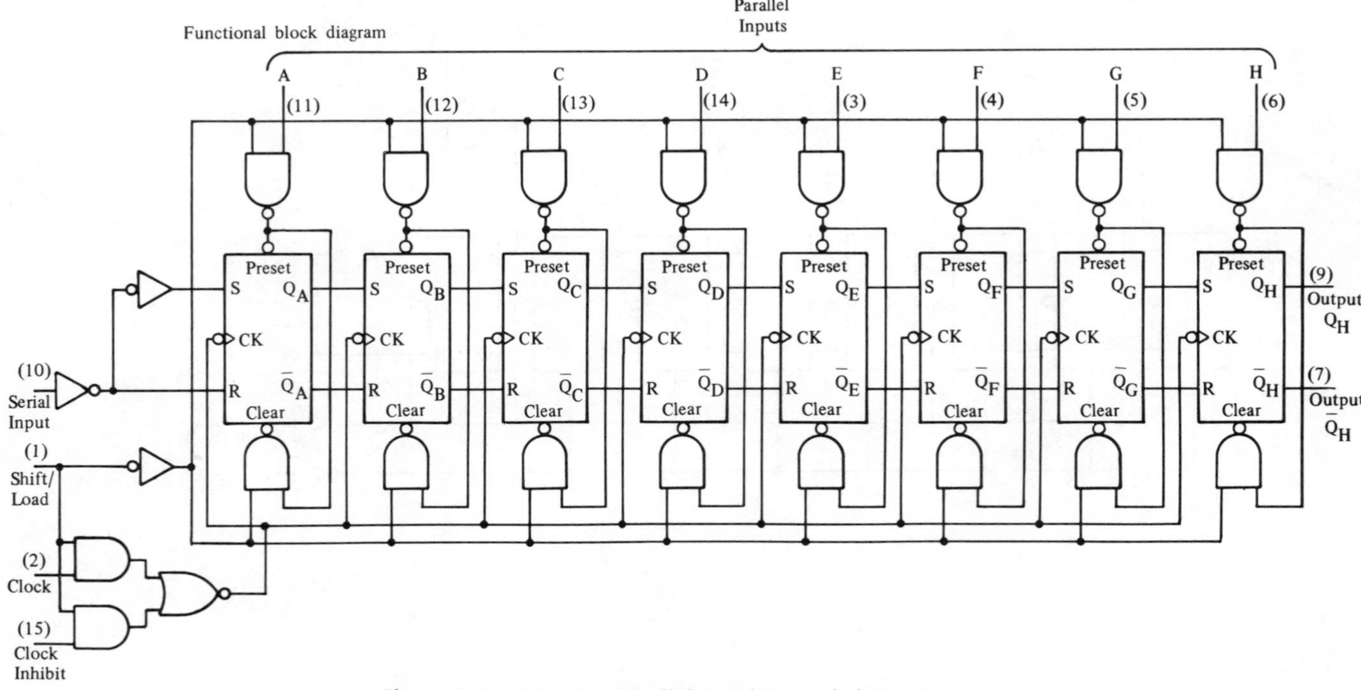

Figure 4.8 SN74165 Parallel-Load 8-Bit Shift Register

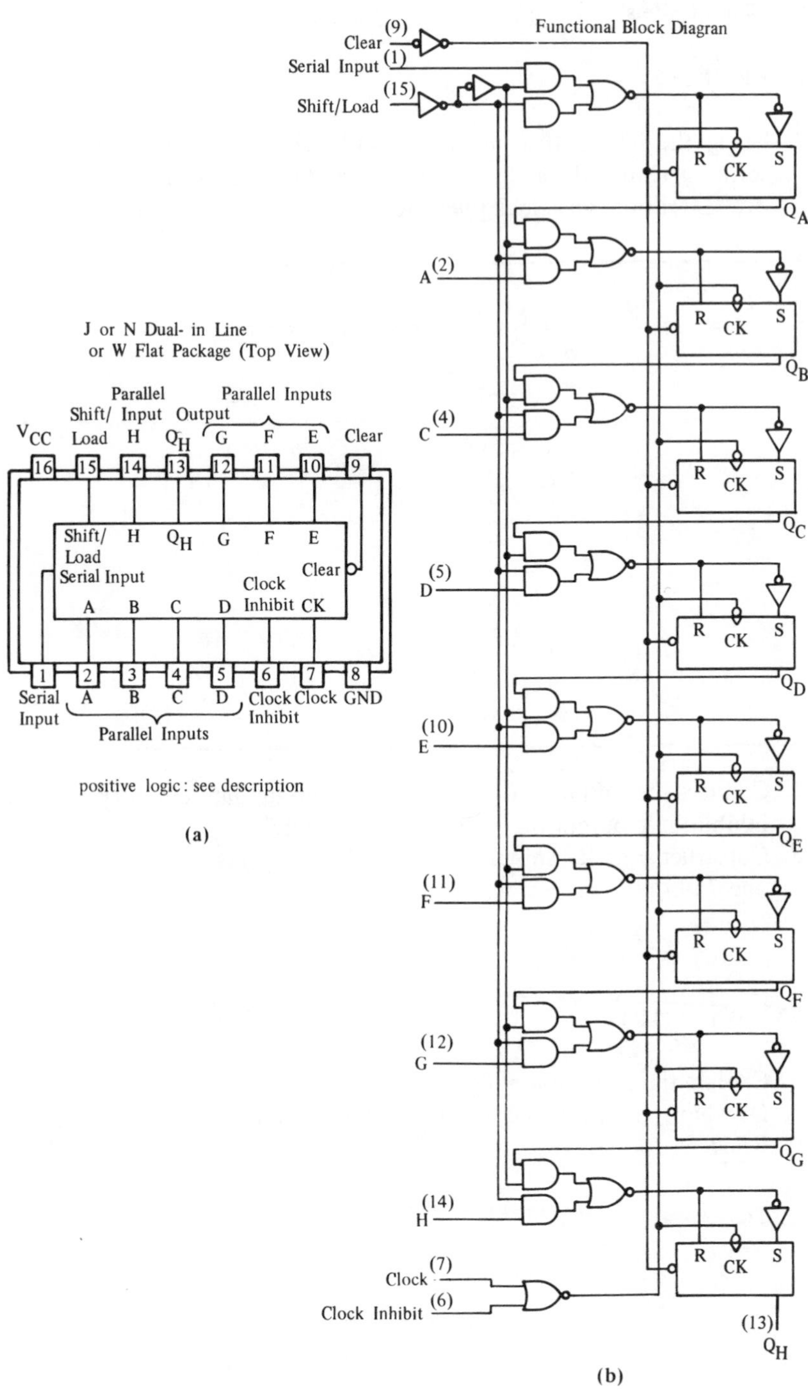

Figure 4.9 SN74166 8-Bit Shift Register. **(a)** Pin-Out. **(b)** Functional Block Diagram

EXERCISES

1. Using the Texas Instruments SN7495 (or equivalent) construct a serial-in shift-right shift register. Enter the data as shown in Table 4.2 (D_S). For each clock pulse, record the outputs Q_3, Q_2, Q_1, and Q_0.

TABLE 4.2

Clock Pulse	D_s	Q_3	Q_2	Q_1	Q_0
0	0	0	0	0	0
1	1				
2	0				
3	1				
4	1				
5	0				
6	0				
7	0				
8	0				
9	0				

2. Connect the outputs of part A to appropriate gate circuitry so as to inhibit these outputs until they are called for.
3. Construct a serial shift register of X bits by connecting more than one 7495 (or equivalent).

STUDY TOPICS

1. Explain the operation of figure 4.10.
2. Of what practical value might a circuit such as 4.10 be in the digital field?

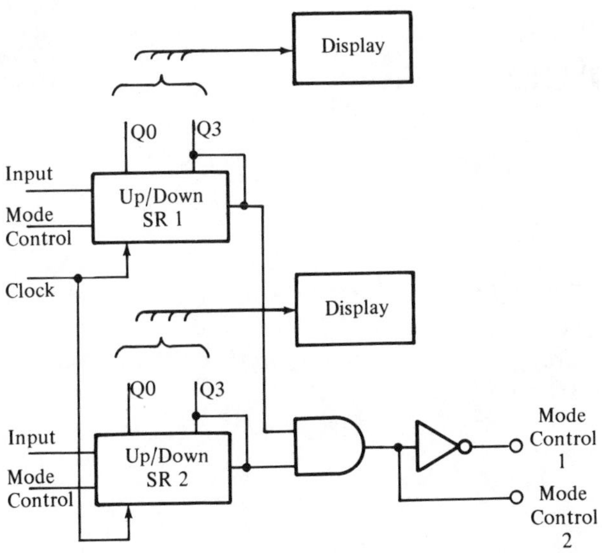

Figure 4.10 Sample Circuit Using Shift Registers

QUESTIONS AND PROBLEMS

4.1 What is a shift register?

4.2 A shift register is a series of ____________.

4.3 Why is a D-type flip-flop used as the input flip-flop of a serial shift register?

4.4 Explain the differences between serial and parallel shift registers.

4.5 Explain the operation (with logic levels) of the shift-left, shift-right shift register.

4.6 What is the maximum delay time that could be expected in shifting one bit through a 7495?

4.7 What is a recirculating shift register?

4.8 What might be one application of this type register?

4.9 Using a gate and a 7495, draw a diagram of a recirculating shift register.

Counters

5

OBJECTIVE: To familiarize the student with operation of binary counters and modulus counters. One method of determining the feedback required for these counters is also discussed.

Introduction: Binary counters are registers that advance through a prescribed sequence of different states each time a pulse appears at the input. They are called *counting registers* or simply *counters.* Modulus counters are counters that complete one count cycle after a specified number of input pulses. The modulus of the counter is the number of input pulses for a complete count cycle.

5.1 BINARY COUNTERS

The register of a binary counter stores information about the accumulated number of input pulses as a binary number as shown in figure 5.1. Notice the J-K flip-flops are connected in such a manner that they behave as T-type flip-flops, where (J and K) are at a logic state 1. This configuration is very similar to the shift registers, which were also constructed from a series of J-K flip-flops. The major difference in the two circuits is the interconnection of the flip-flops. In the binary counter of figure 5.1, each clock input is triggered from the preceding Q output. This means, for example, if the clock triggers on the negative-going pulse it will see such a pulse when the Q out of the preceding FF changes from a logic 1 to a logic 0. Using the waveforms of figure 5.1,

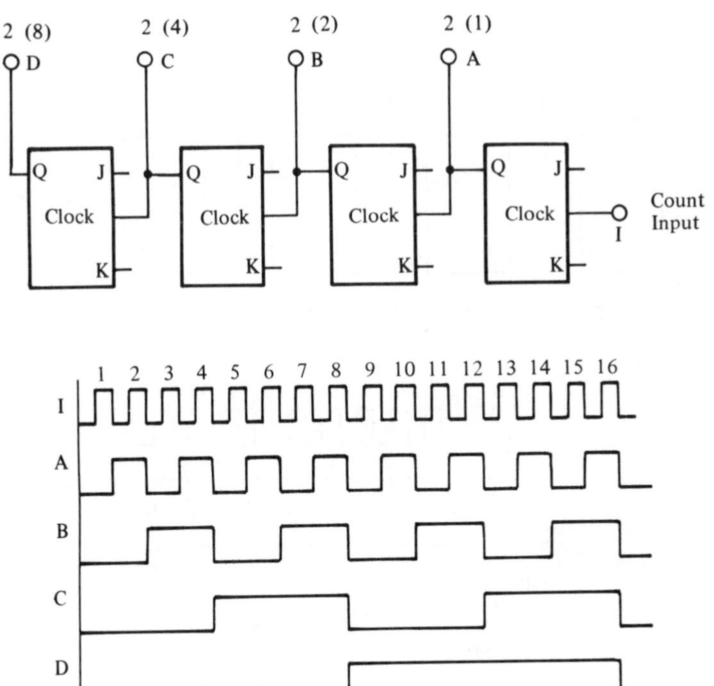

Figure 5.1 Basic Binary Counter and Waveforms

note the effect on each FF as the input pulses arrive at the count input. Starting with each FF having a 0 level at its Q output, the first pulse to enter FF1 causes it to change its Q output from a low to a high. This change is seen at the clock of FF2. However, FF2 will not toggle since the clock pulse has the wrong polarity. The second pulse to arrive at the clock input of FF1 causes it to once again toggle, this time from a high to a low. The clock input of FF2 sees this and it also toggles. The third clock pulse (input to FF1) again causes FF1 to change states. FF2, however, does not. After three input pulses to this counter, the first and second FFs are at logic 1, while FFs 3 and 4 are at logic 0. Reading these levels as a binary number from right to left, yields 0011, or the binary 3. Continuing the input pulses six more counts for a total of nine inputs gives 1001, or the binary 9. Figure 5.1 is a four-stage binary counter. After fifteen input pulses, the counter will have as its output 1111, the binary 15. The sixteenth pulse will cause the counter to return to 0000. The maximum count, then, for this counter is 15 before it resets. Since $2^4 = 16$, this seems logical.

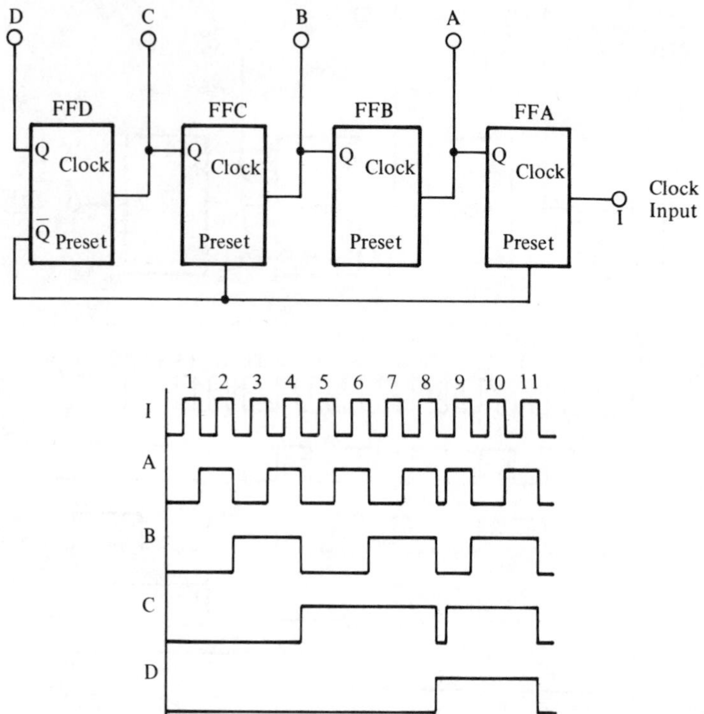

Figure 5.2 Nonsequential Mod-11 Counter and Waveforms

5.2 MODULUS COUNTERS

In a strict sense, most counters are modulus counters, that is, they complete one count cycle after a specified number of input pulses.

Example The simple binary counter in figure 5.1 has a modulus of 16. On the sixteenth input pulse the counter resets to 0, thus completing one complete cycle. To determine how many flip-flops are needed to construct a certain modulus counter the equation below is used.

$M - 1 < 2^n$
M = modulus of counter
n = number of flip-flops needed

Example How many flip-flops are needed to construct a mod-11 counter?

$$\begin{aligned} M - 1 &< 2^n \\ 11 - 1 &< 2^n \\ 10 &< 2^4 \\ 10 &< 16 \end{aligned} \qquad \begin{aligned} 2^0 &= 1 \\ 2^1 &= 2 \\ 2^2 &= 4 \\ 2^3 &= 8 \\ 2^4 &= 16 \end{aligned}$$

Therefore, at least four flip-flops are needed.

A modulus counter uses feedback to accomplish recycling at a specified count. One method of feedback occurs from the output stage of the last flip-flop to the preset inputs of as many flip-flops as are required to preadvance the count cycle. Feedback is determined by subtracting the modulus number from 2^n. For instance, a mod-11 counter requires four flip-flops, so the advance needed is found by:

$$\begin{aligned} \text{Advance needed} &= 2^n - M \\ &= 2^4 - 11 \\ &= 16 - 11 \\ &= 5 \end{aligned}$$

Therefore, feedback is required at flip-flop A and at flip-flop C (binary 1 and 4). A mod-11 counter is shown in figure 5.2. Notice that $\bar{Q}$ is fed back to FF A and FF C's preset input, since a 0 is required at the preset input to set Q to logic 1.

Notice that after feedback occurs, the count shown at the parallel outputs A, B, C, and D is not correct. For instance, after clock pulse 9 the counter has a count of 14. A mod counter that does not display the proper count on every clock pulse is called a *nonsequential* mod counter (as in figure 5.2). By the use of proper gating, a modulus counter can be made to maintain the correct count and still recycle after the proper number of clock pulses. A mod counter that maintains the proper count on every clock pulse is called a *sequential* mod counter (as in figure 5.3). The mod-10 counter shown in figure 5.3 counts sequentially and completes a cycle after ten input pulses. Notice the formula to determine which stages need feedback does not apply when gating is used.

5.3 SYNCHRONOUS COUNTERS

The term *synchronous,* when applied to the counters we are discussing, means simply that all of the FF's that are supposed to change states at a given time will do so almost instantaneously.

When applied to counters, asynchronous means that each flip-flop that is going to change state at a given time (input pulse count) will

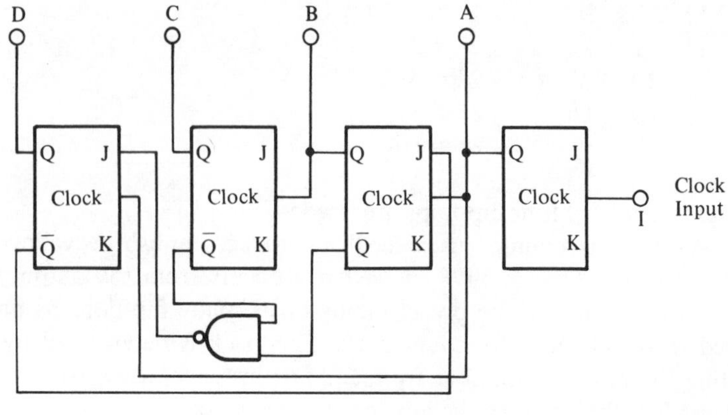

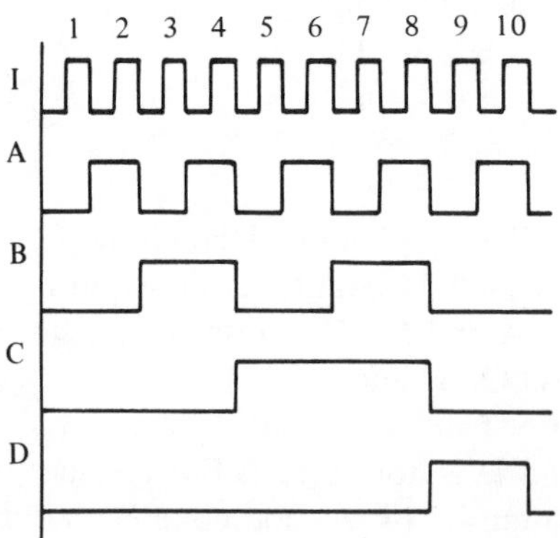

Figure 5.3 Sequential Mod-10 Counter and Waveforms

have to wait until the preceding flip-flop has changed its state. For counters consisting of relatively few flip-flops, the asynchronous counters pose no problem. When there are a larger number of flip-flops in the counter, the delay encountered may cause problems. A counter of four flip-flops, for example, when changing from a count of 7 (0111) to a count of 8 (1000) could have a delay of 80 nanoseconds or greater from the time the eighth pulse enters the counter until the counter had the binary 8 represented. This would be true, for example, if each flip-flop had a delay of 20 nanoseconds from input pulse to change of state of its Q out.

Explain the major difference between synchronous and asynchronous counters.

In the synchronous, all flip-flops that are to change state on any given input pulse will do so almost simultaneously. In the asynchronous, there will be a propagation delay from flip-flop to flip-flop.

5.4 OPERATION OF THE SYNCHRONOUS COUNTER

Figure 5.4 shows a four-bit synchronous counter. Notice the difference between this circuit and the counter shown in figure 5.1. The counter of figure 5.1 was an asynchronous counter. The operation of this circuit is as follows:

1. With all of the Q outputs set at 0 and a logic level 1 applied to the Hi line, the first clock pulse will
2. Cause the Q of FF1 to go to a 1 state. This now puts a one at the J and K inputs of FF2 and to the a input of A gate 1. The second clock pulse will
3. Cause FF1 and FF2 to again change state. FF1 is now a 0 at Q and FF2 is a 1 at its Q. This puts a 1 at b of AND 1 and a 0 at a of AND 1. The third clock pulse will
4. Again cause FF1 to change to a 1 at its Q output. FF2 remains at the same state. Therefore inputs a and b of AND 1 are now both Hi. The fourth clock pulse will
5. Cause FF1 to again change states; FF2 will change at the same instant since its J and K inputs are both Hi. FF3 will also change states at the same time since its J and K inputs are both Hi due to AND gate 1.

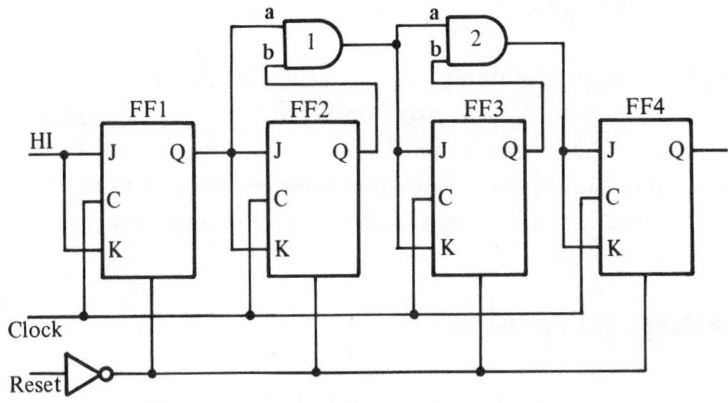

Figure 5.4 Four-Bit Synchronous Counter

5.5 THE SYNCHRONOUS DOWN COUNTER

A down counter is very similar to the types of counters we have been discussing. The counters we have looked at so far start with 0 and give a binary representative of the number of the pulses being fed into the counter in sequence from 0 to 15. The down counter on the other hand starts at 15 and counts down to 0. Figure 5.5 is a synchronous down counter. The main differences between the synchronous down counter and up counter are these:

1. Outputs are taken from the $\overline{Q}$ of the J-K flip-flops to feed the AND gates.
2. The J-K inputs are not tied to a Hi line. **Note:** A reset can be used if desired.

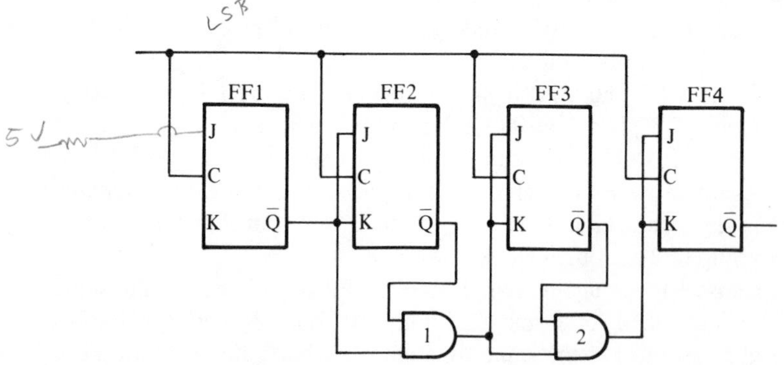

Figure 5.5 Synchronous Down Counter

5.6 RING COUNTERS

Figure 5.6 is a ring counter. This counter is essentially a shift register and will continue to recycle itself until the input is changed or the reset line is pulsed. One application of the ring counter would be as a timing device for a digital system. For the circuit of figure 5.6, the counter (or register) will return to the same output state on every fourth pulse.

5.7 SWITCH-TAIL COUNTERS

This counter is really just another serial shift register with feedback. It is almost the same as the ring counter of Figure 5.7, except the output is

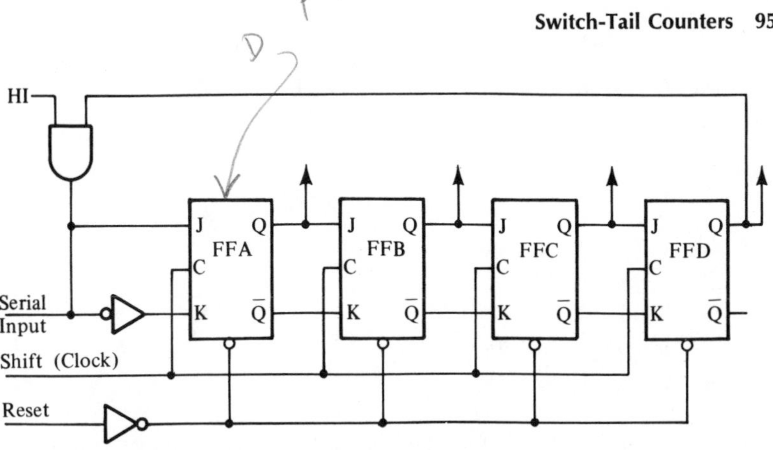

Figure 5.6 Four-Bit J-K Shift Register (Ring Counter)

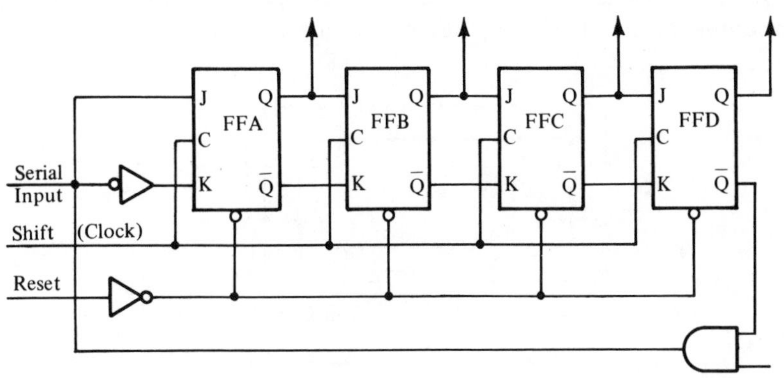

Figure 5.7 Four-Bit J-K Shift Register (Switch-Tail Counter)

Johnson Counter

taken from the $\overline{Q}$ of the most-significant digit. A switch-tail counter is shown in Figure 5.7. Let's see how this counter works.

Example Suppose we have the following in the register.

A	B	C	D
0	1	0	1

Solution What happens for several clock pulses is shown.

cp	A	B	C	D	$\overline{D}$
0	0	1	0	1	0
1	0	0	1	0	1
2	1	0	0	1	0
3	0	1	0	0	1
4	1	0	1	0	1
5	1	1	0	1	0
6	0	1	1	0	1
7	1	0	1	1	0
8	0	1	0	1	0

($\overline{D} = \overline{Q}$ of D)

So the switch-tail counter has the ability to return to its original condition in twice the number of pulses as flip-flops. The switch-tail counter may also be called a *Johnson counter,* a *walking ring counter,* or a *Gray Code Ring.* The switch-tail counter is shown by the circuit of figure 5.8. Here the basic operation of the counter is really shown. The outputs of the last stage are cross-coupled back to the inputs of the first stage. By using the inverter in the line of figure 5.8 and only taking one output back to the input, the same results are achieved.

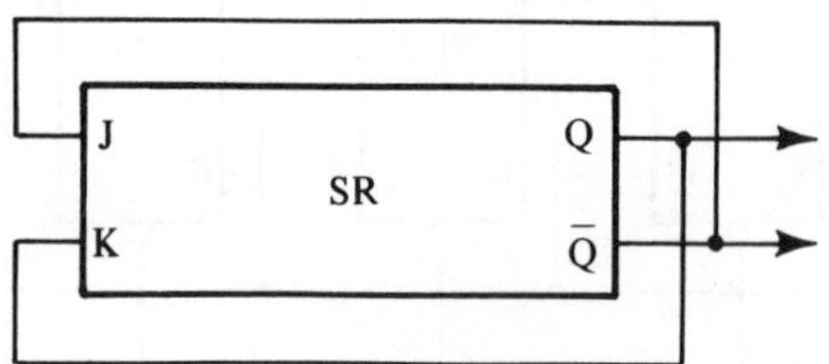

Figure 5.8 Switch-Tail Counter

5.8 SELF-STOPPING ASYNCHRONOUS COUNTERS

A special type of modulus counter is the self-stopping asynchronous counter. Here is a counter as shown in figure 5.9 that counts to a specified number (2^7) and then stops. The final stage in the string does not present an output for the counter. Its function is to turn off the counter when the desired count is reached. When this FF is clocked, its $\overline{Q}$ will go to a logic 0 level. This 0 will be felt at the J-K inputs of the first flip-flop. The first flip-flop is then put in the inhibit mode. This circuit is not very useful, since it can only stop at numbers which are powers of two (2^n). A different approach to the problem is shown in figure 5.10, which shows a self-stopping asynchronous counter capable of stopping on any number. The A, B, C, D inputs to the NAND gate are

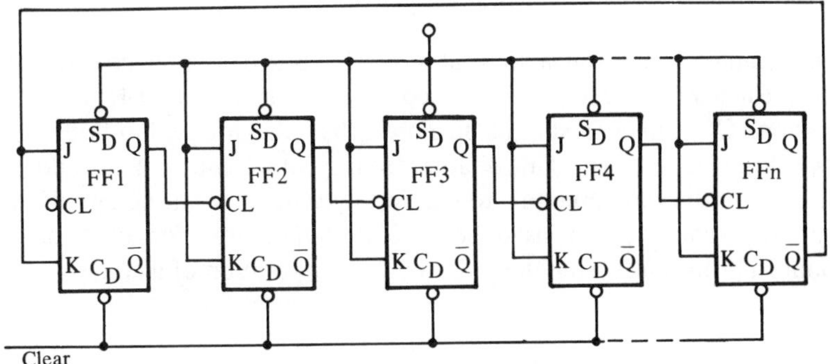

Figure 5.9 Self-Stopping Asynchronous Counter

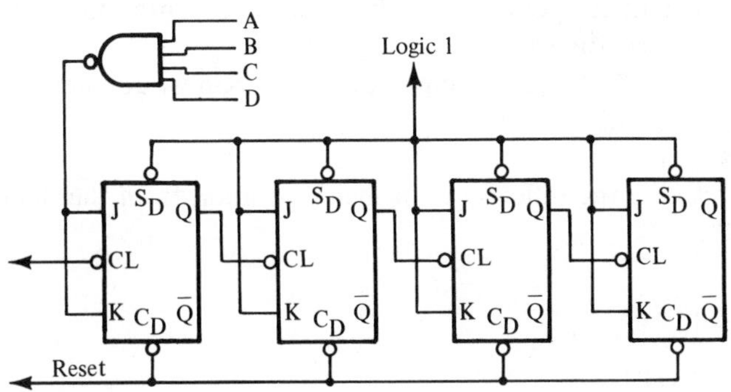

Figure 5.10 Self-Stopping Asynchronous Counter

connected to the Q outputs that will give the binary count of the number the counter is to stop at.

For example, we wish the counter of figure 5.10 to count and stop on the sixth pulse. Therefore Q_2 and Q_4 are connected to the NAND gate inputs A and B. Inputs C and D are connected high to prevent a false output. On the sixth pulse the inputs to the gate will be all 1s. This will cause the output of the gate to be a logic 0. This will inhibit the first flip-flop.

How would you connect the NAND gate for the counter to stop after three pulses?

Connect FF1 to A, FF2 to B, C and D to a high.

SUMMARY

This chapter has taken a look at several types of digital counting circuits. Each of these circuits was constructed using the J-K (or R-S) flip-flop. Of course it should be noted that any toggle flip-flop would have done just as well. This chapter, along with Chapter 4, should give the student the fundamentals to design any type of counting circuit. The four-stage counter was used because its length was found to be just about right for discussing the pulse-by-pulse operation of a counter.

EXERCISES

1. **(a)** Using J-K flip-flops, construct the circuits of figure 3.1 and figure 3.4.
   (b) Verify their operation, through the use of truth tables and timing waveform diagrams.
2. **(a)** Using J-K flip-flops, construct the modulus counters of figures 3.2 and 3.3.
   (b) Repeat step 2a.
   (c) Describe the differences in their operation from the data of step 2b.

STUDY TOPIC

1. Construct a nonsequential and a sequential modulus counter to count to nine pulses.

QUESTIONS AND PROBLEMS

5.1 Which of the following is true for a synchronous counter?
- **(a)** All flip-flops change state at the same time.
- **(b)** Every other flip-flop changes on every other pulse.
- **(c)** Counts faster than the asynchronous counter.
- **(d)** Counts in only one direction.

5.2 Synchronous counters:
- **(a)** Use additional gates to improve their operation.

(b) Use more flip-flops for a given count than do asynchronous counters.
(c) Cannot use J-K flip-flops.
(d) Are not true binary counters.

5.3 In a binary counter, when each flip-flop that is going to change state on a given pulse, change simultaneously that counter is said to be asynchronous.
(a) True
(b) False

5.4 The major difference between an up counter and a down counter (both synchronous) is:
(a) The up needs more flip-flops.
(b) The output is taken from $\overline{Q}$ for down counter.
(c) They will read different (have a different binary state) after six input pulses.
(d) The output is taken from the $\overline{Q}$ for the up counter.

5.5 Ring counters are really shift registers.
(a) True
(b) False

5.6 A ring counter can be used to
(a) Set the timing sequence of a digital system.
(b) Give an accurate binary count.
(c) Count up or down in octal.
(d) Count up or down in binary.

5.7 A ring counter recycles itself
(a) Continually.
(b) On every third clock pulse.
(c) On every other clock pulse.
(d) On every other input pulse.

5.8 A switch tail counter
(a) Is the same as a ring counter.
(b) Is the same as a ring counter except for feedback firing.
(c) Is the same as a hexidecimal ring counter.
(d) Switches counts every other cycle.

5.9 A switch-tail counter is really a shift register.
(a) True
(b) False

5.10 If a J-K flip-flop has a propagation delay of 20 nanoseconds, how long will it take an asynchronous counter to change from 31 to 32?
(a) 20 nanoseconds
(b) 120 nanoseconds

(c) 100 nanoseconds
(d) 60 nanoseconds

5.11 What is a ripple counter?

5.12 Compare the advantages and disadvantages of asynchronous and synchronous counters.

5.13 Using a 74193, draw the logic diagram for a modulus-7 counter. Use the preset method. Draw a timing diagram for all outputs and the input and clock pulses.

5.14 What problems might occur using the preset method of wiring for modulo counters?

5.15 Wire two 7490 devices as a count-down decade counter. Show reset 99 conditions as well as run mode.

5.16 What is the maximum clock rate at which four 7490 devices may be cascaded as a counter circuit?

5.17 Draw a four-bit Johnson counter using J-K flip-flops. Starting count is to be 0.

5.18 Show the timing of the Johnson counter of problem 5.17 for four clock pulses.

5.19 Show the wiring you would use to obtain a modulo-7 counter from a 7490. Use no external gates.

5.20 How many flip-flops are needed to construct a mod-20 modulo counter?

6

Integrated Circuit Counters

OBJECTIVE: To become familiar with the various counters as self-contained integrated circuit devices.

Introduction: In chapter 5 several types of counting circuits were shown. These circuits were for the most part constructed from individual flip-flops and gate circuits. In this unit we will view the counters as "black box" devices having only inputs and outputs. It is hoped that through the combination of chapters 5 and 6 the student will have a thorough understanding of digital counters.

6.1 7490 DECADE BINARY COUNTERS

Figure 6.1 is a page from the Texas Instruments Data Book for TTL integrated circuits. It shows the pin configuration and functional block diagrams for three types of integrated circuit counters. As will be discussed later, there is much more in the way of important data for the operation of the devices contained on that page. We will first turn our attention to the SN7490 (SN5490A, SN7490A, SN54L90, SN74L90). This particular device is a decade counter. That is, the 7490 is a symmetrical divide-by-ten counter when certain external pin connections are made. As seen by the functional block diagram, the counter has a divide-by-five count cycle length. For a further understanding of this device, figure 6.2 is needed. Figure 6.2 is also a page from the Texas Instrument TTL Data Book. Here are found the truth tables and input-output circuits of the counters described in figure 6.1. Again, we will turn our attention to the truth tables for the SN7490. The first table we will look at is labeled 90A, L90 BCD count sequence. This table gives the output logic conditions for the 7490 for each of ten count input conditions. (Q_D is pin 11 and Q_A is pin 12.) This truth table is for a count sequence of 0 to 9. The L (low) corresponds to a logic 0 and the H (high) corresponds to a logic 1.

For the SN7490 to operate as a decade counter, output Q_A (pin 12) must be externally connected to input B (pin 1). This then will give the BCD count as shown in the truth table. Figure 6.3 shows this connection. Also note that the input to the counter is to pin 14 (input A).

The table labeled " '90A, 'L90 Bi-Quinary (5-2)" is a truth table showing the count and output conditions of the 7490 when used as a bi-quinary counter. Note the major difference here is in the conditions of the four outputs during the count sequence. For a count of 0 to 4 the output responds in reverse BCD fashion. At the input of the fifth pulse (count of 5) the device responds with a BCD, A condition. The bi-quinary code is then read as 1 000 is 5 and 1 010 is 7, and so on. To operate the 7490 as a bi-quinary counter requires the Q_D output (pin 11) to be externally connected to input A (pin 14).

The 7490 has three modes of operation. These are shown in the truth table labeled " '90A, 'L90 RESET/COUNT FUNCTION TABLE." This table gives the logic conditions needed at pins 2 and 3 ($R_0(1)$ and $R_0(2)$ and pins 6 and 7 $R_9(1)$ and $R_9(2)$) to set the output of the counter to either 0, 9, or to enable the counter to count. This table has been reproduced in figure 6.4 for ease of explanation. There are three possible operating conditions for the 7490 that may be achieved through seven different settings of the logic levels on pins 2, 3 and 6, 7. The L represents a low logic level (preferably ground) and the H is a high logic level (preferably $+V_{CC}$). The X is a don't-care condition. This means that the particular pin can be at either the high (logic 1) or the low (logic 0) level. Settings 4, 5, 6, or 7 are needed for the device to be in the count mode. If any one of these four sets is met, the device will count the input pulses (BCD or bi-quinary depending on external connections as stated above). To set the counter to 0 (or reset it) the sets of conditions 1 and 2 may be used. To set the 7490 to an output of BCD 9, conditions of set number 3 are needed. The R_0 pins then control the 0 reset while the R_9 pins control the 9 set.

'90A, 'L90 . . . Decade Counters
'92A . . . Divide-by Twelve Counter
'92A, 'L93 . . . 4-Bit Binary Counters

Types	Typical Power Dissipation
'90A	145 mW
'L90	20 mW
'92A, '93A	130 mW
'L93	16 mW

description

Each of these monolithic counters contains four master-slave flip-flops and additional gating to provide a divide-by-two counter and a three-stage binary counter for which the count cycle lengh is divide-by-five for the '90A, 'L90, and 'LS90, divide-by-six for the '92A and 'LS92, and divide-by-eight for the '93A, 'L93, and 'LS93.

All of these counters have a gated zero reset and the '90A, 'L90, and 'LS90 also have gated set-to-nine inputs for use in BCD nine's complement applications.

To use their maximum count length (decade, divide-by-twelve, or four-bit binary) of these counters, the **B** input is connected to the Q_A output. The input count pulses are applied to input **A** and the outputs are as described in the appropriate function table. A symmetrical divide-by-ten count can be obtained from the '90A, or 'LS90 counters by connecting the Q_D output to the **A** input and applying the input count to the **B** input which gives a divide-by-ten square wave at output Q_A.

NC – No internal connection

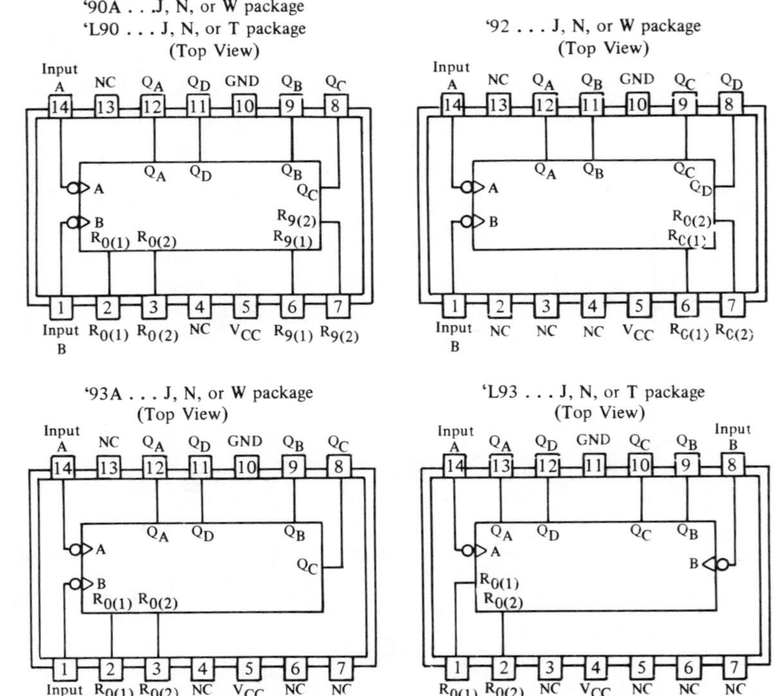

Figure 6.1 IC Counters

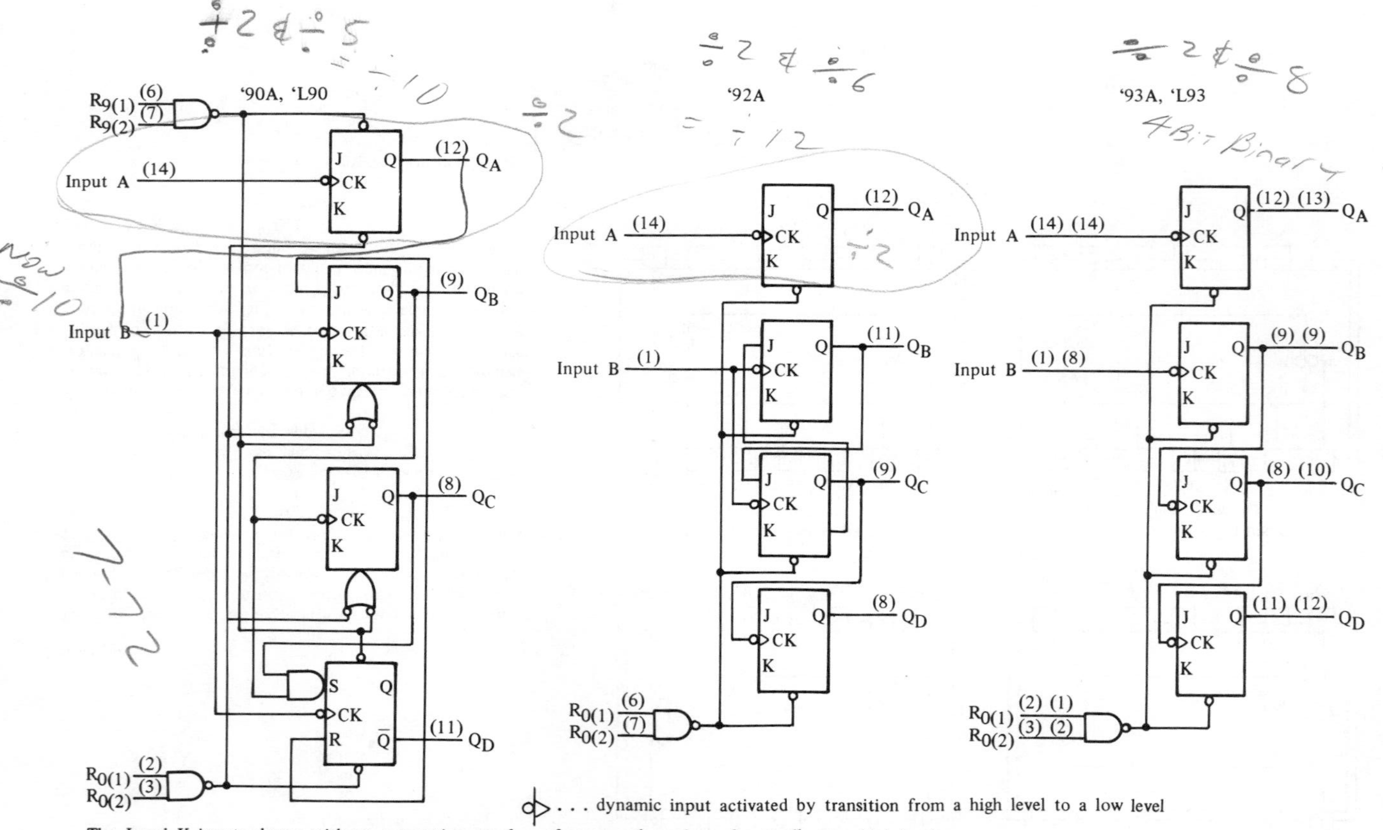

. . . dynamic input activated by transition from a high level to a low level

The J and K inputs shown without connection are for reference only and are functionally at a high level.

Figure 6.1 *continued*

'90A, 'L90
BCD Count Sequence
(See Note A)

Count	Output			
	Q_D	Q_C	Q_B	Q_A
0	L	L	L	L
1	L	L	L	H
2	L	L	H	L
3	L	L	H	H
4	L	H	L	L
5	L	H	L	H
6	L	H	H	L
7	L	H	H	H
8	H	L	L	L
9	H	L	L	H

'90A, 'L90
BI-Quinary (5-2)
(See Note B)

Count	Output			
	Q_A	Q_D	Q_C	Q_B
0	L	L	L	L
1	L	L	L	H
2	L	L	H	L
3	L	L	H	H
4	L	H	L	L
5	H	L	L	L
6	H	L	L	H
7	H	L	H	L
8	H	L	H	H
9	H	H	L	L

'92A
Count Sequence
(See Note C)

Count	Output			
	Q_D	Q_C	Q_B	Q_A
0	L	L	L	L
1	L	L	L	H
2	L	L	H	L
3	L	L	H	H
4	L	H	L	L
5	L	H	L	H
6	H	L	L	L
7	H	L	L	H
8	H	L	H	L
9	H	L	H	H
10	H	H	L	L
11	H	H	L	H

'90A, 'L90
Reset/Count Function Table

Reset Inputs				Output			
$R_{0(1)}$	$R_{0(2)}$	$R_{9(1)}$	$R_{9(2)}$	Q_D	Q_C	Q_B	Q_A
H	H	L	X	L	L	L	L
H	H	X	L	L	L	L	L
X	X	H	H	H	L	L	H
X	L	X	L	Count			
L	X	L	X	Count			
L	X	X	L	Count			
X	L	L	X	Count			

'93A, 'L93
Count Seqence
(See Note C)

Count	Output			
	Q_D	Q_C	Q_B	Q_A
0	L	L	L	L
1	L	L	L	H
2	L	L	H	L
3	L	L	H	H
4	L	H	L	L
5	L	H	L	H
6	L	H	H	L
7	L	H	H	H
8	H	L	L	L
9	H	L	L	H
10	H	L	H	L
11	H	L	H	H
12	H	H	L	L
13	H	H	L	H
14	H	H	H	L
15	H	H	H	H

'92A, '93A, 'L93
Reset/Count Function Table

Reset Inputs		Output			
$R_{0(1)}$	$R_{0(2)}$	Q_D	Q_B	Q_C	Q_A
H	H	L	L	L	L
L	X	Count			
X	L	Count			

Notes: A. Output Q_A is connected to input B for BCD count.
B. Output Q_D is connected to input A for bi-quinary count.
C. Output Q_A is connected to input B.
D. H = high level, L = low level, X = irrlevant

Figure 6.2 IC Counter Truth Tables

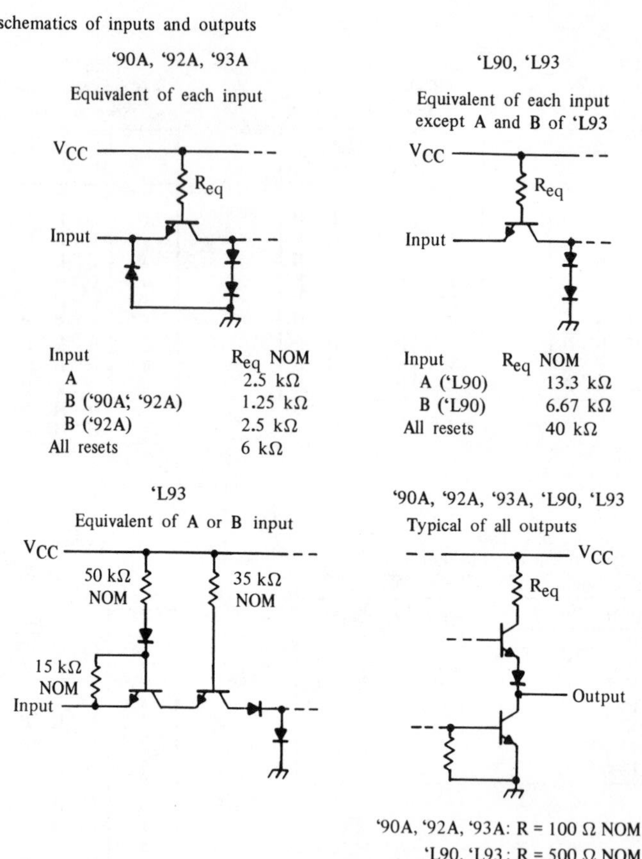

Figure 6.2 *continued*

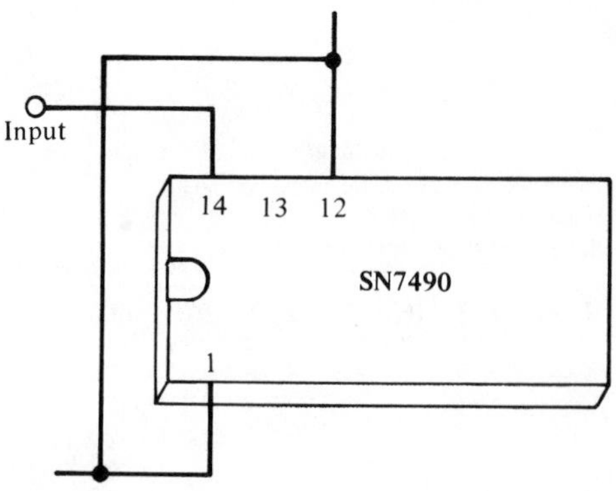

Figure 6.3 External Connections for BCD Counter

RESET	INPUTS			OUTPUT			
$R_0(1)$	$R_0(2)$	$R_9(1)$	$R_9(2)$	Q_D	Q_C	Q_B	Q_A
H	H	L	X	L	L	L	L
H	H	X	L	L	L	L	L
X	X	H	H	H	L	L	H
X	L	X	L	COUNT			
L	X	L	X	COUNT			
L	X	X	L	COUNT			
X	L	L	X	COUNT			

Figure 6.4 Reset/Count Function Table

Figure 6.5 shows a simple switching arrangement for the switching of the 7490 from its count mode to its reset 0 mode. Only one pin (pin 3, $R_0(2)$) needs to be switched. Both the R_9 inputs are held at ground. This insures that the 9 output cannot be falsely achieved. There is no way for the R_9 connections to accidentally go high via switching. When pin 3 is grounded, the counter will count. Conditions of setting 4 have been met. Switching pin 3 to V_{CC} will give the set of conditions for setting 1. This is one of the two reset 0 conditions.

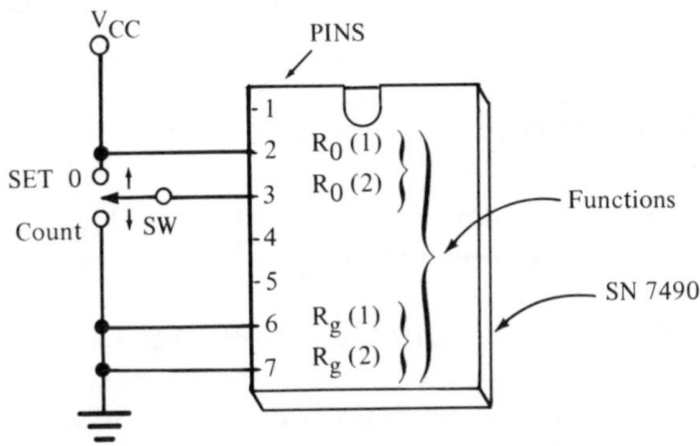

Figure 6.5 Set Zero/Count Switching

6.2 7492 DIVIDE-BY-TWELVE COUNTER

One of the two other devices shown in figures 6.1 and 6.2 is the SN7492A (SN5492A). This device is similar to the 7490 when the 7490 is operated as a bi-quinary counter. Note the differences of the count as shown in figure 6.6. With the bi-quinary, the Q_A output was the most-significant bit with the Q_B the least-significant. The 7492, however, counts through 5 as a typical BCD counter (A is least-significant bit). At the input of the sixth pulse, the counter responds with 1000, similar now to the bi-quinary counter output. The difference is that now the D is the most-significant bit and A is still least-significant. Again, for the 7492 to function properly, the Q_A output must be externally connected to the B input (pin 12 to pin 1).

The 7492 also has a reset/count function table as shown in figure 6.2. Here, however, there are only three sets of conditions available through two inputs. The inputs are $R_0(1)$ (pin 6) and $R_0(2)$ (pin 7). The 7492 then can be reset to 0 by placing both of these pins at the logic 1 level. Placing either or both in the low condition will put the 7492 in the count mode of operation.

					Count
Bi - quinary	A	D	C	B	
	1	0	0	0	5
	1	0	1	0	7
÷ 12	D	C	B	A	
	1	0	0	0	6
	1	0	0	1	7

Figure 6.6 Comparison of Divide-by-Twelve with Bi-Quinary Coding

6.3 74190 SYNCHRONOUS UP/DOWN COUNTER

If the 7490 can be called the workhorse of the decade counter ICs, the 74190 is its slick city cousin. Here is a device, as shown in figure 6.7, that is not only a decade counter, but one that will count up or down and do it from a presettable count. The 74190 has the same four data outputs as the 7490 (Q_A to Q_B). It too is a four-bit counter. Pin 14 is the clock input to the counter. Pin 5 is the up/down mode control. A logic 1 for count down and a logic 0 for count up. A high level at pin 4, the enable input, inhibits the counter from counting.

Figure 6.8 is a logic diagram of the 74190. The reader should be able to verify the logic behavior of the circuit when, for instance, the enable (pin 4) is at either a logic 1 or a logic 0. Figure 6.9 shows the input and output waveforms for a typical sequence of operation of this counter. Pin 11 is the load input; information present at this input works in conjunction with the data inputs A, B, C, and D. The information (logic levels) present at the data inputs will be transferred to the corresponding A, B, C, D outputs when the load input changes from a high to a low logic level.

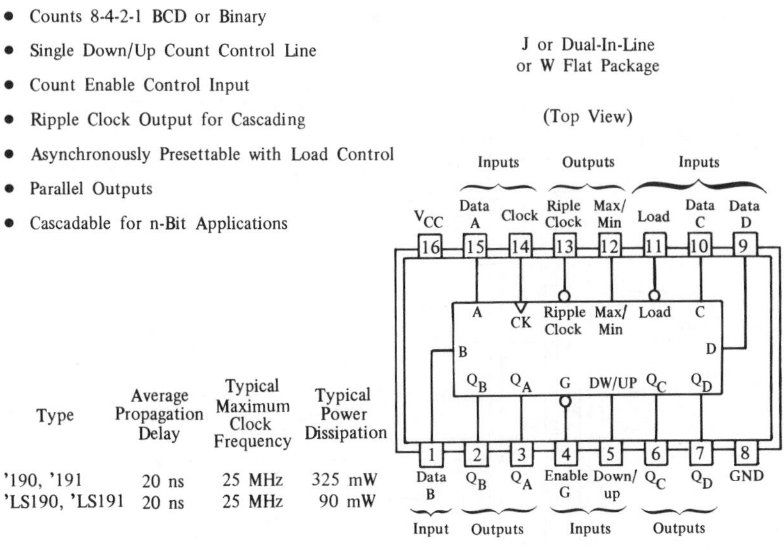

Type	Average Propagation Delay	Typical Maximum Clock Frequency	Typical Power Dissipation
'190, '191	20 ns	25 MHz	325 mW
'LS190, 'LS191	20 ns	25 MHz	90 mW

asynchronous inputs : Low input to load seats Q_A = A, Q_B @ B' Q_C @ C' and Q_B = B, Q_C = C, and Q_D = D

Figure 6.7 SN74190

With loading accomplished, and with the down/up control at a logic 0 and the enable input at a logic 0, the counter will start a count sequence in time with the input clock pulses. The counter is in its count up mode so it continues to count from its preset value to a maximum, a count of 9 in this case. When the counter reaches its maximum value, it will return to 0 and start to count up to 9 again. Decade-counter action will then continue until the counter receives a different set of instructions at its "coding" inputs. Should the enable input, for example, go to a high at this time, the counter will simply stop counting. Its outputs will remain at their last count. Returning the enable to 0 will allow the counter to count once again. If during the time the enable was high, the down/up control went to a logic 1, the counter would start to count down when it was again enabled. This means it would count down to 0, its minimum value, then return to 9 and start to count down again.

As if the fact that the 74190 can count in either direction, start from a preset count and start or stop its counting on command were not enough, it has two other features. Notice in figure 6.7 that pins 12 and 13 are also labeled as outputs. We shall look at pin 12 first. This is called the max/min output. It is normally at a low level. At the time the

'190, 'LS190 Decade Counters

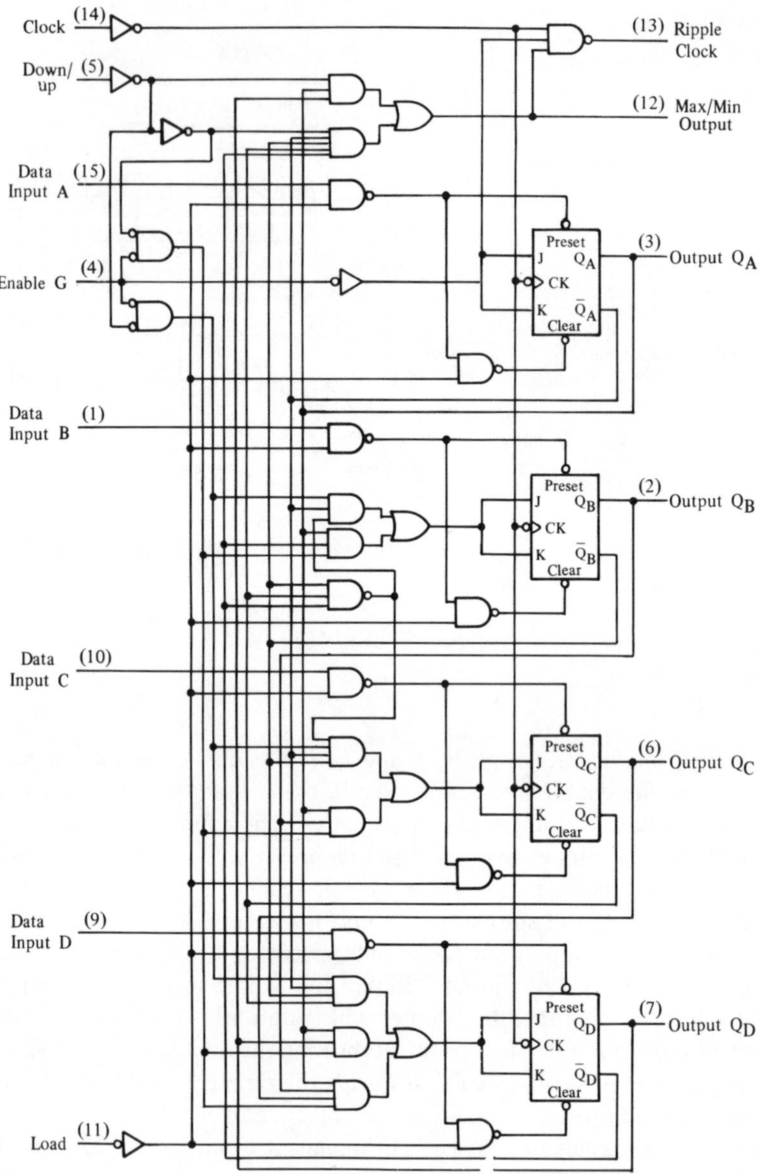

... Dynamic input activated by a transition from a high level to a low level.

Figure 6.8 SN74190 Functional Block Diagrams

'190, 'LS190 Decade Counters

Typical load, count, and inhibit sequences

Illustrated below is the following sequence:

1. Load (preset) to BCD seven.
2. Count up to eight, nine (maximam), zero, one, and two.
3. Inhibit.
4. Count down to one, zero (minimum), nine, eight, and seven.

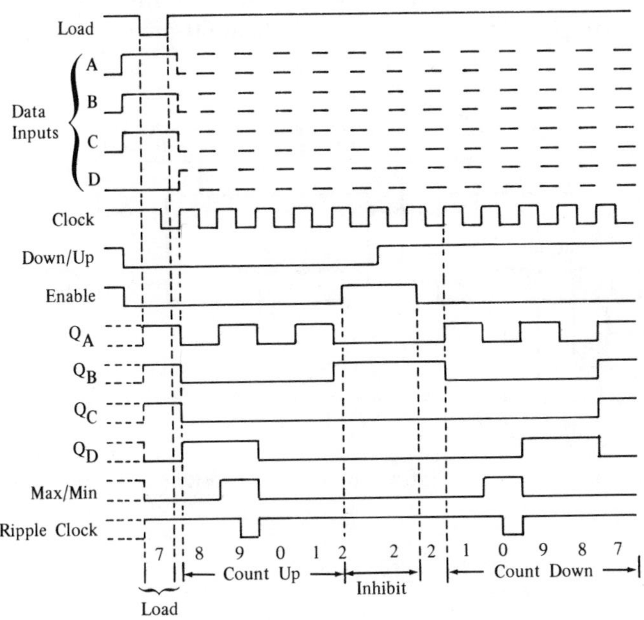

Figure 6.9 SN74190 Sequence of Operation

counter reaches its maximum or minimum count, depending on whether or not it is counting up or down, the output at pin 12 will go high and remain there for a full cycle of the input clock pulse. This could be particularly useful for obtaining a signal out when the counter has reached its count.

Example Suppose we wish to have an output after six pulses have entered the counter. Simply set the counter to a preset of 6. Put it in the count down mode. On the sixth pulse then, the output level of pin 12 will go high. This could then be gated to wherever the signal needed to go to signify a count of 6 had been obtained.

The ripple clock output gives a similar output. This output, however, goes from a high to a low and remains low only for the negative duration of the clock input pulse. By using these two outputs, the 74190 becomes a very versatile counting and timing device.

Figure 6.10 shows how three 74190's might be connected to produce a counter which can count seconds and minutes. The first counter (1) is simply a decade counter. Its ripple output will clock counter 2 every tenth pulse. Counter 2 will produce a ripple output pulse for every sixth pulse it receives. It would take 60 input pulses to counter 1 to produce a pulse out of 2 to counter 3.

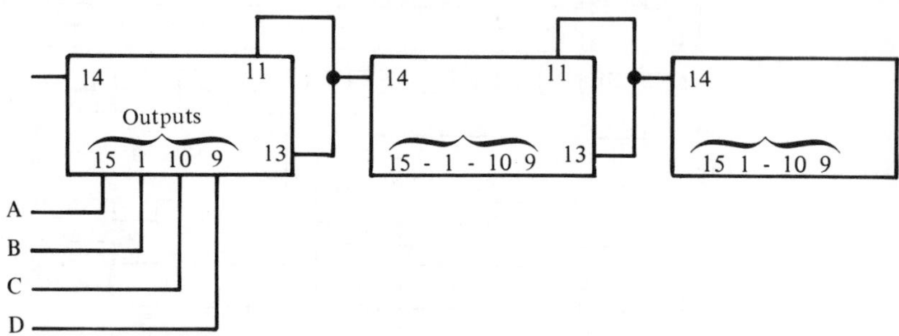

Figure 6.10 Three 74190 Connected to Produce a Three-Digit Counter

6.4 74192 SYNCHRONOUS UP/DOWN COUNTER

A device that is very similar to the 74190 is shown in figures 6.11, 6.12, and 6.13. The 74192 is also a four-bit synchronous counter. The major differences between the 74190 and 74192 are as follows:

1. The 74192 has no mode control. It has instead two (or dual) clocks. One clock input will cause the counter to count up, the other will cause it to count down.
2. In addition to the load input (for presetting the counters) the 74192 has a clear input. It will set the counter to 0 at its output regardless of the data inputs.
3. In place of the max/min and ripple outputs, the 74192 has a carry output and a borrow output. These outputs will each produce a negative pulse equal in duration to the negative por-

tion of the clock input. The borrow output will be produced when the output count reaches a minimum during the count down cycle. The carry output will be produced when the maximum count is reached during the count up cycle.

The 74192 may be preset to any count from zero to nine and made to count up or down from that preset value.

- Cascading Circuitry Provided Internally
- Synchronous Operation
- Individual Preset to Each Flip-Flop
- Fully Independent Clear Input

SN54', SN54LS' . . . J OR W Package
SN54L' . . . J Package
SN74', SN74L', SN74LS' . . . J OR N Package

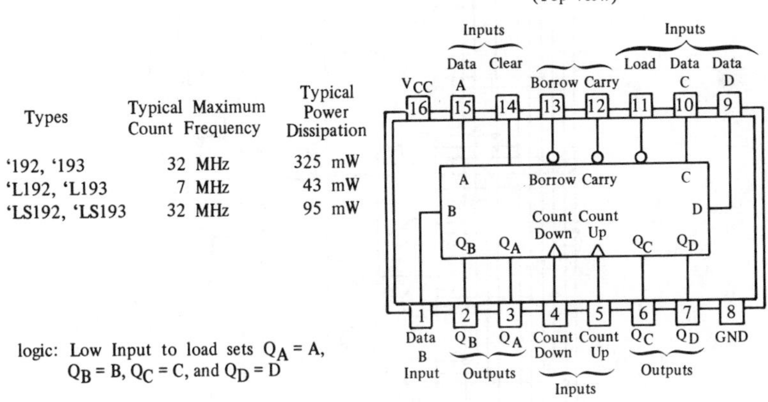

Types	Typical Maximum Count Frequency	Typical Power Dissipation
'192, '193	32 MHz	325 mW
'L192, 'L193	7 MHz	43 mW
'LS192, 'LS193	32 MHz	95 mW

logic: Low Input to load sets $Q_A = A$, $Q_B = B$, $Q_C = C$, and $Q_D = D$

absolute maximum ratings over operating free-air temperature range (unless otherwise noted)

	SN54'	SN54L'	SN54LS'	SN74'	SN74L'	SN74LS'	Unit
Supply voltage, V_{CC} (see Note 1)	7	8	7	7	8	7	V
Input voltage	5.5	5.5	7	5.5	5.5	7	V
Operating free-air temperature range	-55 to 125			0 to 70			°C
Storage temperature range	-65 to 150			-65 to 150			°C

Note1 : Voltage values are with respect to network ground terminal.

Figure 6.11 74192

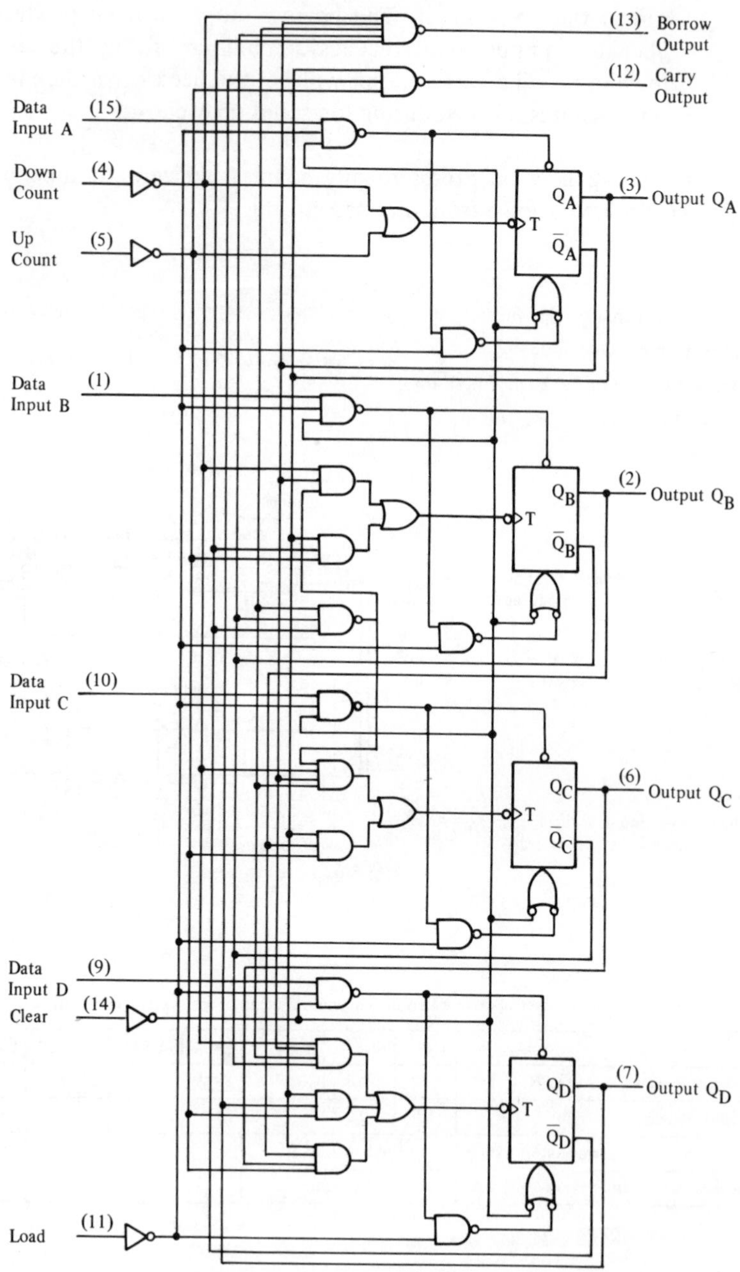

Figure 6.12 74192 Functional Block Diagram

typical clear, load, and count sequences

Illustrated below is the following sequence:

1. Clear outputs to zero.
2. Load (preset) to BCD seven.
3. Count up to eight, nine, carry, zero, one, and two.
4. Count down to one, zero, borrow, nine, eight, and seven.

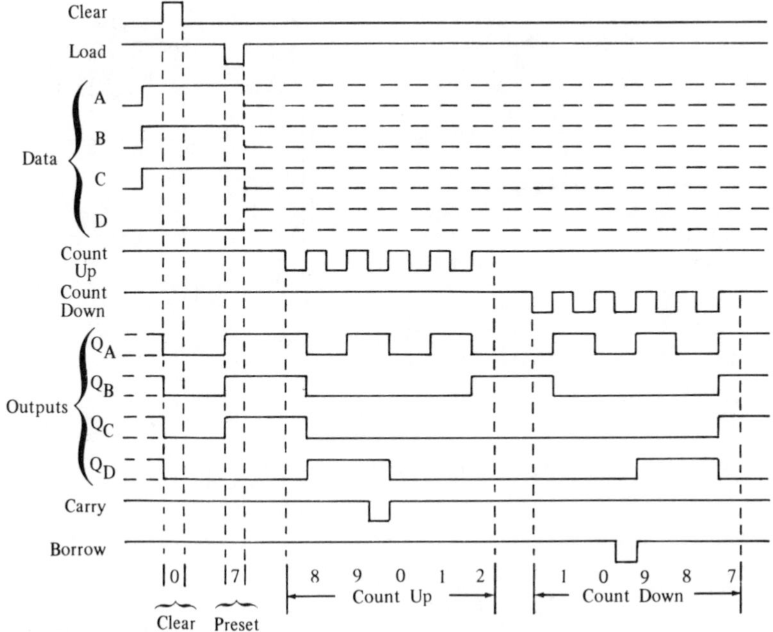

Notes: A. Clear overrides load, data, and count inputs.
B. When counting up, count-down input must be high; when counting down, count-up input must be high.

Figure 6.13 Timing Diagram for SN74192 Operation

6.5 74176 PRESETTABLE DECADE COUNTER

The 74176 is presented here to illustrate an asynchronous type of a presettable decade counter. This device is shown in figures 6.14 and 6.15. Another reason for showing the 74176 at this time is shown in figure 6.15. Here, instead of showing the waveforms for a particular count or load sequence, the truth (or function) tables for the 74176 are given. The operation of this device should be apparent from its logic diagram, as shown in figure 6.16. The 74176, it should be noted, has a logic diagram very similar to the 7490. The changes or differences between the devices should give the reader an insight to the need for so many different types of counters constructed on integrated circuit chips.

SN54176, SN54177 . . . J or W Package
SN74176, SN74177 . . . J or N Package
(Top View)

Reduce-Power versions of SN54196, SN54197, SN74196, and SN74197 50-MHz counters

D-C coupled counters designed to replace signetics 8280, 8281, 8290, and 8291 counters in most applications

Performs BCD, Bi-quinary, or Binary counting

Fully programmable

Fully independent clear input

Guaranteed to count at input frequencies from 0 to 35 MHz

Input clamping diodes simplify system design

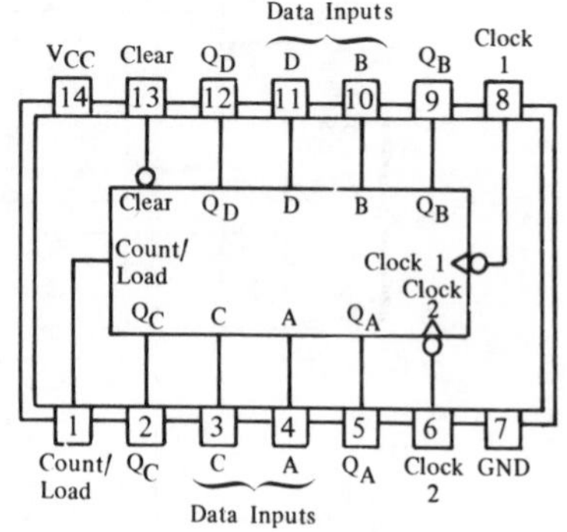

asynchronous input: Low input to clear sets Q_A, Q_B, Q_C, and Q_D, low

Figure 6.14 Pin-Out for SN74176/177

SN54176, SN74176
Function Tables

Decade (BCD)
(See Note A)

Count	Output			
	Q_D	Q_C	Q_B	Q_A
0	L	L	L	L
1	L	L	L	H
2	L	L	H	L
3	L	L	H	H
4	L	H	L	L
5	L	H	L	L
6	L	H	H	L
7	L	H	H	H
8	H	L	L	L
9	H	L	L	H

BI-Quinary (5-2)
(See Note B)

Count	Output			
	Q_A	Q_D	Q_C	Q_B
0	L	L	L	L
1	L	L	L	H
2	L	L	H	L
3	L	L	H	H
4	L	H	L	L
5	H	L	L	L
6	H	L	L	H
7	H	L	H	L
8	H	L	H	H
9	H	H	L	L

H = high level, L = low level

Notes: A. Output Q_A connected to clock-2 input.
B. Output Q_D connected to clock-1 input.

SN54177, SN74177
Funtion Table

(See Note A)

Count	Output			
	Q_D	Q_C	Q_B	Q_A
0	L	L	L	L
1	L	L	L	H
2	L	L	H	L
3	L	L	H	H
4	L	H	L	L
5	L	H	L	H
6	L	H	H	L
7	L	H	H	H
8	H	L	L	L
9	H	L	L	H
10	H	L	H	L
11	H	L	H	H
12	H	H	L	L
13	H	H	L	H
14	H	H	H	L
15	H	H	H	H

Figure 6.15 Function Tables for SN74176/177

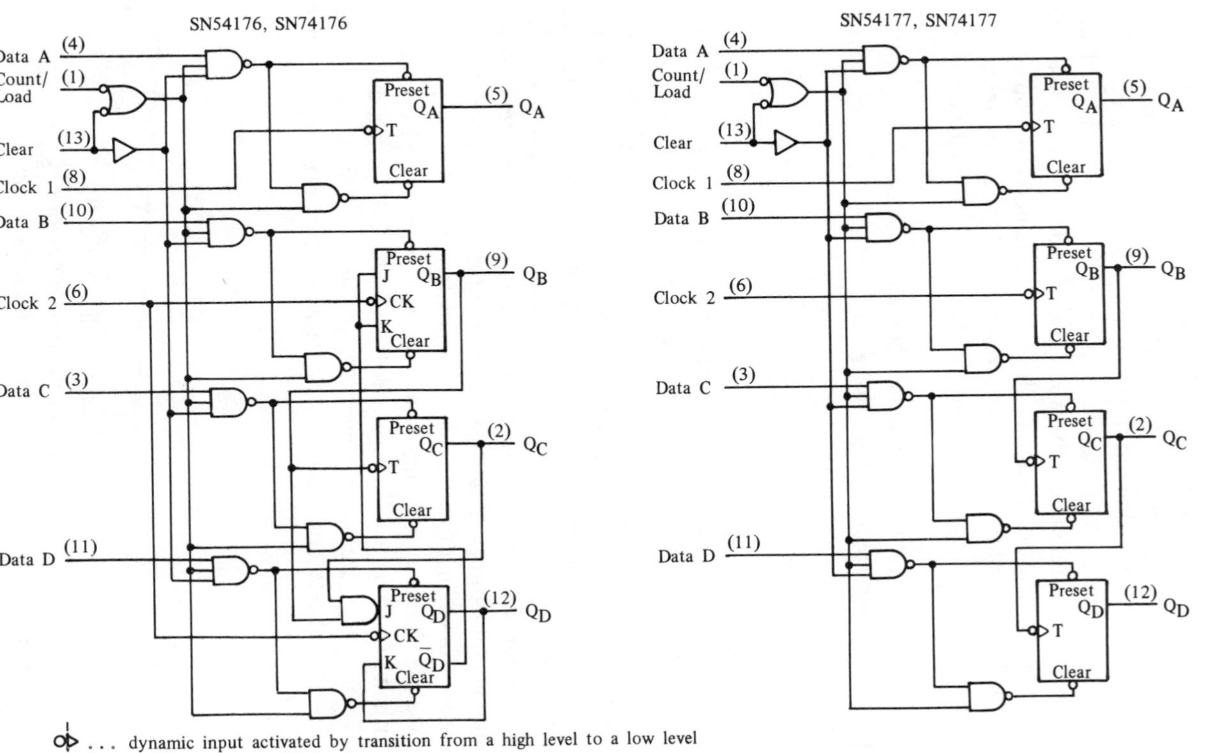

. . . dynamic input activated by transition from a high level to a low level

Figure 6.16 Functional Block Diagram for SN74176/177

6.6 74177 BINARY COUNTER

In figure 6.1, a very similar device to the 7490 was shown. This was the 7493, which is a binary counter. The binary counter is the general case of counters constructed using flip-flops. Figures 6.14, 6.15 and 6.16 also show the 74177 device. This device is a binary counter version of the 74176. The major differences between the decade counters and the binary counters are internal gating. The binary counters are allowed to count to their full binary limit before restarting their count sequence. A binary four-bit counter (as shown in figure 6.16) then can count from 0 to 15.

The 74193 was also the binary equivalent of the 74192 as the 74191 was the binary equivalent of the 74190.

SUMMARY

This chapter has taken a look at several types of binary and decade counters available as single integrated circuit units. The TI series was used for explanation simply because staying with one type of data sheet eases the readers' chores. The appendices give several other types of devices from several other manufacturers, along with the appropriate data sheets. This chapter has by no means tried to cover the field of devices used as counters. It is not the intent of this chapter, or this text, to replace any manufacturer's data sheets. By the time this book is read, the data sheets shown in this unit may be outdated. It is hoped that the reader will be able to use the information presented here and in chapter 5 to go hand in hand with any set of data sheets. All of the counters shown in chapter 5 were not shown in this unit. The reader is left to seek the data sheets and attempt to construct the counting circuits needed.

EXERCISES

1. Wire a 7490 and verify the truth table of figure 6.2.
   a. Use a bounceless switch for pulsing the input clock.
   b. Use LED displays to monitor the four outputs of the 7490.

2. Wire the circuit of figure 6.5 for a count/reset circuit to the 7490.

3. Restructure and wire figure 6.5 to enable the operator to set the counter to a 0, a 9, or the count mode.

   a. Use the simplest switching arrangement. Have your instructor verify this arrangement.

4. Construct and verify the operation of either a 74192 or 74176 counter.

   a. Use a bounceless switch for input pulses to the counter.

QUESTIONS AND PROBLEMS

6.1 To use the SN 7490 as a BCD counter, which external connections are needed?

6.2 Why must a bounceless switch be used to provide input pulses for the 7490 counter?

6.3 What is the difference between an SN 7490 and an SN 5490?

6.4 Show the connections needed to use the SN 7490 as a bi-quinary counter.

6.5 Of what value is a bi-quinary counter?

6.6 Show all connections needed for the SN 74190 to be used as an up/down counter. Show simple switching for mode control, etc.

6.7 Show a timing diagram for the 7490 used as a bi-quinary counter.

6.8 Show a circuit that would automatically have a 74190 count from 0 to 9 and then back down to 0, then back to 9, etc.

6.9 Of what value might a circuit such as the one of problem 6.8 be in a digital system?

6.10 Draw a complete system using the circuit of problem 6.8.

Encoding and Decoding

7

OBJECTIVE: To understand the definitions of encoding and decoding and to see the application and function of various circuits used for encoding and decoding.

Introduction: In this chapter the circuits of encoders and decoders will be shown. These circuits are very useful in the study of digital computer electronics. Simply, these circuits are the means by which information is changed into the binary system and is brought out of the binary system. You will not only see the function of these devices, but also will be shown their purpose. This chapter will focus its attention on several integrated circuit devices whose functions are encoding and decoding.

7.1 ENCODERS

By definition, the encoder is a circuit that changes information or data from any number system to the binary number system. Figure 7.1 shows a logic circuit diagram for the conversion of a decimal number (0–9) to its BCD equivalent. One use of this type of circuit is in the converting of information from a device such as a 10-key keyboard to a binary number. The small pocket calculators are a good example of the use of this type of circuitry.

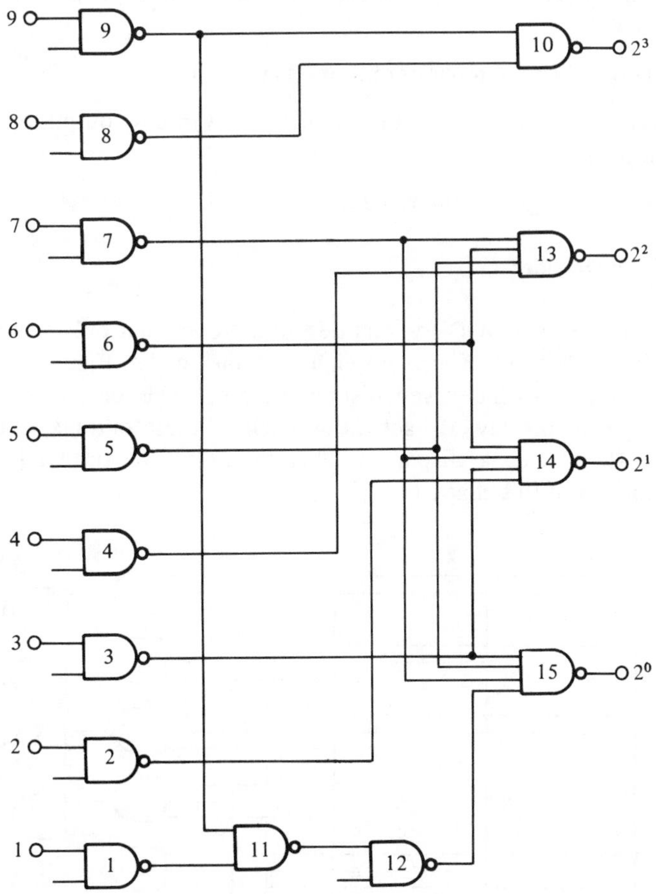

Figure 7.1 Decimal-to-BCD Conversion Logic Diagram

7.2 DECODERS

The definition of a decoder (or decoding) is the conversion from the binary number system to any other number system.

An encoder converts some numbering system into a binary system. True or False?

Answer: True

What determines the number of output gates needed by an encoder?

Answer: The number of output bits needed.

When encoding an octal number, how many output gates are needed?

Answer: Eight. Octal, remember is base 8.

BCD-Octal Decoder

Figure 7.2 shows a BCD-to-octal decoder. Notice that only three inputs are needed ($2^3 = 8$). There are eight outputs shown. Figure 7.3 shows a device that also has seven distinct outputs. This device is called a seven-segment display. By activating each of these outputs, a decimal number (0–9) can be displayed. This device will be further looked at in section 7.4 of this chapter.

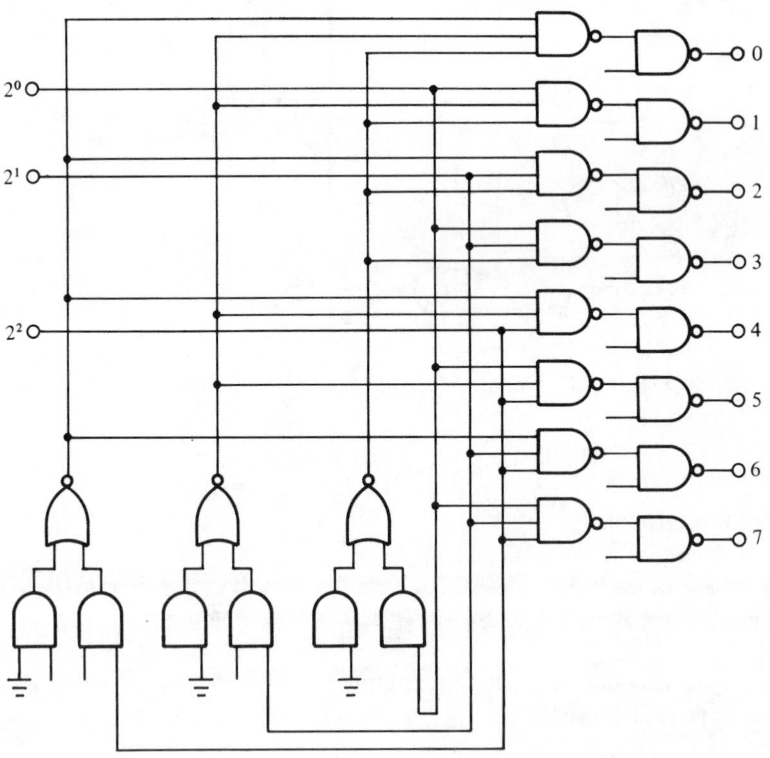

Figure 7.2 BCD-to-Octal Decoder

Figure 7.3 Seven-Segment Display

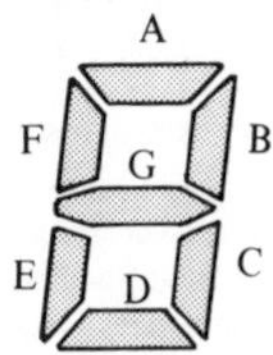

7.3 INTEGRATED CIRCUIT BCD—DECIMAL DECODER/DRIVES

Figure 7.4 shows the data page for a device that will decode a BCD to a decimal number. This device is the 74141. By coding the inputs, pins 3, 6, 7, 4 (A, B, C, D as shown in the function table), one of the outputs (0–9) will be on. This device is used to drive gas-filled cold-cathode tubes directly. Figure 7.5 shows a type 74145 BCD–decimal decoder driver also. This device is similar to the 74141 with the significant difference being in the off state output voltages and the output sink-current capabilities. The 74141 can sink 7 milliamperes and handle output voltages to 60 volts in the off state. This makes it ideal for driving the cold-cathode display tubes. The 74145 on the other hand can handle 80 milliamperes of output sink current. It has a 15-volt off state output voltage capability. Both of these devices are open-collector devices. Figure 7.6 illustrates the open-collector concept. The load must be placed between V_{CC} and the open-collector output. Remember for the 74141 the V_{CC} can be up to 60 volts dc.

7.4 INTEGRATED CIRCUIT BCD—SEVEN SEGMENT DECODER/DRIVERS

One of the most widely used display devices now on the market is the seven-segment display shown in figure 7.3. Each segment of this display is a lamp or an LED (light-emitting diode).

Figures 7.7, 7.8, 7.9, 7.10, and 7.11 are the data sheets for four types of BCD–seven-segment decoder/drivers. These are the 7446A, 7447A, 7448, and 7449. Each of these units is basically the same with minor variations. They are all positive-logic devices and all except the 7448A have open-collector outputs. The function tables should be studied and followed for specific operation of these devices. Figure 5.8 shows the segment identification and the displays for various numerical inputs. The inputs to each of these devices is BCD to the ABCD inputs. Figure 7.12 shows the block diagram connections for the general connection to a seven-segment decoder driver and a seven-segment display.

- Drives gas-filled cold-cathode indicator tubes directly
- Fully decoded inputs ensure all outputs are off for invalid codes
- Input clamping diodes minimize transmission-line effects

description

The SN74141 is a second-generation BCD-to-decimal decoder designed specifically to drive cold-cathode indicator tubes. This decoder demonstrates an improved capability to minimize switching transients in order to maintain a stable display.

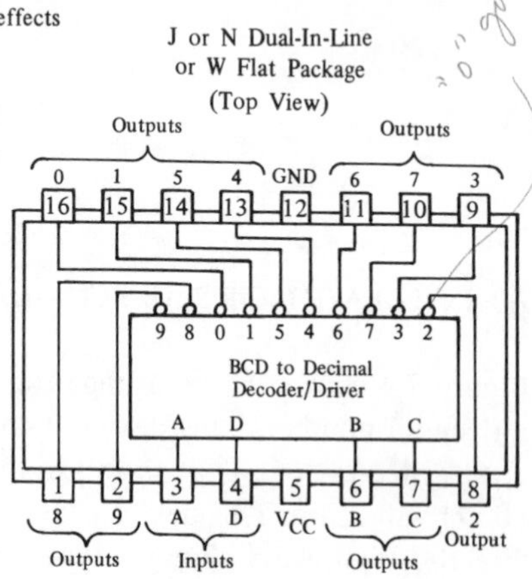

Positive logic: see truth table

Full decoding is provided for all possible input states. For binary inputs 10 through 15, all the outputs are off. Therefore the SN74141, combined with a minimum of external circuitry, can use these invalid codes in blanking leading- and/or trailing-edge zeros in a display. The ten high-performance, n-p-n output transistors have a maximum reverse current of 50 microamperes at 55 volts.

Low-forward-impedance diodes are also provided for each input to clamp negative-voltage transitions in order to minimize transmission-line effects. Power dissipation is typically 80 milliwatts. The SN74141 is characterized for operation over the temperature range of 0°C to 70°C.

functional block diagram

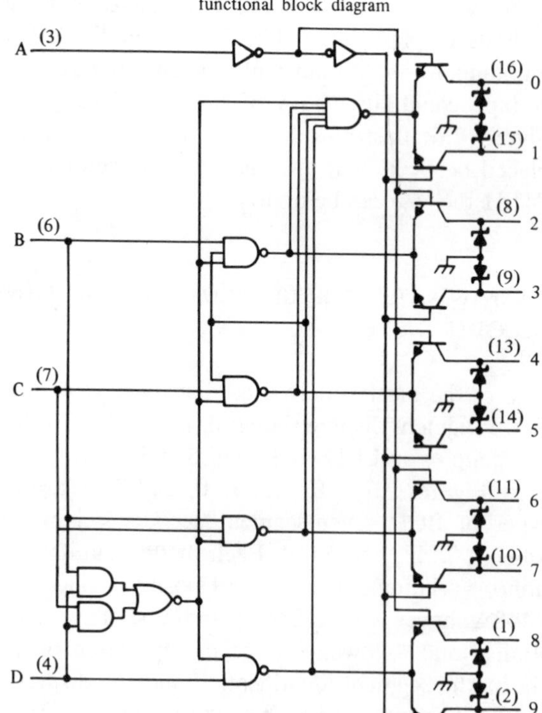

Function Table

Input				Output
D	C	B	A	On†
L	L	L	L	0
L	L	L	H	1
L	L	H	L	2
L	L	H	H	3
L	H	L	L	4
L	H	L	H	5
L	H	H	L	6
L	H	H	H	7
H	L	L	L	8
H	L	L	H	9
H	L	H	L	None
H	L	H	H	None
H	H	L	L	None
H	H	L	H	None
H	H	H	L	None
H	H	H	H	None

H = high level, L = low level
† All other outputs are off

Figure 7.4 Type SN74141 BCD-to-Decimal Decoder/Driver

- Full Decoding of Input Logic
- 80-mA Sink-Current Capability
- All Outputs Are Off for Invalid BCD Input Conditions

J or N Dual-in-Line or
W Flat Package

(Top View)

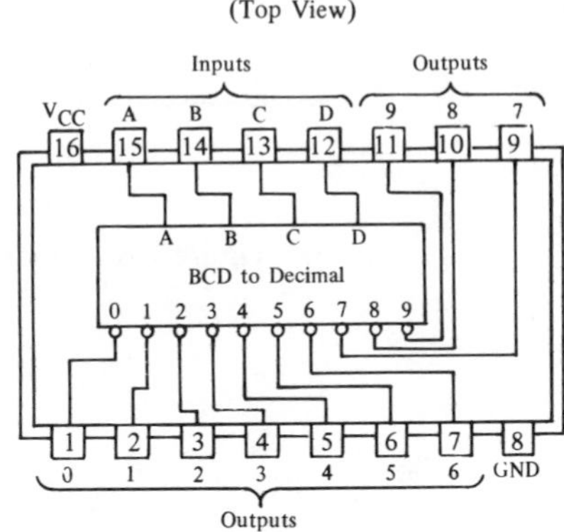

Positive logic: see function table

description

These monolithic BCD-to-decimal decoder/drivers consist of eight inverters and ten four-input NAND gates. The inverters are connected in pairs to make BCD input data avaible for decoding by the NAND gates. Full decoding of valid BCD input logic ensures that all outputs remain off for all invalid binary input conditions. These decoders feature TTL inputs and high-performance, n-p-n output transistors designed for use as indicator/relay drivers or as open-collector logic-circuit drivers. Each of the high-breakdown output transistors (15 volts) will sink up to 80 milliamperes of current. Each input is one normalized Series 54/74 load. Inputs and outputs are entirely compatible for use with TTL or DTL logic circuits, and the outputs are compatible for interfacing with most MOS integrated circuits. Power dissipation is typically 215 milliwatts.

logic

Function Table

NO.	Inputs				Outputs									
	D	C	B		0	1	2	3	4	5	6	7	8	9
0	L	L	L	L	L	H	H	H	H	H	H	H	H	H
1	L	L	L	H	H	L	H	H	H	H	H	H	H	H
2	L	L	H	L	H	H	L	H	H	H	H	H	H	H
3	L	L	H	H	H	H	H	L	H	H	H	H	H	H
4	L	H	L	L	H	H	H	H	L	H	H	H	H	H
5	L	H	L	H	H	H	H	H	H	L	H	H	H	H
6	L	H	H	L	H	H	H	H	H	H	L	H	H	H
7	L	H	H	H	H	H	H	H	H	H	H	L	H	H
8	H	L	L	L	H	H	H	H	H	H	H	H	L	H
9	H	L	L	H	H	H	H	H	H	H	H	H	H	L
Invalid	H	L	H	L	H	H	H	H	H	H	H	H	H	H
	H	L	H	H	H	H	H	H	H	H	H	H	H	H
	H	H	L	L	H	H	H	H	H	H	H	H	H	H
	H	H	L	H	H	H	H	H	H	H	H	H	H	H
	H	H	H	L	H	H	H	H	H	H	H	H	H	H
	H	H	H	H	H	H	H	H	H	H	H	H	H	H

H = high level (off), L = low level (on)

Functional block diagram

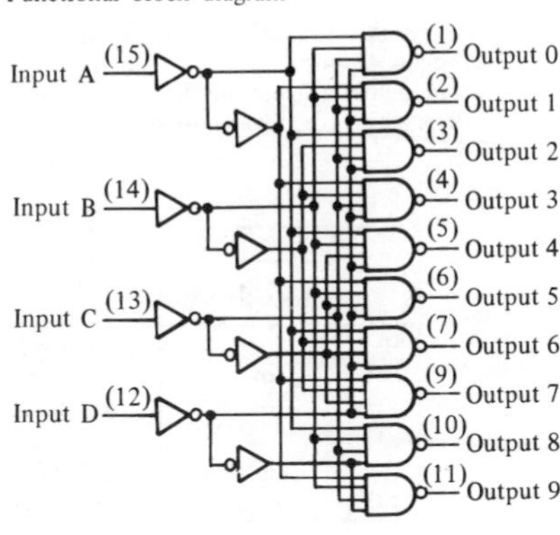

Figure 7.5 Types SN54145, SN74145 BCD-to-Decimal Decoder/Drivers

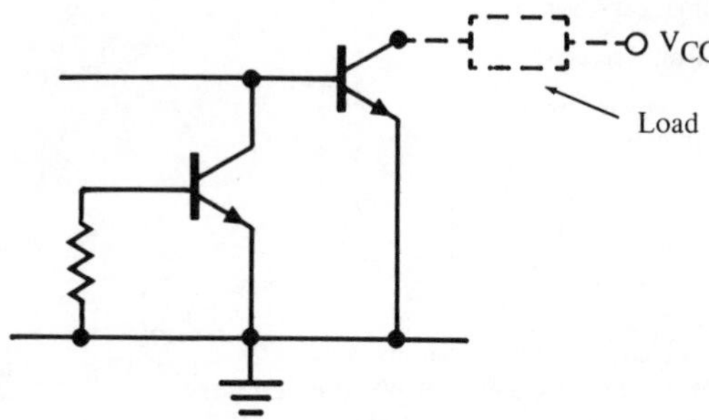

Figure 7.6 Open-Collector Concept

description

Of these BCD-to-seven-segment decoder/driver circuits, the '46A, 'L46A,'47A, and 'L47 feature active-low outputs designed for driving indicators directly, and the other two, '48 and '49, feature active-high outputs for driving lamp buffers. The following table summarizes the differences in the driver outputs and gives the typical power dissipation.

Type	Driver Outputs				Typical Power Dissipation
	Active-Level	Output Configuration	I_{OL} Sink Current	Max Voltage	
'46A	low	open-collector	40 mA	30 V	320 mW
'L46	low	open-collector	20 mA	30 V	160 mW
'47A	low	open-collector	40 mA	15 V	320 mW
'L47	low	open-collector	20 mA	15 V	160 mW
'48A	high	2- kΩ pull-up	6.4 mA	5.5 V	265 mW
'49A	high	open-collector	10 mA	5.5 V	165 mW

All of the circuits except '49 have ripple-blanking input/output controls and a lamp test input. The '49 circuit incorporates a direct blanking input. Segment identification with resultant displays are shown on the following page.

Figure 7.7 BCD-to-Seven-Segment Decoder/Drivers

• All Circuits Types Feature Lamp Intensity Modulation Capability

'46A, '47A, 'L46, 'L47 feature

• Open- Collector Outputs Drive Indicators Directly
• Lamp-Test Provision
• Leading/trailing Zero Suppression

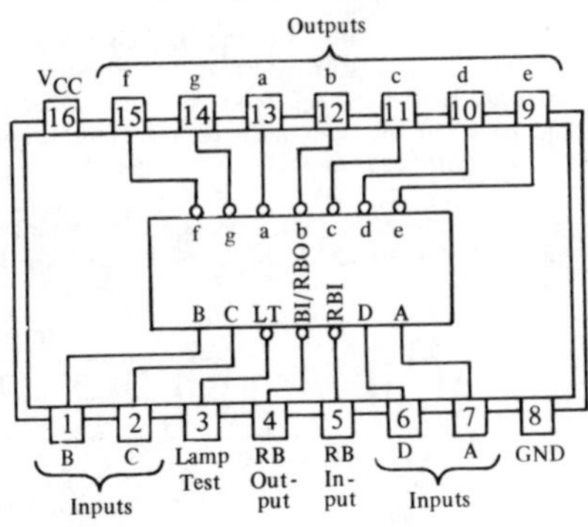

'48 features

• Internal Pull-Ups Eliminate Need for External Resistors
• Lamp-Test Provision
• Leading/Trailing Zero Suppression

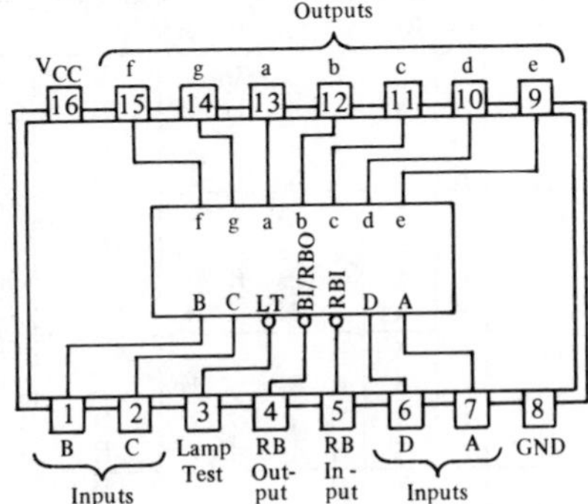

'49 features

• Open-Collector Outputs
• Blanking Input

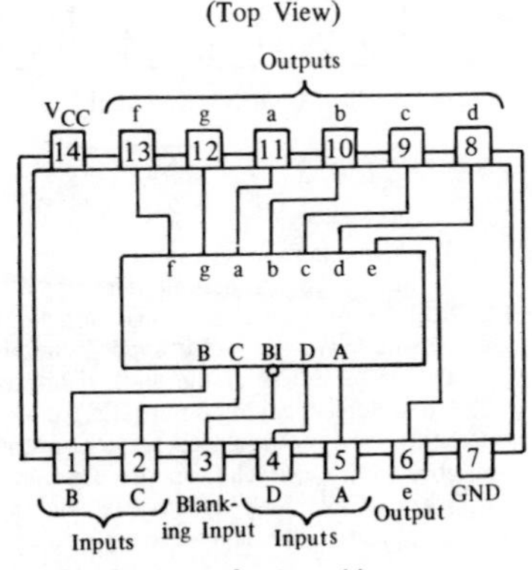

positive logic: see funtion tables

Figure 7.7 *continued*

description

The '46A, '47A, '48, 'L46, and 'L47 circuits incorporate automatic leading and/or trailing-edge zero-blanking control (RBI and RBO). Lamp test (LT) of types may be performed at any time when the BI/RBO node is at a high level. All types contain an overriding input (B) which can be used to control the lamp intensity by pulsing or to inhibit the, outputs. Inputs and outputs are entirely compatible for use with TTL or DTL logic outputs.

Series 54 and Series 54L devices are characterized for operation over the full military temperature range of-55°C to 125°C; Series 74 and Series 74L devices are characterized for operation from 0°C

a
f g b
e c
d

Segment Identification

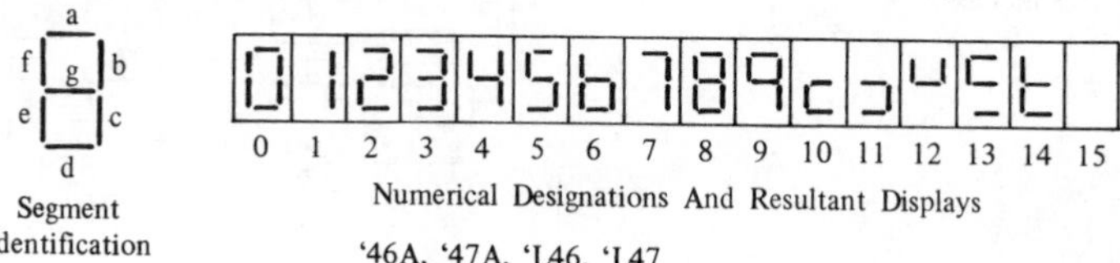

Numerical Designations And Resultant Displays

'46A, '47A, 'L46, 'L47

Function Table

Decimal or Function	Inputs						BI/RBO†	Outputs							Note
	LT	RBI	D	C	B	A		a	b	c	d	e	f	g	
0	H	H	L	L	L	L	H	ON	ON	ON	ON	ON	ON	OFF	1
1	H	X	L	L	L	H	H	OFF	ON	ON	OFF	OFF	OFF	OFF	1
2	H	X	L	L	H	L	H	ON	ON	OFF	ON	ON	OFF	ON	
3	H	X	L	L	H	H	H	ON	ON	ON	ON	OFF	OFF	ON	
4	H	X	L	H	L	L	H	OFF	ON	ON	OFF	OFF	ON	ON	
5	H	X	L	H	L	H	H	ON	OFF	ON	ON	OFF	ON	ON	
6	H	X	L	H	H	L	H	OFF	OFF	ON	ON	ON	ON	ON	
7	H	X	L	H	H	H	H	ON	ON	ON	OFF	OFF	OFF	OFF	
8	H	X	H	L	L	L	H	ON	ON	ON	ON	ON	ON	ON	
9	H	X	H	L	L	H	H	ON	ON	ON	OFF	OFF	ON	ON	
10	H	X	H	L	H	L	H	OFF	OFF	OFF	ON	ON	OFF	ON	
11	H	X	H	L	H	H	H	OFF	OFF	ON	ON	OFF	OFF	ON	
12	H	X	H	H	L	L	H	OFF	ON	OFF	OFF	OFF	ON	ON	
13	H	X	H	H	L	H	H	ON	OFF	OFF	ON	OFF	ON	ON	
14	H	X	H	H	H	L	H	OFF	OFF	OFF	ON	ON	ON	ON	
15	H	X	H	H	H	H	H	OFF	OFF	OFF	OFF	OFF	OFF	OFF	
BI	X	X	X	X	X	X	L	OFF	OFF	OFF	OFF	OFF	OFF	OFF	2
RBI	H	L	L	L	L	L	L	OFF	OFF	OFF	OFF	OFF	OFF	OFF	3
LT	L	X	X	X	X	X	H	ON	ON	ON	ON	ON	ON	ON	4

H = high level, L = low level, X = irrelevant

Notes:
1. The blanking input (B) must be open or held at a high logic level when output functions 0 through 15 are desired. The ripple-blanking input (RBI) must be open or high if blanking of a decimal zero is not desired.
2. When a low logic level is applied directly to the blanking input (BI), all segment outputs are off regardless of the level of any other input.
3. When ripple-blanking input (RBI) and inputs A, B, C, and D are at a low level with the lamp test input high, all segment outputs go off and the ripple-blanking output (RBO) goes to a low level (response condition).
4. When the blanking input/ripple blanking output (BI/RBO) is open or held high and a low is applied to the lamp-test input, all segment outputs are on.

†BI/RBO is wire-AND logic serving as blanking input (BI) and/or ripple–blanking output

Figure 7.8 BCD-to-Seven-Segment Decoder/Drivers

'48 Function Table

Decimal or Function	Inputs						BI/RBO†	Outputs							Note
	LT	RBI	D	C	B	A		a	b	c	d	e	f	g	
0	H	H	L	L	L	L	H	H	H	H	H	H	H	L	1
1	H	X	L	L	L	H	H	L	H	H	L	L	L	L	1
2	H	X	L	L	H	L	H	H	H	L	H	H	L	H	
3	H	X	L	L	H	H	H	H	H	H	H	L	L	H	
4	H	X	L	H	L	L	H	L	H	H	L	L	H	H	
5	H	X	L	H	L	H	H	H	L	H	H	L	H	H	
6	H	X	L	H	H	L	H	L	L	H	H	H	H	H	
7	H	X	L	H	H	H	H	H	H	H	L	L	L	L	
8	H	X	H	L	L	L	H	H	H	H	H	H	H	H	
9	H	X	H	L	L	H	H	H	H	H	L	L	H	H	
10	H	X	H	L	H	L	H	L	L	L	H	H	L	H	
11	H	X	H	L	H	H	H	L	L	H	H	L	L	H	
12	H	X	H	H	L	L	H	L	H	L	L	L	H	H	
13	H	X	H	H	L	H	H	H	L	L	H	L	H	H	
14	H	X	H	H	H	L	H	L	L	L	H	H	H	H	
15	H	X	H	H	H	H	H	L	L	L	L	L	L	L	
BI	X	X	X	X	X	X	L	L	L	L	L	L	L	L	2
RBI	H	L	L	L	L	L	L	L	L	L	L	L	L	L	3
LT	L	X	X	X	X	X	H	H	H	H	H	H	H	H	4

Notes: 1. The blanking input (BI) must be open or held at a high logic level when output functions 0 through 15 are desired. The ripple-blanking input (RB) must be open or high, if blanking of a decimal zero is not desired.

2. When a low logic level is applied directly to the blanking input (BI), all segment outputs are low regardless of the level of any other input.

3. When ripple-blanking input (RB) and inputs A, B, C, and D are at a low level with the lamp-test input high, all segment outputs go low and the ripple-blanking output (RBO) goes to a low level (response condition)'

4. When the blanking input/ripple-blanking output (B1/RBO) is open or held high and a low is applied to the lamp-test input, allsegment outputs are high.

†BI/RBO is wire-AND logic serving as blanking input (B) and/or ripple-blanking output (RBO)

'49 Function Table

Decimal or Function	Inputs					Outputs							Note
	D	C	B	A	BI	a	b	c	d	e	f	g	
0	L	L	L	L	H	H	H	H	H	H	H	L	1
1	L	L	L	H	H	L	H	H	L	L	L	L	
2	L	L	H	L	H	H	H	L	H	H	L	H	
3	L	L	H	H	H	H	H	H	H	L	L	H	
4	L	H	L	L	H	L	H	H	L	L	H	H	
5	L	H	L	H	H	H	L	H	H	L	H	H	
6	L	H	H	L	H	L	L	H	H	H	H	H	
7	L	H	H	H	H	H	H	H	L	L	L	L	
8	H	L	L	L	H	H	H	H	H	H	H	H	
9	H	L	L	H	H	H	H	H	L	L	H	H	
10	H	L	H	L	H	L	L	L	H	H	L	H	
11	H	L	H	H	H	L	L	H	H	L	L	H	
12	H	H	L	L	H	L	H	L	L	L	H	H	
13	H	H	L	H	H	H	L	L	H	L	H	H	
14	H	H	H	L	H	L	L	L	H	H	H	H	
15	H	H	H	H	H	L	L	L	L	L	L	L	
BI	X	X	X	X	L	L	L	L	L	L	L	L	2

H = high level, L= low level, X = irrelevant

Notes: 1. The blanking input (B) must be open or held at a high logic level when output functions 0 through 15 are desired.

2. When a low logic level is applied directly to the blanking input (B), all segment outputs are low regardless of the level of any other input.

Figure 7.9 BCD-to-Seven-Segment Decoder/Drivers

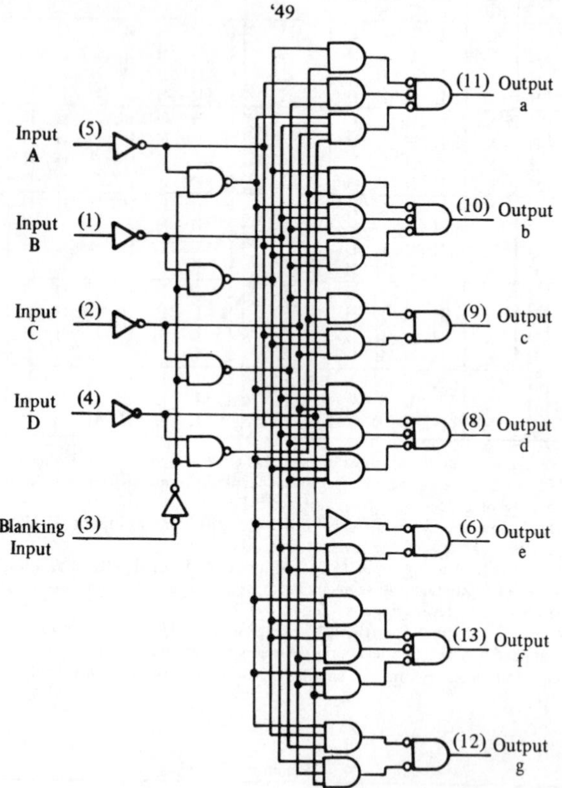

Figure 7.10 BCD-to-Seven-Segment Decoder/Drivers

Figure 7.10 *continued*

functional block diagrams

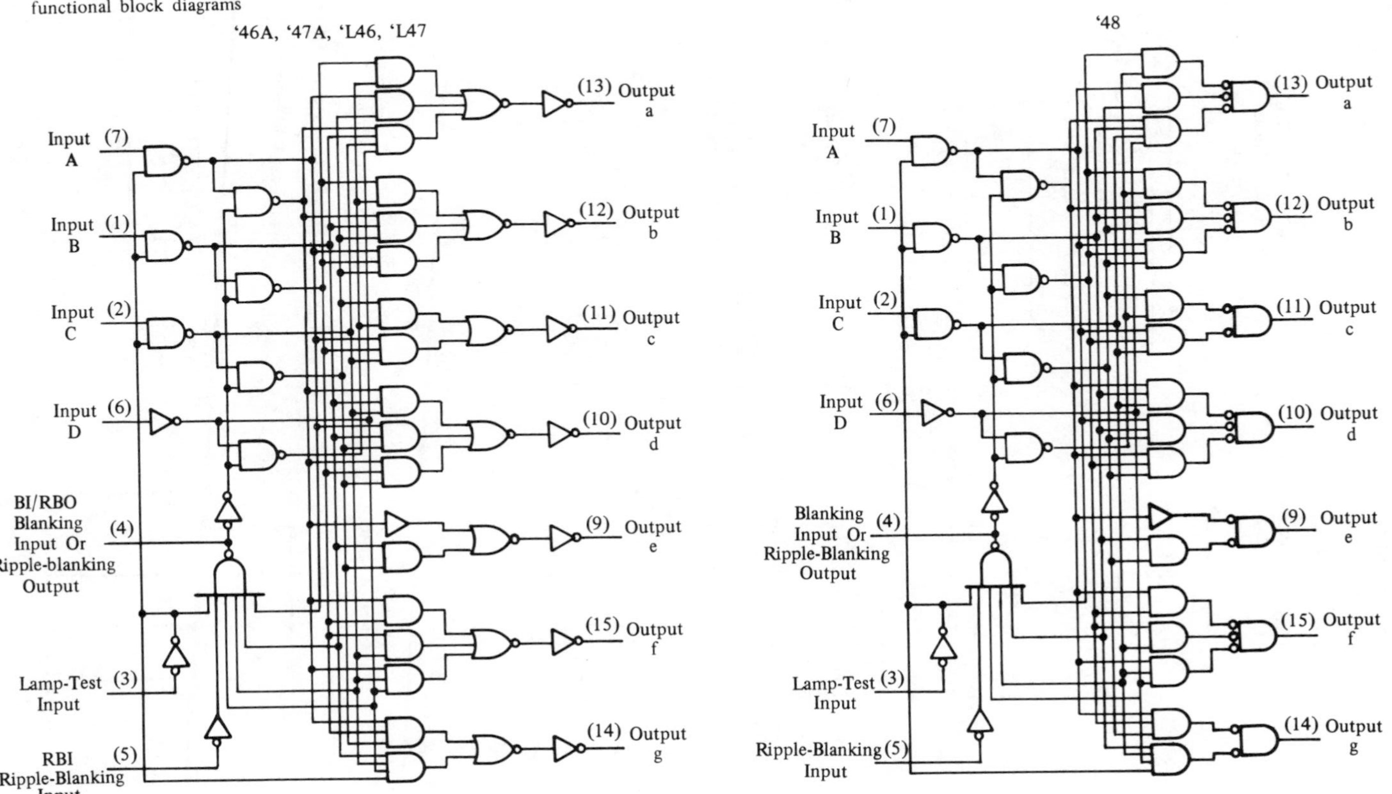

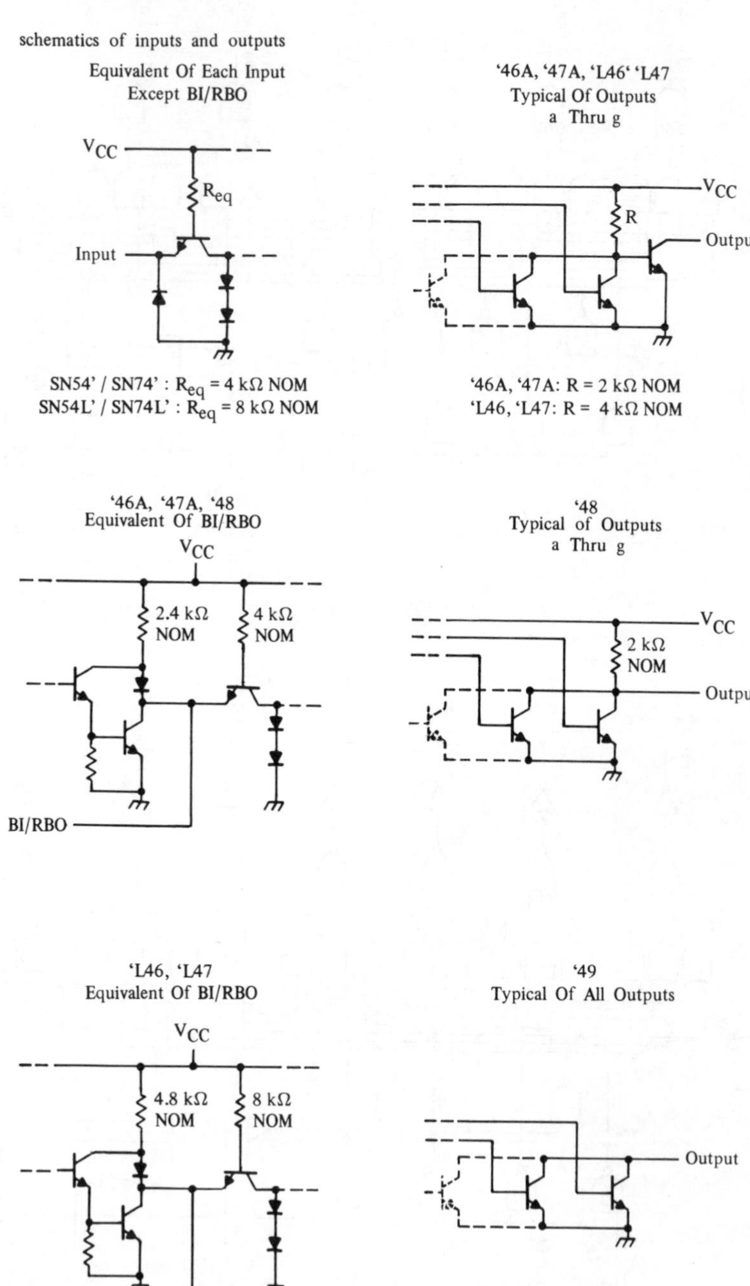

Figure 7.11 BCD-to-Seven-Segment Decoder/Drivers

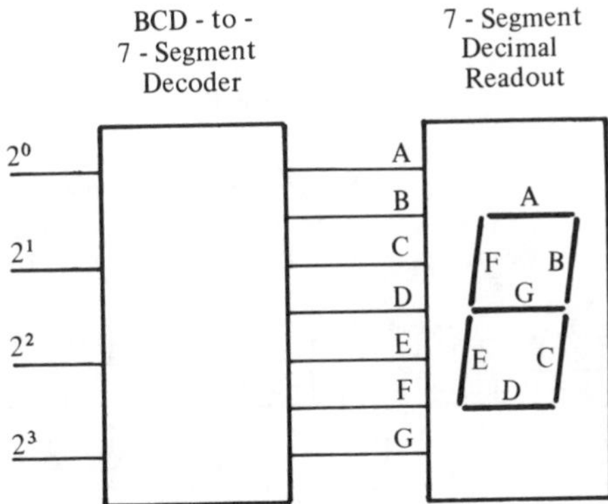

Figure 7.12 Block Diagram of Interconnections—Seven-Segment Decoder/ Driver to Readout

7.5 PRIORITY ENCODERS

Figure 7.13 shows two types of encoders. The 74147 is a ten-line to four-line priority encoder. The 74148 is an eight-line to three-line priority encoder. The function of these two devices is similar. The 74147 converts a ten-digit decimal to a four-bit binary-coded decimal. The 74148 converts an eight-digit decimal to a three-bit binary coded decimal. The data input lines are checked with the data (low) at the highest input being translated to the proper BCD output.

7.6 PARITY

Parity is the name given to one method of error detection during the transmission or reception of binary data. The parity method of error detection is quite simple. The detection device simply counts the number of logic highs (1's) contained in the binary data. For example, even parity would mean the binary data had an even number of 1's while odd parity would denote an odd number of 1's contained in the binary word. To use the parity method for error detection, an extra bit of data is included with the actual binary data. This bit is called the *parity bit*. This extra bit is added by a device called a *parity generator*. Let's see how the system actually works. Suppose we wish to send the two hexadecimal characters 77. When written in binary form this would be 01110111. Now let's suppose we want to use odd parity. This means we want to

SN54147, SN74147

- Encodes 10-Line Decimal to 4-Line BCD
- Applications Include:

  Keyboard Encoding
  Range Selection

- Typical Data Delay . . . 10 ns
- Typical Power Dissipation . . . 225 mW

J or N Dual-In-line or
W Flat Package (Top View)

SN54148, SN74148

- Encodes 8 Data Lines to 3-Line Binary (Octal)
- Application Include:

  N-Bit Encoding
  Code Converters and Generators

- Typica Data Delay . . . 10 ns
- Typical Power Dissipation . . . 190 mW

J or N Dual-In-Line or
W Flat Package (Top View)

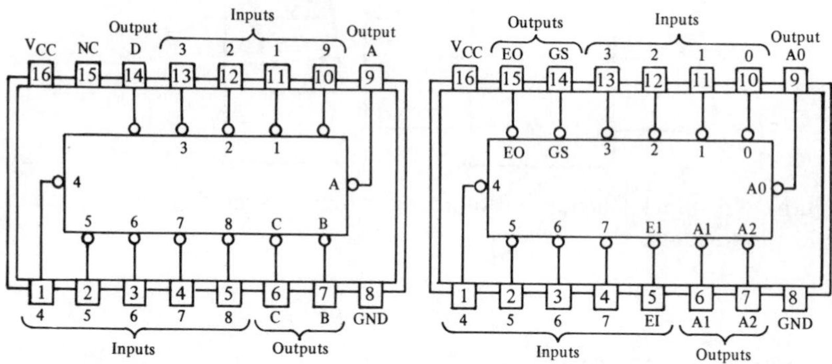

positive logic: seefunction table

NC – No internal connection

description

These TTL encoders feature priority decoding of the inputs to ensure that only the highest-order data line is encoded. The SN54147 and SN74147 encode nine data lines to four-line (8-4-2-1) BCD. The implied decimal zero condition requires no input condition as zero is encoded when all nine data lines are at a high logic level. All inputs are buffered to represent one normalized Series 54/74 load. The SN54148 and SN74148 encode eight data lines to three-line (4-2-1) binary (octal). Cascading circuitry (enable input EL and enable output EO) has been provided to allow octal expansion without the need for external circuitry. For all types, data inputs and outputs are active at the low logic level.

SN54147, SN74147

Function Table

Inputs									Outputs			
1	2	3	4	5	6	7	8	9	D	C	B	A
H	H	H	H	H	H	H	H	H	H	H	H	H
X	X	X	X	X	X	X	X	L	L	H	H	L
X	X	X	X	X	X	X	L	H	L	H	H	H
X	X	X	X	X	X	L	H	H	H	L	L	L
X	X	X	X	X	L	H	H	H	H	L	L	H
X	X	X	X	L	H	H	H	H	H	L	H	L
X	X	X	L	H	H	H	H	H	H	L	H	H
X	X	L	H	H	H	H	H	H	H	H	L	L
X	L	H	H	H	H	H	H	H	H	H	L	H
L	H	H	H	H	H	H	H	H	H	H	H	L

SN54148, SN74148

Function Table

Inputs									Outputs				
EI	0	1	2	3	4	5	6	7	A2	A1	A0	GS	EO
H	X	X	X	X	X	X	X	X	H	H	H	H	H
L	H	H	H	H	H	H	H	H	H	H	H	H	L
L	X	X	X	X	X	X	X	L	L	L	L	L	H
L	X	X	X	X	X	X	L	H	L	L	H	L	H
L	X	X	X	X	X	L	H	H	L	H	L	L	H
L	X	X	X	X	L	H	H	H	L	H	H	L	H
L	X	X	X	L	H	H	H	H	H	L	L	L	H
L	X	X	L	H	H	H	H	H	H	L	H	L	H
L	X	L	H	H	H	H	H	H	H	H	L	L	H
L	L	H	H	H	H	H	H	H	H	H	H	L	H

H = high logic level, L = low logic level, X = irrelevant

Figure 7.13 Ten-Line-to-Four-Line and Eight-Line-to-Three-Line Priority Encoders

have our binary data contain an odd number of 1's. Our word, however, contains six 1's, an even number. This means we need to add an extra 1 to the word. This extra bit, when added, will not change the data word. So an extra bit position is added to the data word:

__01110111
Parity bit space

Now, we will simply insert the extra 1 in the parity-bit space, giving us

101110111

which is the data 77 with an odd parity bit added. If on the other hand we wanted to send the data 73, it would appear as

01110011

This data word contains an odd number of 1's, therefore we will not add any. We will, however, add a logic 0 to give us

001110011

This is now the data word 73 with correct odd-parity.

Figure 7.14 illustrates a system to transmit data with parity. Here we see the elements of a system that will transmit two hex words with odd parity. The system is also capable of receiving the data with error detection.

The two hexadecimal words are transmitted from A to B. The two hexadecimal words are also sent to a unit called a *parity generator*. Here, the two words are checked and the number of 1's is counted. If the number is even, a parity bit of 1 is sent out on the parity-bit line. If the number is odd, a 0 is sent out on the parity-bit line. At the receiving end of the system, the two hexadecimal words are sent to a receiver and to a parity checker unit. The parity checker checks the two words and the extra parity bit. If the number of 1's is odd, the unit signals OK. If the number of 1's is even, an error signal is given.

Parity Generators/Checkers

To examine and understand how the system of figure 6.14 operates more fully, we will discuss a device that is used as both a parity-bit generator and parity checker. One such device is the 74280. The pin out for this device is given in figure 7.15. Notice that there are nine inputs

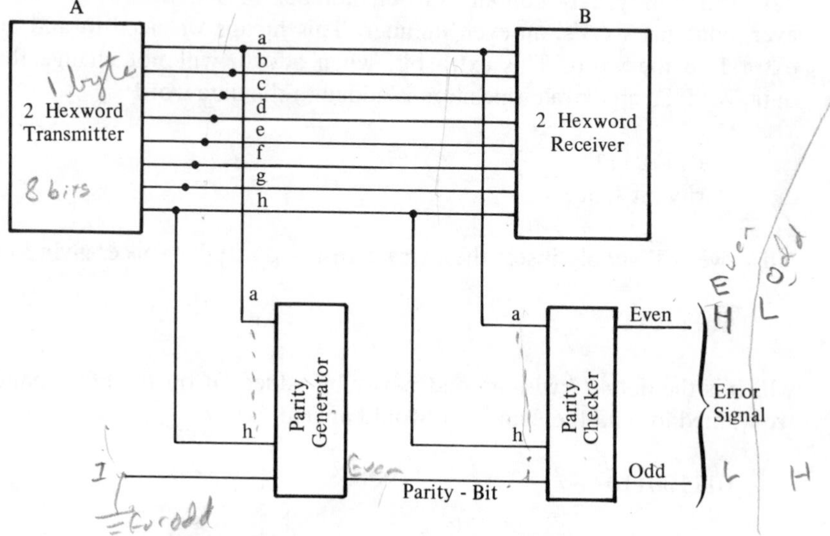

Figure 7.14 Parity Generator/Checker System

- Generates either odd or even parity for nine data lines
- Cascadable for n-bits
- Can be used to upgrade existing Systems using MSI parity circuits
- Typical data-to-output delay of only 14ns for 'S280 and 33 ns for 'LS280
- Typical power dissipation:
  'LS280 . . . 80 mW
  'S280 . . . 335 mW

Function Table

Number of Inputs A Thru 1 that are high	Outputs	
	Σ even	Σ odd
0, 2, 4, 6, 8	H	L
1, 3, 5, 7, 9	L	H

H = high level, L = low level

SN54LS280, SN54S280 . . . J or W package
SN74LS280, SN74S280 . . . J or N package
(Top View)

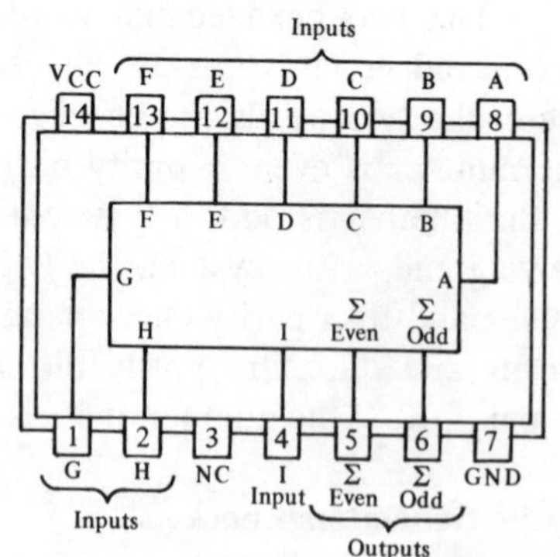

Logic: see function table

NC - No internal connection

Figure 7.15 The 74280

to this device and two outputs. If the number of inputs that are at a logic 1 are even, pin 5 (Σ EVEN) will be a logic 1 and pin 6 (Σ ODD) will be low. If the number of inputs at a logic 1 is odd then pin 5 will be high (logic 1).

So, now we see that the parity generator and parity checker of our system (figure 7.14) are actually the same device. Figure 7.16 now shows the actual connections of two 74280 devices as used in the system. If the hexadecimal 77 were transmitted, the 74280 used as a parity generator would count the number of inputs at logic 1. This as we saw earlier would be 6. The ninth input line is physically tied to 0, therefore the sum is still 6. Pin 5 (Σ EVEN) will go high. At the receiving end of the system, the two hex-words will be input to the checker along with the input from the extra parity bit. This 74280 will now see seven lines that are high (logic 1). This will produce a logic 1 at pin 6 (Σ ODD) and light the signal which shows odd.

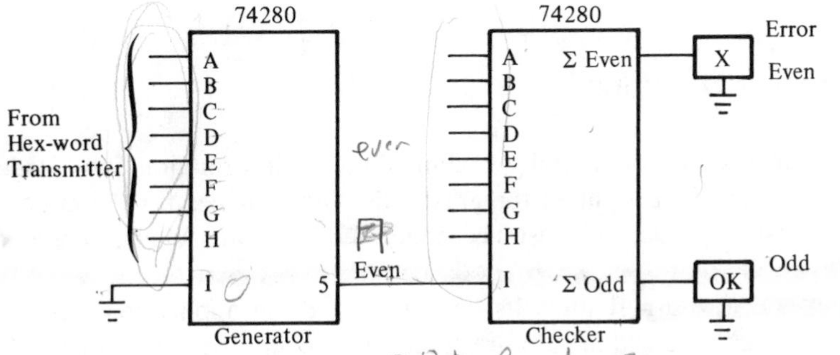

Figure 7.16 System Using 74280 Devices

It must be emphasized at this point that this check in no way guarantees error-free transmission, reception, or detection. The parity checker will only catch one type of error, the number of bits that are at a logic 1. It will, however, reduce the chance of overall error in the system and parity checking is used extensively in digital systems and computers.

Figure 7.17 shows a system for parity detection using exclusive-OR gates. Of course, several 74280 devices could be cascaded to enable detection of larger data-word groups. Figure 7.17 is sometimes called a *parity tree*. This method requires two packages, whereas the 74280 requires one package. A designer might need to make a cost versus package-count decision when doing an actual design project, so it is important that more than one way to implement a design be understood.

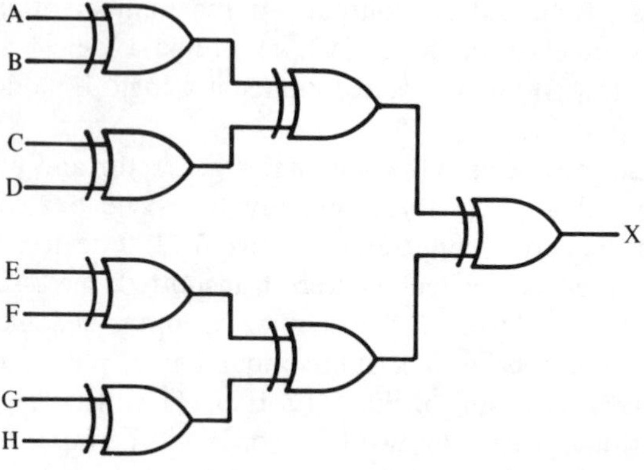

X = A⊕B⊕C⊕D⊕E⊕F⊕G⊕H

Figure 7.17 Eight-Bit Parity Checker Using Exclusive-OR Gates

7.7 THE GRAY CODE

In our systems of digitally coding data or information, one code is rather unique. It is called the gray code. Other names for the gray code are position code, unit-distance code, and minimum change code. The gray code is an unweighted code, and therefore is not suited for arithmetic operations. It finds its best uses in those applications where an error might occur due to more than bit changing per time. Figure 7.18 shows the conversion from decimal to binary to gray code. To further illustrate the use of the gray code as a positional code, figure 7.19 shows a pattern that can be used as a shaft position indicator. Note that the shaft is capable of rotation with one of sixteen different bit patterns being selected for a position indication. More bits could be selected by decreasing the angular displacement between each sector. For any change from one sector to the next adjoining sector, only one bit changes. For example, as shown in figure 7.18, when going from seven to eight decimal we must go from 0111 to 1000 binary. This requires a change in all four bits of the binary information. To make this change in gray code is a change from 0100 to 1100. Only the most-significant bit was changed. This is very important. The fewer bits we have to change, the more we decrease the possibility of error. The time it takes for a system to change all of its bits can sometimes be long, thus while all of the bits are changing, there is the possibility of a wrong reading of the data.

Decimal	Code: Binary	Code: Gray
0	0000	0000
1	0001	0001
2	0010	0011
3	0011	0010
4	0100	0110
5	0101	0111
6	0110	0101
7	0111	0100
8	1000	1100
9	1001	1101
10	1010	1111
11	1011	1110
12	1100	1010
13	1101	1011
14	1110	1001
15	1111	1000

Figure 7.18 Gray-Code Conversion

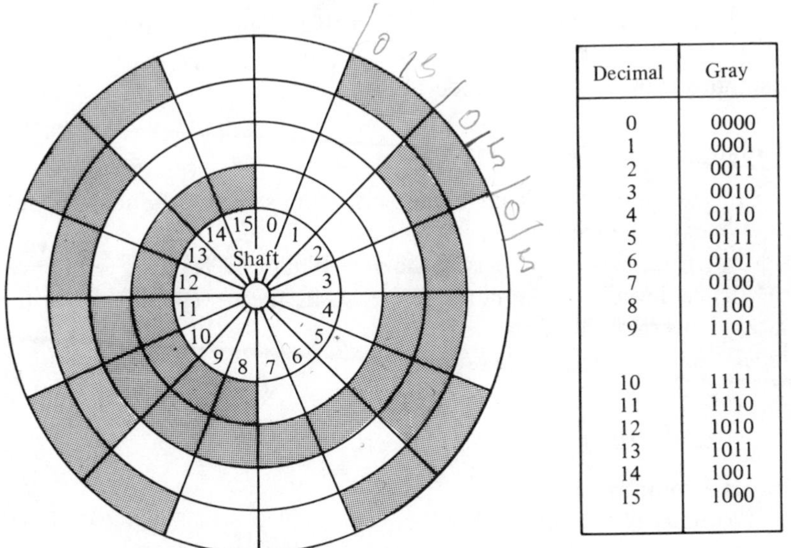

Decimal	Gray
0	0000
1	0001
2	0011
3	0010
4	0110
5	0111
6	0101
7	0100
8	1100
9	1101
10	1111
11	1110
12	1010
13	1011
14	1001
15	1000

Figure 7.19 Gray Code and Coding Wheel

Gray Code Conversion

It is important to our study of codes that we be able to convert from one to another easily and accurately. Before we build a conversion circuit, let's examine the steps in converting from binary to gray and from gray to binary codes.

To convert from binary to gray requires the following simple steps.

Example Convert the binary 1010 to its gray equivalent.

Solution

Step 1 The most-significant bit of the gray code is the same as the most-significant bit of the binary.

Step 2 The second bit (to the right of MSB) of the gray is the result of the exclusive-OR of the MSB and second bit of the binary.

Step 3 The third gray-code bit is the exclusive-OR of the second and third bits of the binary.

Step 4 The fourth gray-code bit is the exclusive-OR of the third and fourth binary. Therefore,

Binary	1010	10 decimal
Gray	1111	10 decimal

To convert from gray to binary is also a straightforward procedure.

Example Convert the gray code 1111 to binary.

Solution

Step 1 The most-significant binary bit is the same as the most-significant gray-code bit.

Step 2 If the next gray bit is 0, the next binary bit is equal to the preceding bit. If the next gray bit is 1, the next binary bit is the inverted preceding binary bit.

Step 3 Same as step 2. Therefore,

Gray	1111	decimal 10
Binary	1010	decimal 10

Gray is 1 so invert.

Another illustration is given here:

Gray	1001	decimal 9
	1110	

Circuit Conversion of Gray Code

Although it is certainly useful to know how to do the conversions of gray code by hand, it is useful and practical to be able to construct and implement circuitry that will do it for us. A clue to the method by which such conversion may be done was given in the example showing binary-to-gray-code conversion. Remember, the conversion mode use of the OR operation.

Figure 7.20 shows the use of three exclusive-OR gates to implement the binary-to-gray-code conversion. The three gates actually comprise three-fourths of one IC package, the 7486.

In a similar manner, the gray-to-binary conversion can be implemented. Figure 7.21 illustrates this procedure. Again, the exclusive-OR gate is used. Now, however, since it was necessary to know the state of the previous bit, the output of each higher value gate is fed to the next descending value gate. The values in parentheses are for the second example of the discussion.

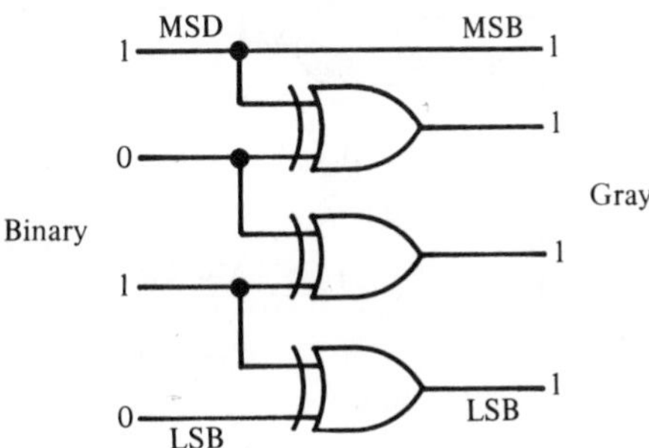

Figure 7.20 Binary-to-Gray-Code Conversion Using Exclusive-OR Gates

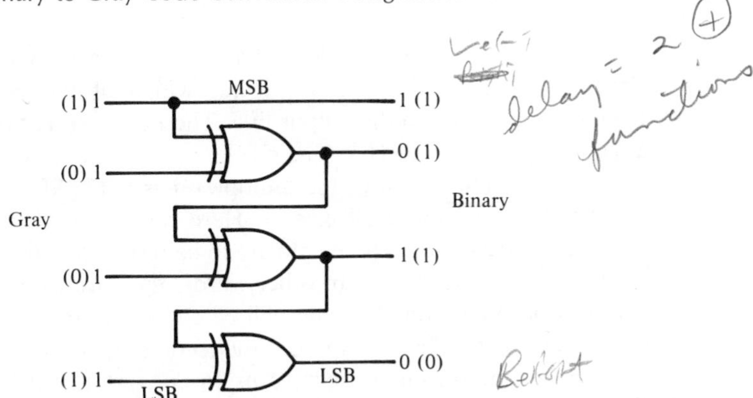

Figure 7.21 Gray-Code-to-Binary-Conversion Using Exclusive-OR Gates

7.8 MULTIPLEXING (DATA SELECTIONS)

Data selection is simply the process of selecting only one set of data from several sets of data. An example of data selection is the television channel selector switch. By use of this switch, one set of data (one channel) is selected for presentation. This is also multiplexing. The term multiplexing will be used for data selection in the remainder of this unit. Electronic multiplexers play an important part of data transfer and examination. Figure 7.22 shows a simple two-input multiplexer. The truth table of figure 7.22 really tells the story. The level of the input S determines which data (A or B) is passed to the output Y. The input S may be thought of as a switch as shown in 7.22 connecting either line A or line B to the line Y.

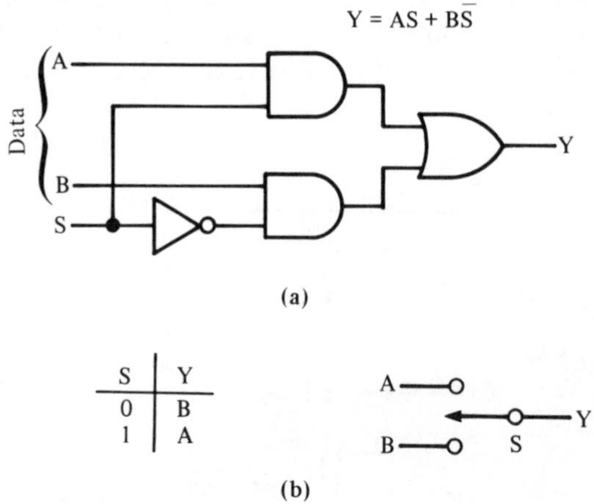

S	Y
0	B
1	A

A
B
S
Y

(b)

Figure 7.22 Two-Input Multiplexer

Figure 7.23 shows a block diagram approach to the multiplexer circuit. The select lines simply decide which of the input data lines (D0–D7) is output to the output line. The next step is to add a counter as shown in figure 7.23(b). The clock frequency will not determine the rate at which the output of the multiplexer is changed. Figure 7.24 will explain the operation a bit clearer. Here we see the data are stored at the multiplexer inputs. These data are to be transferred to the multiplexer output at some clock rate. In other words, the parallel data at the input will now become serial data and will be constantly transferred at a given clock frequency. The multiplexer then not only selects a given bit of data to be output, but it enables us to convert parallel data to serial data. The ability to do this will allow us to send digital information via radio waves or via microwave links.

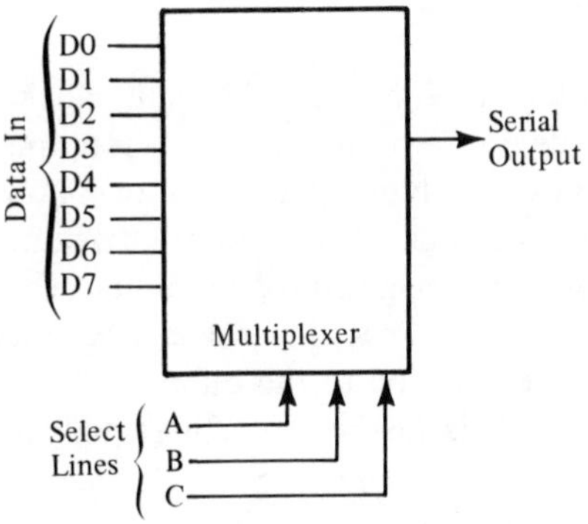

Data In: D0, D1, D2, D3, D4, D5, D6, D7 → Multiplexer → Serial Output

Select Lines: A, B, C

Select A	Select B	Select C	Output Signal
0	0	0	D0
0	0	1	D1
0	1	0	D2
0	1	1	D3
1	0	0	D4
1	0	1	D5
1	1	0	D6
1	1	1	D7

(a)

Multiplexer

Select

Clock

Counter

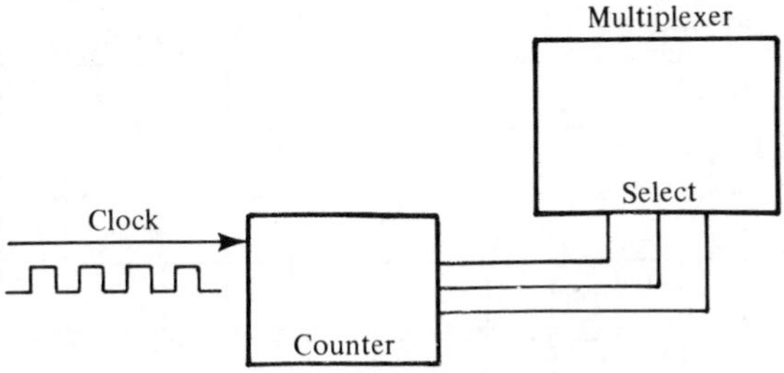

(b)

Figure 7.23 Parallel-to-Serial Data Transmission

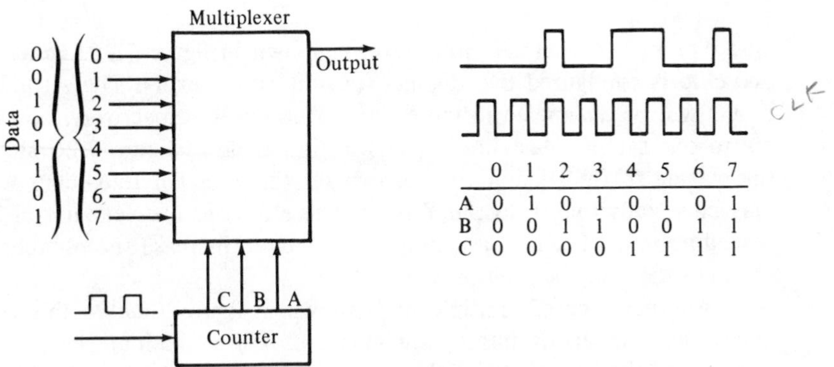

Figure 7.24 Multiplexer Operation with Timing Sequence

The 74150 Data Selector/Multiplexer

Figure 7.25 shows the pin-out diagram function table and block diagram for the 74150 device. This device is a one-of-sixteen multiplexer. That is, there are sixteen input lines. The block diagram shows how the device may be constructed using gate logic. One important feature to note is that the output line will yield the negative of the input. The device itself is otherwise self-explanatory. The four select lines decide which of the sixteen inputs will be channeled to the output. The single strobe line enables the output line. This insures that there will be no erratic data on the output line.

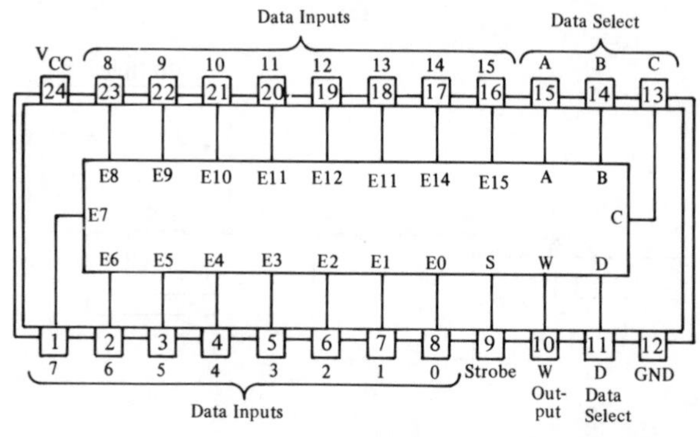

Figure 7.25 Pin-Out for SN74150

Two-Channel Multiplexers

One type of two-channel multiplexer is shown in figure 7.26. Here, on one chip is configured two distinct four-bit multiplexers. The A and B inputs will be channeled to the Y_1 and Y_2 outputs respectively, according to one set of select lines. The function table of figure 7.26 shows this. Using this particular device, two sets of data (of four-bits each) may be serially output to the Y lines. These devices are very useful for cascading multiplexers. Cascading enables us to increase the number of channels that may be multiplexed.

Another type of multiplexer is shown in figure 7.27. In this configuration, two sets of data of four-bits each are at the input leads. Depending on the logic level of the select line, the complete set of data is sent to four output lines.

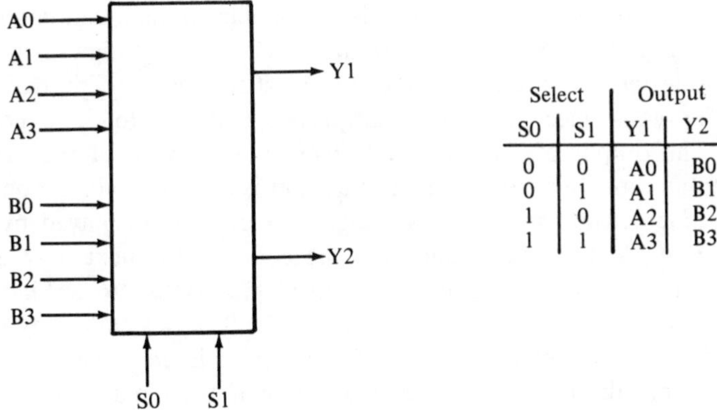

Select		Output	
S0	S1	Y1	Y2
0	0	A0	B0
0	1	A1	B1
1	0	A2	B2
1	1	A3	B3

Figure 7.26 Two-Channel Multiplexer

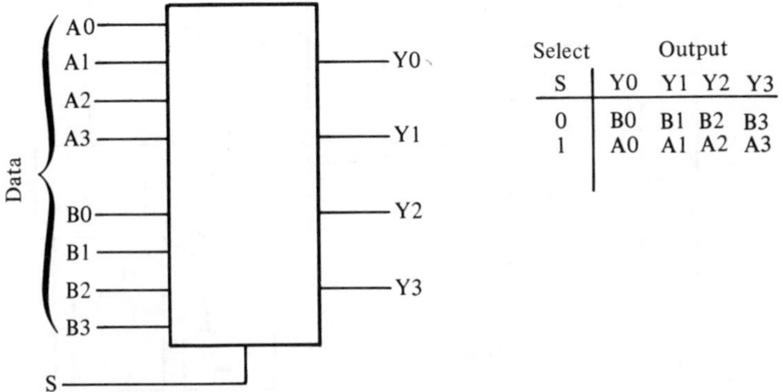

Select	Output			
S	Y0	Y1	Y2	Y3
0	B0	B1	B2	B3
1	A0	A1	A2	A3

Figure 7.27 Two-Channel Four-Bit Multiplexer

Multiplexer Applications

One of the more common uses of multiplexing is in seven-segment display circuits. The reason for this is two-fold. First, there is, as you will see, an obvious savings in chip count. Second, through multiplexing, all displays will not be lit at once, therefore, a savings in power consumption. Figure 7.28 is a block diagram of a system for a three-digit multiplexed display.

The counter is connected to a clock that determines the rate at which data is transferred. The output of the counter is tied to both the multiplexers and the display units. It is tied to the display units through a one-of-four decoder. The counter is a modular counter that counts 0, 1, 2, 0, 1, 2 . . . . At a count of 0, the least-significant digit of the display is selected. At the same time, the least-significant digit from the binary inputs is selected by the multiplexers and sent to the decoder/driver. The output of the decoder/driver is sent to all of the display units at one time. Since only one display unit is selected (via the one-of-four decoder), the least-significant digit will only be displayed by one display unit. At the next count from the counter, the next digit is selected by the multiplexers and sent to all of the display units. Since the count is now 1, only the second display unit will be selected. Any number of displays may be connected in this manner. Even if the input data are changing, the display can be flicker-free if the clock frequency is above 100 hertz.

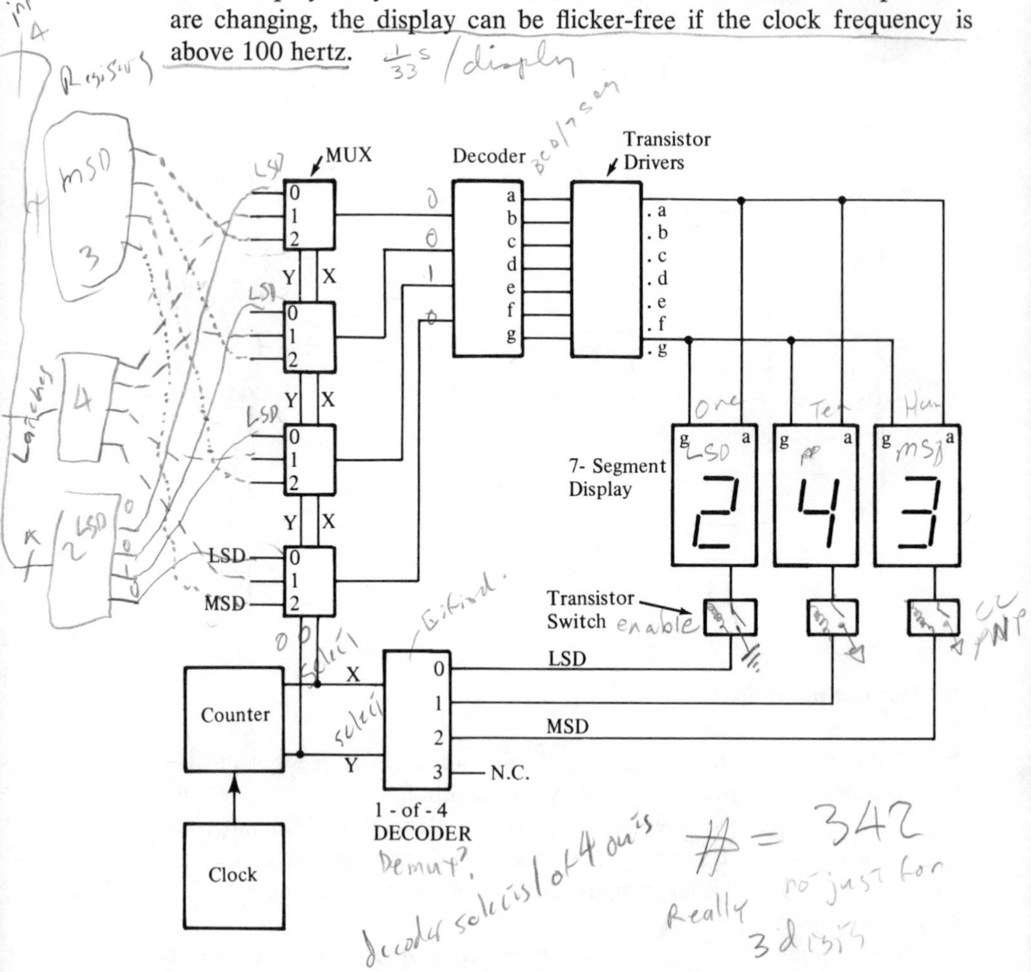

Figure 7.28 3 Digit Multiplexing System

Advantages of this type system are lower chip count; only one decoder/driver is needed regardless of the number of digits to be displayed; less power consumption, the displays are not all on at one time; less wiring; only one set of data lines are needed from the single display decoder/driver.

To increase the size of the display to eight digits, only 4 eight-input multiplexers are needed. A mod-counter that counts to eight and a one-of-eight decoder are used to complete the circuitry. It is left as a student exercise to draw this circuit.

Demultiplexing

Just as we may take parallel data and convert it to serial data (figure 7.24) by the use of a multiplexer, so we may do the reverse. The circuitry by which this is accomplished is called *demultiplexing.* Figure 7.29 shows a block diagram and timing diagram representation of a four-output demultiplexer. Here we see the serial data being input to the multiplexer, probably from a multiplexer unit. The serial data *must* be synchronized to the input clock and the count on the select lines. As the count at the select lines changes, each output is selected (tied to the serial input line). Figure 7.29(b) shows a typical bit of serial data and the corresponding output levels. So we can see then that the serial data is "recaptured" as 1011 and placed at the parallel outputs of the demultiplexer.

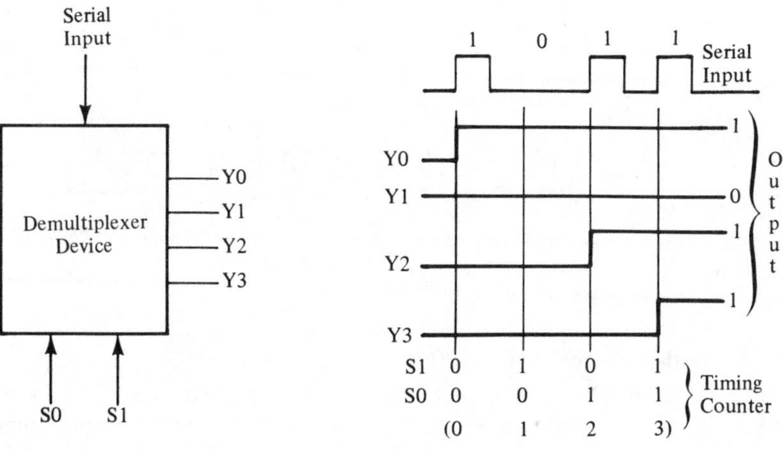

Figure 7.29 Timing for Demultiplexer Signals

IC Decoder/Demultiplexer

The 74154 is an example of a demultiplexer circuit on a single chip. The 74154 is called a *four-line-to-sixteen-line demultiplexer.* This means it is capable of recapturing a full sixteen-bit digital word. A look at figure 7.30 will show how this is accomplished. When used as demultiplexer, the 74154 has two data input lines. These are G1 and G2 (pins 18 and 19). Inputs A, B, C and D (pins 20, 21, 22, 23) will be used to synchronize the input data to a clock frequency. Serial data will be input at either G1 or G2. The level of this data signal (logic 1 or logic 0) will be sent to one of the sixteen output lines in accordance with the binary code at the A, B, C, D inputs. ~~The function table (figure 7.30(b)) shows this.~~ Also notice that a high on either of the G inputs will cause *all* of the output lines to go high. This will occur regardless of the state of the A, B, C, D lines. One last item on the 74154; notice that if the input lines G1 and G2 are held a constant low, the output line going low will be an indication of the binary count at the A, B, C, D inputs. When used in this manner, the device may be used as binary decoder.

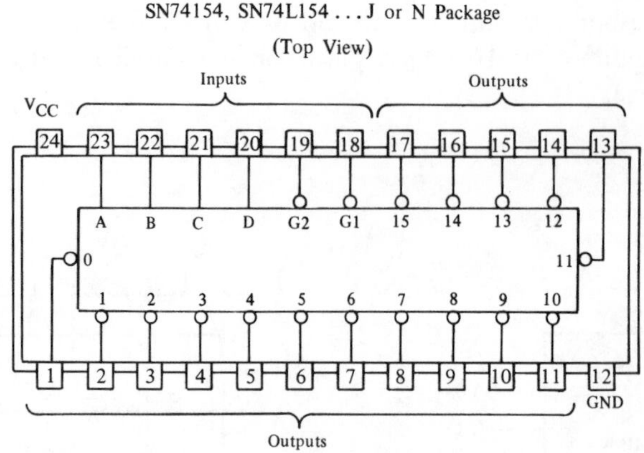

Figure 7.30 Pin-Out for SN74154

Another device that is very similar to the demultiplexer is the *addressable latch.* One such device, the 74259, is shown here. The 74259 is an eight-bit addressable latch. This means it has eight output lines. These lines are not only selectable, they are latchable. One advantage of this device over the 74154 is that data selected by this device may be held at the output for a user-specified length of time. You remember

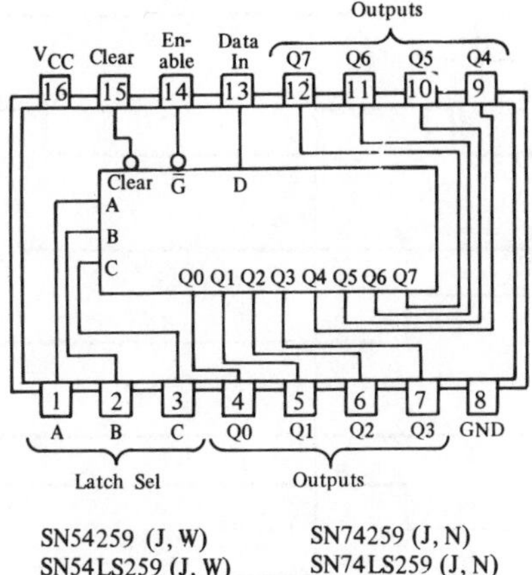

Figure 7.31 Pin-Out for SN74259

that with the 74154, data (output levels) were constantly changing with the input codes at A, B, C, D. As shown in figure 7.31, the 74259 has three select input lines. These lines determine which of the eight output latches is addressed. By use of two other input lines, clear and enable ($\overline{G}$) one of four modes of operation of the device may be selected. When used as an addressable latch, the timing of the enable and address (select input) lines is very important. Figures 7.32 and 7.33 illustrate this important point. Figure 7.32 shows a timing sequence that must be followed in general for error-free operation of the 74259. Notice that the enable line ($\overline{G}$) is held high while the address (select input) lines are changing. This insures that the outputs will not change during a change of address. Figure 7.33 is a more complete diagram that illustrates in detail the input timing signals required to latch two-bits of data.

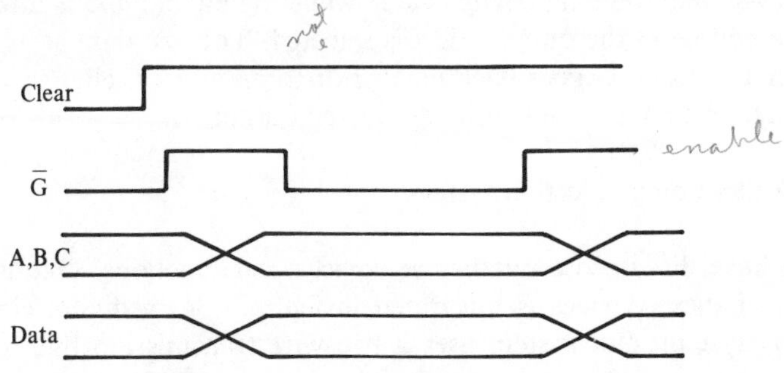

Figure 7.32 Signal Timing for 74259

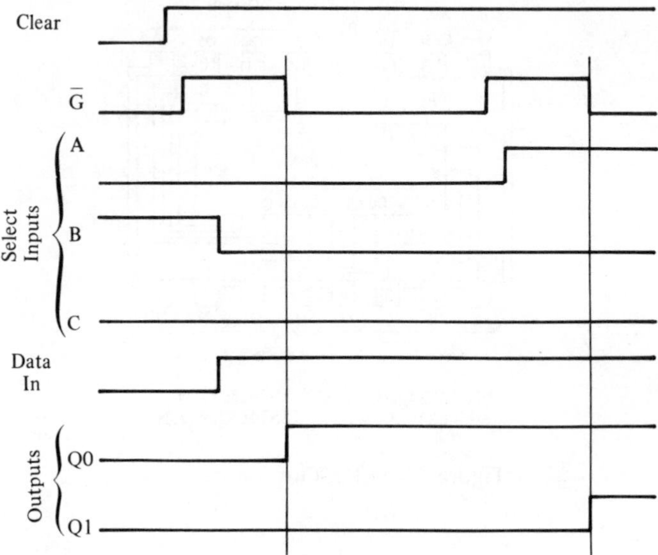

Figure 7.33 Detail Timing for 74259

Here we see the clear line held high. Prior to a change of address (to address 0) the enable line is taken high. Data are already on the line when the address changes to 0. The outputs (Q_0 and Q_1) are 0. When the enable line goes low, the data at the data-in are transferred and latched at Q_0. Enable once again goes high prior to an address change. The address lines change to address change. The address lines change to address latch 1. Enable goes low latching the data (1) into location 1. Notice that unlike the 74154, the status of Q_0 did not change with the change in address.

Should the data change value while the enable line is low, this new data will be at the output address selected. The last data seen at data-in (pin 13) when $\overline{G}$ goes back high prior to an address change will be the data latched at the previously addressed output.

A Digital Communication System

We have discussed a number of devices and functions. One use of several of these devices is illustrated in figure 7.34, a digital communications system. Our system uses a two-wire transmission line to transmit data from point A to point B. The system could also have been de-

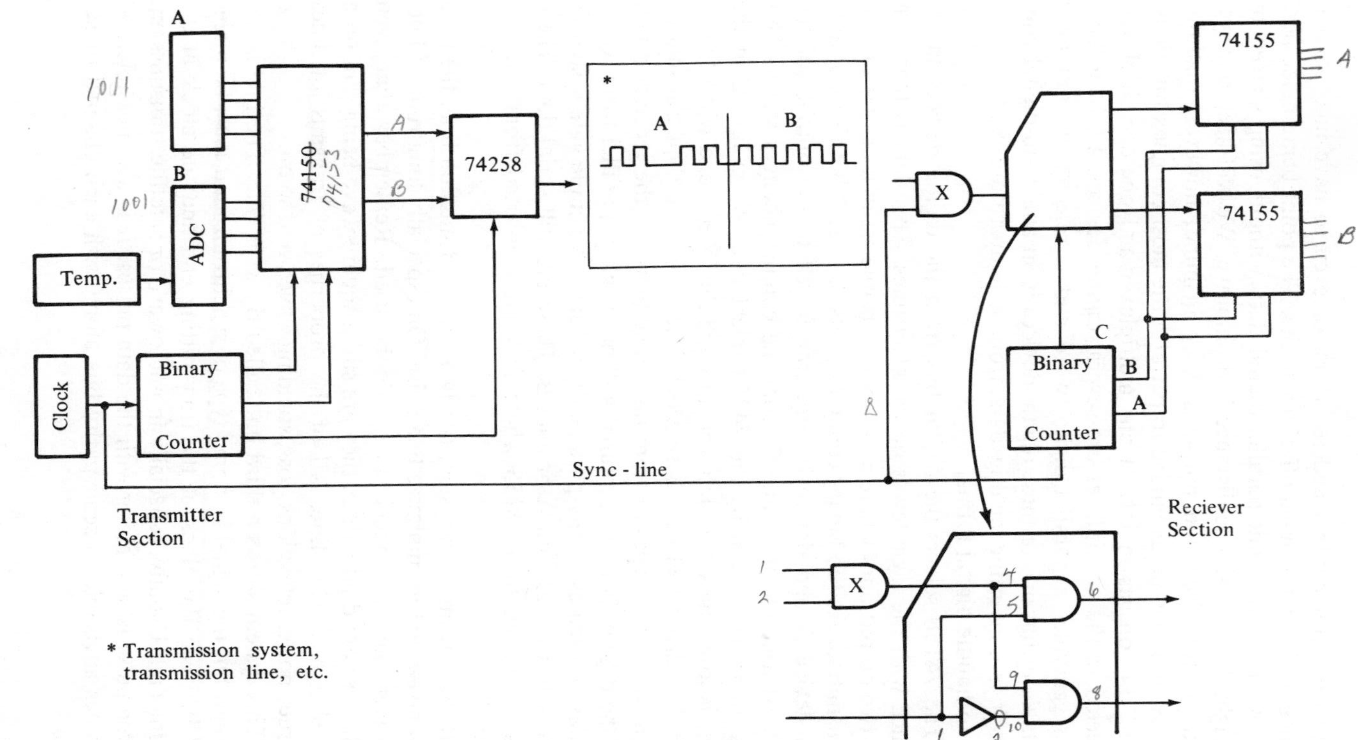

Figure 7.34 Complete Transmission Circuitry

signed to use transmitters and receivers, or even to have incorporated a bounce of the radio signal off of a satellite. The point being made here is that we must transmit parallel digital data without using sixteen or even eight data lines. The distance is prohibitive. We will keep our system simple for discussion purposes; keep in mind, however, that this simple system may be expanded to perform the most complex functions. Figure 7.34 consists mainly of one multiplexer and one demultiplexer. Connected to the multiplexer are several digital devices. Each of these devices generates a parallel digital word and is selected in turn by a 74150. Note that one of these digital devices is an analog-to-digital converter. It is converting analog data from a temperature sensor to an easier-to-handle digital format.

The 74150 selects the digital bits from the digital devices in accordance with the logic levels on its select lines. This data is then sent to the two output lines where a 74258 selects one set of data at a time to be transmitted. As the binary counter counts 0, 1, 2, 3 both sets of data (from device A and device B) are transferred to the outputs of the 74150. At a count of 4, line C from the counter changes states. This change in states is seen by the 74258 select line. When line C of the counter is low, one set of data gets through the 74258 and when line C is high the other set of data is selected. These data are then transmitted (via one of several types of transmission systems) to the receiver point. Here, the digital data are captured and put back in parallel form where they may be used as in any parallel digital data. A simple gate system is used to select either data A or data B. These are still serial data. These data are then fed to a 74155 where they are converted from serial to parallel.

If the transmission system is a two-wire transmission line, the lines should be twisted to minimize line noise. On especially long runs of line, line drivers and line receivers should be used. Remember that even though these are digital data, they are still a square wave being sent on a piece of transmission line. All of the properties of transmission lines hold true: noise, reflections, loss, standing waves, and so on.

This system shows a third line called the *sync line*. The line is not necessary, but the signal is. Some type of synchronization must be maintained between the clock at the transmitting end and the clock at the receiving end. Usually, especially in microwave or satellite transmission, this sync pulse is sent along with the data pulses. It is also possible and sometimes advisable to send parity bits along with each channel or set of data.

Multiplexers may also be used to decode a Boolean function. For example, suppose we had the function (or condition) where:

$$F = AB\bar{C} + A\bar{B}C + \bar{A}BC + \bar{A}B\bar{C}$$
$$F = (\ 6 \ + \ 5 \ + \ 3 \ + \ 2\ )$$

We know from our study of gates that we could devise a gating sequence that would produce a high output when one of the four proper input conditions was present. A multiplexer may be used to accomplish the same result at far less cost and much simpler circuitry. Figure 7.35 shows the implementation of such a circuit using one 74153 device. Through the use of a truth table, verify the accuracy of this circuit. Develop a circuit that will implement other Boolean functions.

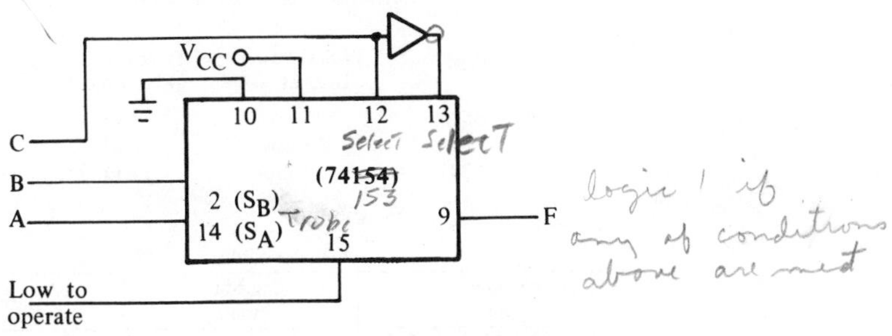

Figure 7.35 Use of Multiplexer as Boolean Decoder

7.9 MAGNITUDE COMPARATORS

One device that should not be overlooked by anyone interested in digital design is the magnitude comparator. One such device available to the circuit designer is the 7485 four-bit magnitude comparator. This device will produce an output based on the results of the comparison of two 4-bit binary digits.

Figure 7.36 shows the pin-out diagram and the function tables for this device. The 7485 will compare these two sets of input data and produce a logic 1 at one of three outputs (A < B, A > B, A = B) that tells the status of the inputs.

Figure 7.37 shows the connections needed to cascade several of these devices to compare two 24-bit words.

SN5485, SN54LS85, SN54S85 . . . J or W Package
SN7485, SN74LS85, SN74S85 . . . J or N Package
(Top View)

SN54L85 . . . J Package
SN74L85 . . . J or N Package
(Top View)

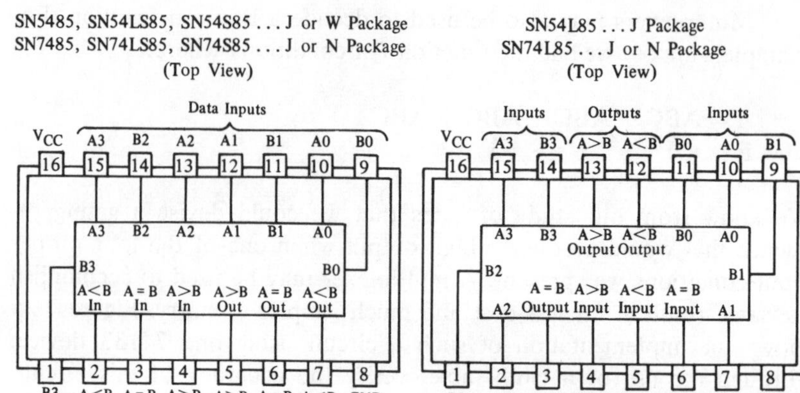

positive logic: see function tables

Type	Typical Power Dissipation	Typical Delay (4-Bit Words)
'85	275 mW	23 ns
'L85	20 mW	90 ns
'LS85	52 mW	24 ns
'S85	365 mW	11 ns

Function Table

Comparing Inputs				Cascading Inputs			Outputs		
A3,B3	A2,B2	A1,B1	A0,B0	A>B	A<B	A=B	A>B	A<B	A=B
A3>B3	X	X	X	X	X	X	H	L	L
A3<B3	X	X	X	X	X	X	L	H	L
A3=B3	A2>B2	X	X	X	X	X	H	L	L
A3=B3	A2<B2	X	X	X	X	X	L	H	L
A3=B3	A2=B2	A1>B1	X	X	X	X	H	L	L
A3=B3	A2=B2	A1<B1	X	X	X	X	L	H	L
A3=B3	A2=B2	A1=B1	A0>B0	X	X	X	H	L	L
A3=B3	A2=B2	A1=B1	A0<B0	X	X	X	L	H	L
A3=B3	A2=B2	A1=B1	A0=B0	H	L	L	H	L	L
A3=B3	A2=B2	A1=B1	A0=B0	L	H	L	L	H	L
A3=B3	A2=B2	A1=B1	A0=B0	L	L	H	L	L	H

'85, 'LS85, 'S85

A3=B3	A2=B2	A1=B1	A0=B0	X	X	H	L	L	H
A3=B3	A2=B2	A1=B1	A0=B0	H	H	L	L	L	L
A3=B3	A2=B2	A1=B1	A0=B0	L	L	L	H	H	L

'L85

A3=B3	A2=B2	A1=B1	A0=B0	L	H	H	L	H	H
A3=B3	A2=B2	A1=B1	A0=B0	H	L	H	H	L	H
A3=B3	A2=B2	A1=B1	A0=B0	H	H	H	H	H	H
A3=B3	A2=B2	A1=B1	A0=B0	H	H	L	H	H	L
A3=B3	A2=B2	A1=B1	A0=B0	L	L	L	L	L	L

H = high level, L = low level, X = I irrelevant

Figure 7.36 Four-Bit Magnitude Comparators

Typical Application Data

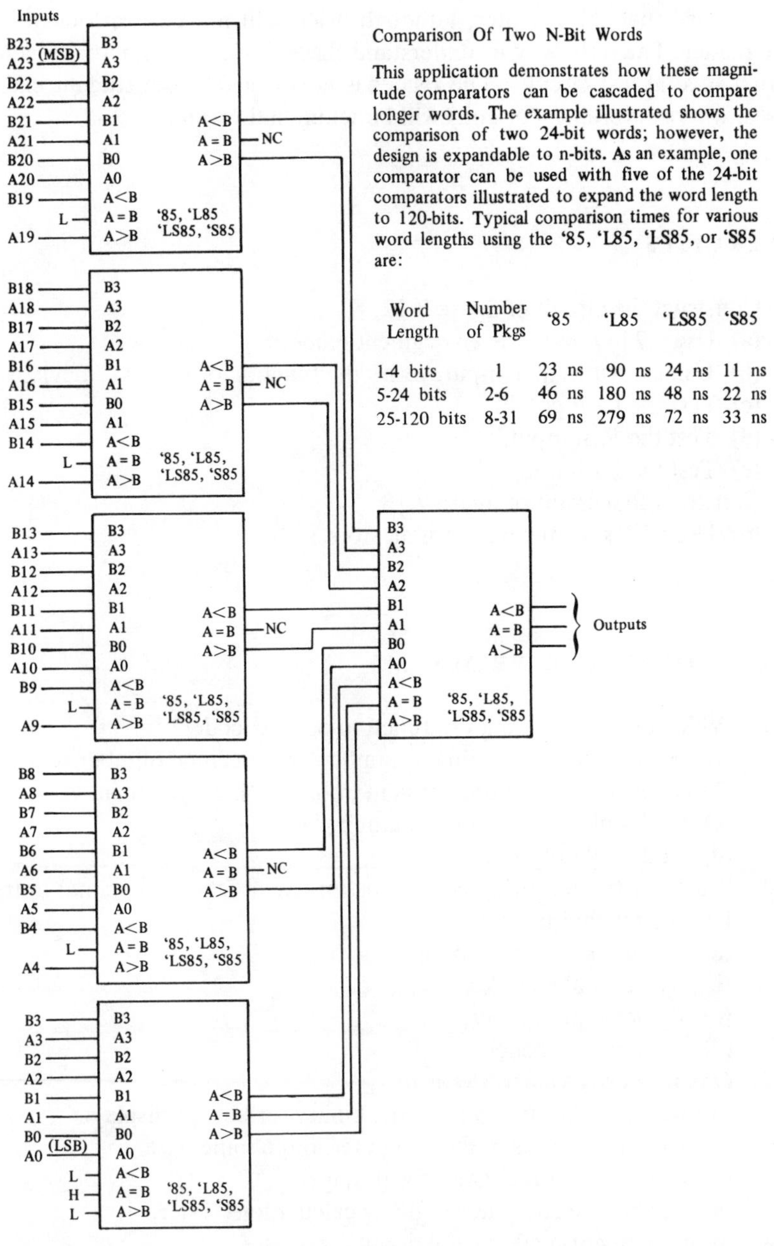

Comparison Of Two N-Bit Words

This application demonstrates how these magnitude comparators can be cascaded to compare longer words. The example illustrated shows the comparison of two 24-bit words; however, the design is expandable to n-bits. As an example, one comparator can be used with five of the 24-bit comparators illustrated to expand the word length to 120-bits. Typical comparison times for various word lengths using the '85, 'L85, 'LS85, or 'S85 are:

Word Length	Number of Pkgs	'85	'L85	'LS85	'S85
1-4 bits	1	23 ns	90 ns	24 ns	11 ns
5-24 bits	2-6	46 ns	180 ns	48 ns	22 ns
25-120 bits	8-31	69 ns	279 ns	72 ns	33 ns

Comparison of Two 24-Bit Words

Figure 7.37 Use of Magnitude Comparator to Compare Two n-Bit Words

SUMMARY

It is hoped that this chapter, although brief, will not be overlooked by the reader. The only way to understand these devices is to actually construct several test circuits. The reader is now urged to the student exercises, where several test circuits will be set up and verified.

EXERCISES

1. Construct the circuit of figure 7.12.
  (a) Use a 7447 and a seven-segment readout.
  (b) Using a truth table input, verify the function of the 7447 device.
  (c) Test the L_T input.
  (d) Test the RB_1 input.
  (e) Test the B1 input.

2. Construct the circuit of figure 7.13.
  (a) Use LEDs for the output indicators.

QUESTIONS AND PROBLEMS

7.1 Which of the following is a function of an encoder?
  (a) Selects the proper binary number for a decimal number.
  (b) Transforms a number system into some binary format.
  (c) Codes all incoming information.
  (d) Codes all output information.

7.2 Which of the following is the most practical form of decimal entry into digital equipment?
  (a) The ten-stroke keyboard.
  (b) By decimal encoder.
  (c) By decimal decoder.
  (d) By binary decoder.

7.3 One purpose of a decoder is to
  (a) Return information from the binary format for display.
  (b) Get all numbers in the computer in the same base.
  (c) Select the correct gates for timing.
  (d) Enable the computer to do its calculations faster.

7.4 Seven-segment readouts are popular because
  (a) They can read binary information.

(b) They can be seen for long distances.
(c) They are easily activated by a simple decode circuit.
(d) They read directly in octal code.

7.5 Decoder input gates are determined by
(a) The bit combination necessary to generate the desired output.
(b) The number of outputs from the encoder.
(c) The type of encoder used.
(d) None of the above.

7.6 By definition, what is the difference between an encoder and a decoder?

7.7 Give an example of where a priority encoder such as the 74147 might be used.

7.8 The following need to be sent using even parity. Add the needed parity bit:
(a) __10100001
(b) __01100011
(c) __11100001
(d) __01000101

7.9 Using odd parity, add the needed parity bit to the following hexadecimal words.
(a) B3
(b) 42
(c) 17
(d) 21
(e) FF

7.10 What type of error will the parity check reduce?

7.11 Will parity assure error-free transmission of the data words?

7.12 What is meant by a parity tree?

7.13 Where might the gray code be used?

7.14 Convert binary 9 to gray code.

7.15 Design a circuit that implements the use of gray code.

7.16 Convert gray 14 to binary and to decimal.

7.17 Redraw figure 7.19 to double the number of discrete shaft positions.

7.18 How does gray code differ from binary or BCD codes?

7.19 Design a system to convert a ten-key keyboard to a digital binary input. Use any device necessary.

7.20 What is the disadvantage of the cold-cathode type of display?

7.21 What are the limitations of the seven-segment display?

7.22 Define *multiplexing*.

7.23 Design a system where a multiplexer may be used to select data from one of four input sources. The data will consist of an eight-bit word. The multiplexer system must let each eight-bit word pass before going to the next input.

7.24 Design a system that will enable the user to know which of three switches is closed first, which is closed second, and which is closed third. *Note:* Any one switch may be closed more than once. *Hint:* Use a ring counter (or a Johnson counter) and seven-segment displays.

8

Counting Systems

OBJECTIVE: To put together the material on counters and introduce complete integrated circuit counting systems.

Introduction: As you have seen in the previous chapters, digital counters are a valuable building block in logic systems. Counters are used for the most part in conjunction with other circuit devices to provide a display of their count states. Usually, the counters are contained on one chip and the decoding device is on another chip. There are times, however, when the total chip count for a particular unit must be reduced. This chapter will present a look at several self-contained counting and decoding circuits.

8.1 BCD COUNTER/BCD DECODER DRIVER

The BCD counting system of earlier chapters is contained on a single integrated circuit. The device is the SN74142 shown in figure 8.1. The SN74142 is a divide-by-ten counter complete with a four-bit latch and a decoder driver. This single unit replaces three separately packaged MS1 chips. The 74142 has all the necessary interconnections for operation of the complete circuit directly on the chip. This improves the reliability and reduces the cost of the entire system.

Figure 8.2 is a function table showing the operational states of the system. Notice from this table that an input of a logic low level to pin 1 (clear) will reset the counter to 0 and hold it there. To operate the counter, a logic high level is needed at the clear input (pin 1). The in-

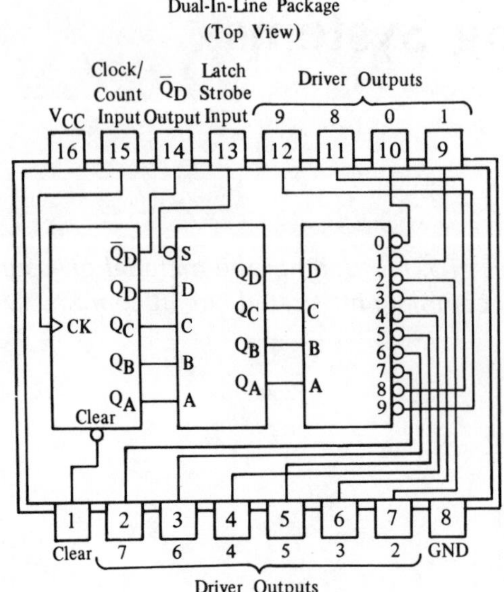

Figure 8.1 Type SN74142 BCD Counter/Four-Bit Latch/BCD Decoder/Driver

Function Table

Inputs			Outputs	
Count Pulse (Clock)	Clear	Latch Strobe	On†	$\overline{Q}_D$
X	L	L	0	H
1	H	L	1	H
2	H	L	2	H
3	H	L	3	H
4	H	L	4	H
5	H	L	5	H
6	H	L	6	H
7	H	L	7	H
8	H	L	8	L
9	H	L	9	L
10	H	L	0	H
11	H	H	0	H

†All other outputs are off.
H = high level, L = low level, X = irrelevant

Figure 8.2 Type SN74142 Function Table

put to the counting circuitry is through pin 15. Each positive-going transition to this input will cause the counter to be incremented.

Pin 13, the latch strobe input serves to activate the four-bit latch wired between the counter and decoder sections. While the input signal to pin 13 is low, the output from the counter section is fed straight through to the decoder. Changing the signal fed to pin 13 from a logic low to a logic high level causes the latch section of the circuit to store the data that were present at the latch input (the counter output) just prior to the low-to-high transition at pin 13.

Notice that the $\bar{Q}D$ output from the counter section is not wired to the latch, but rather is brought out to a pin (pin 14) of the integrated circuit. This output is used to clock a second counter circuit for the counting of larger than single-digit numbers (decimal numbers).

The Decoder/Driver

The decoder/driver section of this device is separated in our discussion from the remainder of the chip for a very good reason. The decoder/driver section of the 74142 accepts the binary-coded decimal output of the counter (by way of the latch) and decodes it. The decoded output is decimal, that is, it has ten separate output lines. One, and only one, of these output lines will be activated for any binary-coded input condition. Again, the function table of figure 8.2 shows the output states from the decoder for the count pulse condition. The decoder section of the 74142 is used not only to decode the binary output of the counter section but to drive the display device. The 74142 is used for driving the cold-cathode high-voltage type of display such as the NIXIE* tube.

Figure 8.3 shows a typical connection to a cold-cathode display tube. Each of the output lines from the decoder/driver section of the 74142 will be connected to a cathode of the display tube. Each cathode is shaped as a decimal number, 0 through 9. As each output line is turned on (decoded), that particular line and its associated cathode will go low. This will place a potential of B^+ from cathode to anode of the tube, causing the gas in the tube to ionize. This ionization will take place very close to the cathode, and the cathode will appear to glow. Since the cathode is shaped as a decimal number, that number will appear as a glow in the tube.

With the off lines appearing as an open circuit, they will have a minimum of 60.0 volts between them and ground. This is dependent upon the value of B^+ and the particular tube being used.

---

* NIXIE is a registered trademark of the Burroughs Corporation.

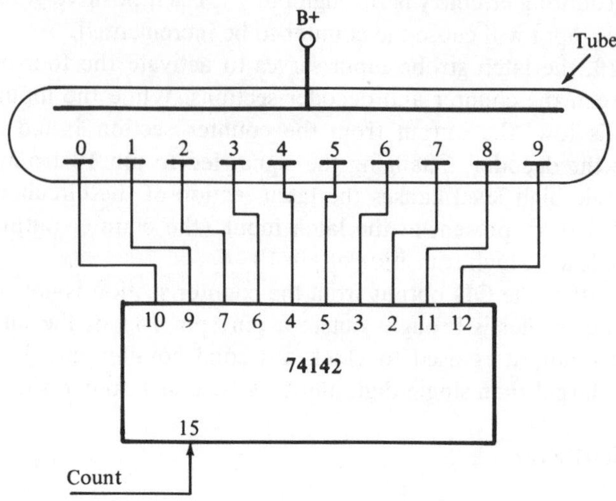

Figure 8.3 Connections for Cold-Cathode Display Tube Operation

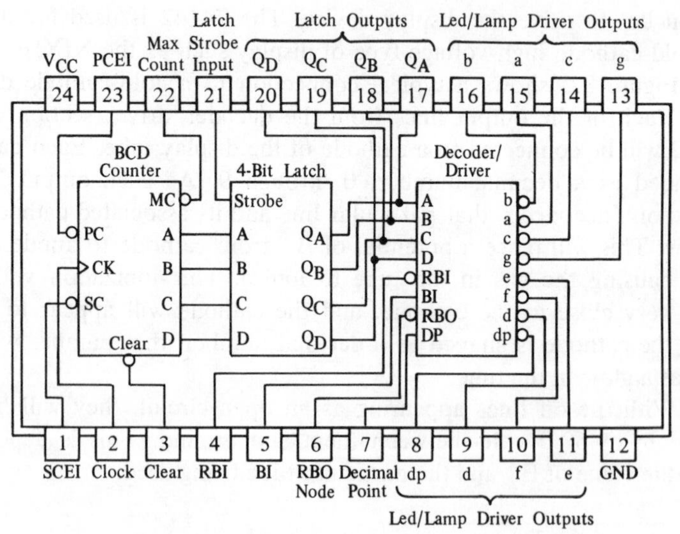

Figure 8.4 74144 Block Diagram

8.2 FOUR-BIT COUNTER/LATCH, SEVEN-SEGMENT DRIVER

Figure 8.4 shows the logic block diagram and pin assignment for a device similar to the 74142. This device is the SN74144. It will perform the same general function as the 74142 with a few minor differences. Notice that this device is a twenty-four pin MS1 chip. Table 8.1 lists the function of each of these pins plus a short description of that function.

TABLE 8.1 Pin Functions for SN74144

Function	*Pin No.*	*Description*
CLEAR INPUT	3	When low, resets and holds counter at 0. Must be high for normal counting.
CLOCK INPUT	2	Each positive-going transition will increment the counter provided that the circuit is in the normal counting mode (serial and parallel count enable inputs low, clear input high).
PARALLEL COUNT ENABLE INPUT (PCEI)	23	Must be low for normal counting mode. When high, counter will be inhibited. Logic level must not be changed when the clock is low.
SERIAL COUNT ENABLE INPUT (SCEI)	1	Must be low for normal counting mode, also must be low to enable maximum count output to go low. When high, counter will be inhibited and maximum count output will be driven high. Logic level must not be changed when the clock is low.
MAXIMUM COUNT OUTPUT	22	Will go low when the counter is at 9 and serial count enable input is low. Will return high when the counter changes to 0 and will remain high during counts 1 through 8. Will remain high (inhibited) as long as serial count enable input is high.
LATCH STROBE INPUT	21	When low, data in latches follow the data in the counter. When high, the data in the latches are held constant, and the counter may be operated independently.

TABLE 8.1 (*Continued*)

Function	*Pin No.*	*Description*
LATCH OUTPUT (Q_A, Q_B, Q_C, Q_D)	17, 18, 19, 20	The BCD data that drive the decoder can be stored in the four-bit latch and are available at these outputs for driving other logic and/or processors. The binary weights of the outputs are: $Q_A = 1$, $Q_B = 2$, $Q_C = 4$, $Q_D = 8$.
DECIMAL POINT INPUT	7	Must be high to display decimal point. The decimal point is not displayed when this input is low or when the display is blanked.
BLANKING INPUT (BI)	5	When high, will blank (turn off) the entire display and force RBO low. Must be low for normal display. May be pulsed to implement intensity control of the display.
RIPPLE-BLANKING INPUT (RBI)	4	When the data in the latches are BCD 0, a low input will blank the entire display and force the RBO low. This input has no effect if the data in the latches are other than 0.
RIPPLE-BLANKING OUTPUT (RBO)	6	Supplies ripple blanking information for the ripple blanking input of the next decade. Provides a low if BI is high, or if RBI is low and the data in the latches are BCD 0; otherwise, this output is high. This pin has a resistive pull-up circuit suitable for performing a wire-AND function with any open-collector output. Whenever this pin is low the entire display will be blanked; therefore, this pin may be used as an active-low blanking input.
LED/LAMP DRIVER OUTPUTS (a, b, c, d, e, f, g, dp)	15, 16, 14, 9 11, 10, 13, 8	Outputs for driving seven-segment LEDs or lamps and their decimal points.

Decoder/Driver

There is one major difference between the 74144 and the 74142. The 74144 is used to drive a different type of display unit. The display driven by the 74144 is a seven-segment display. This means initially that the decoder/driver has eight outputs instead of the ten from a 74142. The voltage requirements are also lower. That is, the output voltage will be much lower than the 60.0 volts at the outputs of the 74142. More than one output line may be activated during the decoding of the binary count. (Figure 8.5.) Refer to the section on displays.

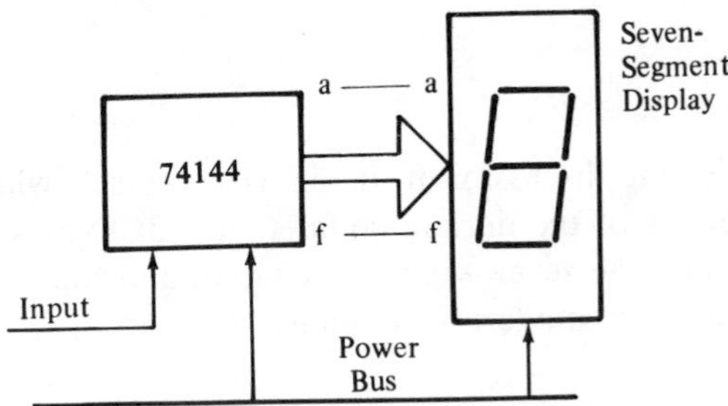

Figure 8.5 Connections for Seven-Segment Display Operation

Latch Outputs

Another difference between the two devices is that the latch outputs are brought out to external pins on the device. This enables the user of the chip to divert the counter output to circuits other than display.

SUMMARY

This unit has shown several integrated circuit devices that are actually small electronic systems. The counting systems shown were used to illustrate a complex functioning integrated circuit. Counting systems were used because the student should have a fair understanding of that system by this time.

A complete four-bit counting system may be purchased on a single chip. There are many advantages to this type of circuit configuration. There are also some disadvantages to using such a ready-made system, and these should be mentioned. When a complete system is packaged on a single chip, many of the interconnections are not brought out to external pin connections. Since the primary purpose of using a single chip is to reduce total chip count, this can usually be overlooked in favor of device economy. If the device is to function purely as a black box input-output device, having access to these interconnections is usually not of great importance. It must be noted here, however, that the inaccessibility of these points can be a hindrance for some types of troubleshooting. Their nonexistence also limits the use of the device.

In conclusion, this chapter introduced the reader to a new family of integrated circuitry, the complete system.

EXERCISES

1. Read the following discussion to familiarize yourself with the operation and capabilities of the decade counter, the BCD-to-seven-segment decoder/driver, and the seven-segment readout, and follow the instructions given to construct and test the circuits.

Decade Counter

The SN7490 decade counter is comprised of a divide-by-two section and a divide-by-five section. These sections may be used separately or externally connected to perform the function of a sequential mod-10 counter. Two sets of direct reset inputs are provided to allow setting the BCD count to a 0 or a 9. The reset/count truth table is given in Table 8.2 and the logic diagram is shown in Figure 8.6.

TABLE 8.2 Reset/Count Truth Table

R0		*R9*		*Output*			
Pin 2	*Pin 3*	*Pin 6*	*Pin 7*	Q_3	Q_2	Q_1	Q_0
1	1	0	X	0	0	0	0
1	1	X	0	0	0	0	0
X	X	1	1	1	0	0	1
X	0	X	0		Count		
0	X	0	X		Count		
0	X	X	0		Count		
X	0	0	X		Count		

X = Don't Care

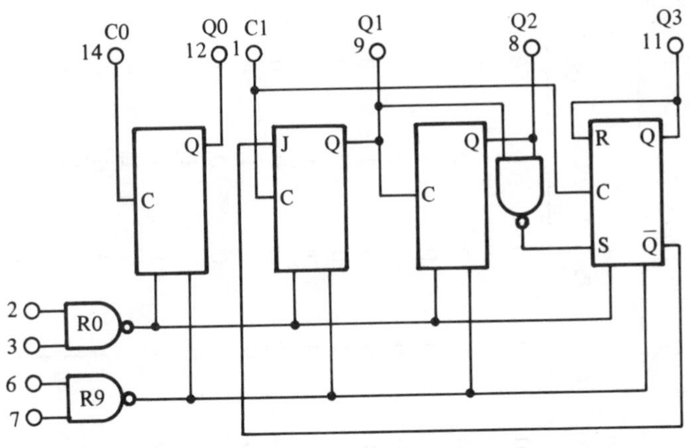

V_{CC} = Pin 5
Ground = Pin 10

Figure 8.6 Logic Diagram for a SN7490

To simplify the testing procedure for the various functions of the SN7490, use the count mode given below.

	Pin 2	Pin 3	Pin 6	Pin 7
Count Mode	0	1	0	1

To use the direct reset (R0) to reset the count to 0, place a logic 1 at pin 2 with the other pins as given above.

To use the direct reset (R9) to preset the count to 9, place a logic 1 at pin 6 with the other pins as given.

Remember to use the SN7490 as a decade counter. An external connection is required between pins 12 and 1.

BCD-To-Seven Segment Decoder/Driver

The SN7447 decoder/driver is used to translate the binary coded output of the decade counter into a suitable format to drive the seven-segment readout. The decoder/driver has seven open-collector outputs that must be connected through a load to V_{CC}. The segments of the readout act as the load. If the decoder/driver is tested separately, load resistors must be provided at all outputs. If an output goes low it will provide a current path to ground for its segment and the segment will light. Maximum output voltage of the SN7447 is 15 volts. The truth table for the SN7447 is given in Table 8.3.

TABLE 8.3 Truth Table for SN7447

	Input							*Output*						
Digit or Function	*LT* *Pin 3*	*RBI* *Pin 5*	*D* *Pin 6*	*C* *Pin 2*	*B* *Pin 1*	*A* *Pin 7*	*BI/RBO* *Pin 4*	*a* *Pin 13*	*b* *Pin 12*	*c* *Pin 11*	*d* *Pin 10*	*e* *Pin 9*	*f* *Pin 15*	*g* *Pin 14*
0	1	1	0	0	0	0	1	0	0	0	0	0	0	1
1	1	X	0	0	0	1	1	1	0	0	1	1	1	1
2	1	X	0	0	1	0	1	0	0	1	0	0	1	0
3	1	X	0	0	1	1	1	0	0	0	0	1	1	0
4	1	X	0	1	0	0	1	1	0	0	1	1	0	0
5	1	X	0	1	0	1	1	0	1	0	0	1	0	0
6	1	X	0	1	1	0	1	1	1	0	0	0	0	0
7	1	X	0	1	1	1	1	0	0	0	1	1	1	1
8	1	X	1	0	0	0	1	0	0	0	0	0	0	0
9	1	X	1	0	0	1	1	0	0	0	1	1	0	0
10	1	X	1	0	1	0	1	1	1	1	0	0	1	0
11	1	X	1	0	1	1	1	1	1	0	0	1	1	0
12	1	X	1	1	0	0	1	1	0	1	1	1	0	0
13	1	X	1	1	0	1	1	0	1	1	0	1	0	0
14	1	X	1	1	1	0	1	1	1	1	0	0	0	0
15	1	X	1	1	1	1	1	1	1	1	1	1	1	1
BI	X	X	X	X	X	X	0	1	1	1	1	1	1	1
RBI	1	0	0	0	0	0	0	1	1	1	1	1	1	1
LT	0	X	X	X	X	X	1	0	0	0	0	0	0	0

X = Don't Care

Three special function inputs are also provided. The lamp test input (pin 3) provides the capability of testing to insure all segments are working properly. In normal operation a logic 1 is maintained at the input; if the lamp test input is grounded (logic 0) all outputs will go to a logic 0 (all segments will light). The ripple-blanking input (pin 5) may be used for suppression of nonsignificant 0's in a display system. The blanking input (pin 4) may be used to change the lamp intensity. In normal operation the blanking input is high. If it is grounded all outputs go high (extinguishing the lamp segments). If a modulated signal (alternating between high and low states) is introduced at the blanking input it will reduce the duty cycle of the segments and thus reduce the intensity. When several readouts are used this may significantly reduce the power-supply current requirements. By varying the duty cycle of the modulated signal the intensity may be changed.

Seven-Segment Readout

This display device has seven incandescent segments that draw approximately 8 milliamperes each. The common inputs may be connected to either V_{CC} or ground depending on the type of driver used. The SN7447 decoder/driver requires that the common inputs be connected to V_{CC}. A logic 0 at the input will cause that segment to light. The device has no ground connection, but instead establishes ground through the outputs of the decoder/driver. The recommended segment voltage range is from 4 to 6.5 volts dc or ac. The segment designation and the pin connections are given in Figure 8.7.

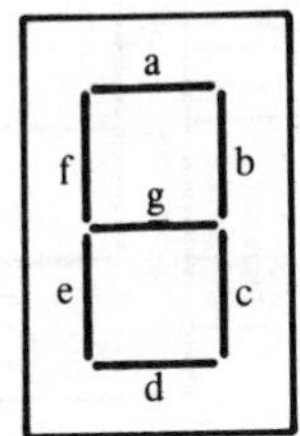

Left pins	Right pins
NC 1	16 NC
Com 2	15 a
f 3	14 b
g 4	13 Com
Com 5	12 Com
e 6	11 c
d 7	10 Com
NC 8	9 Dec Pt

NC – No connection may be left open or tied to common.

Figure 8.7 Segment Designations and Pin Connections

Connect the proper inputs to the seven-segment readout and check to insure each segment lights properly. Now connect the SN7447 BCD-to-seven segment decoder/driver to the readout. Connect the proper inputs (A, B, C, and D) and fill in Table 8.4.

TABLE 8.4 Truth Table

D	C	B	A	a	b	c	d	e	f	g
0	0	0	0							
0	0	0	1							
0	0	1	0							
0	0	1	1							
0	1	0	0							
0	1	0	1							
0	1	1	0							
0	1	1	1							
1	0	0	0							
1	0	0	1							

Now add the SN7490 decade counter to the test circuit (Figure 8.8). Clock the counter ten times and see if the segments light properly each time. Remember to connect the proper inputs to the decade counter so that it is in the count mode. Test the R0 and R9 direct resets of the SN7490. Connect a square-wave signal (*Caution:* use a low-voltage signal of less than 2 volts) to the blanking input of the decoder/driver and observe how the intensity is affected by a duty cycle of 50 percent as compared to a duty cycle of 100 percent.

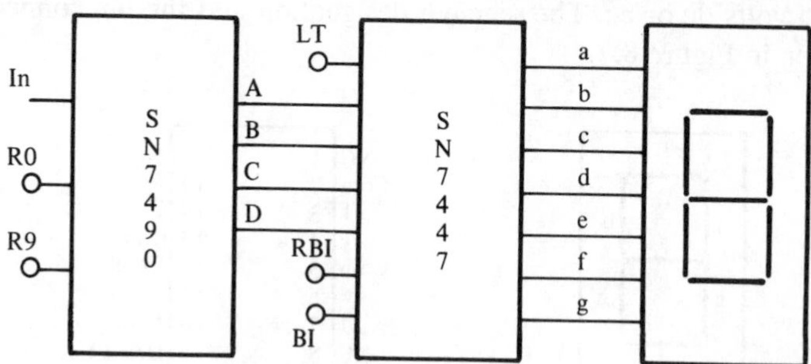

Figure 8.8 Test Circuit

Connect a SN74144 as shown in figure 8.5. Clock the counter while observing the decimal display and verify the function of each of the pins.

QUESTIONS AND PROBLEMS

8.1 Design a frequency counter using the devices of this chapter.

8.2 Design a counting system that will be capable of producing either a decimal readout of its count status or a hexadecimal count of that status.

9

Memory and Memory Devices

OBJECTIVE: To present the topic of memory and the devices used to store information in a digital system.

Introduction: Memory is the means by which a digital computer stores its information. Figure 9.1 shows the memory unit in the overall block diagram of a digital computer. The information stored in memory may be data or instructions. For example, data going to and from the arithmetic unit will generally be temporarily stored in memory. The data coming into the computer or leaving the computer are also placed in memory prior to being manipulated by the computer. Various sets of instructions are also stored in the computer's memory. These tell the computer how to manipulate the data, when to seek new information, and when to print or release the data it has stored. The memory unit also holds repetitive information such as mathematical tables, constants, printed messages, and more.

9.1 TYPES OF MEMORY

There are primarily two major types of memory devices, permanent memory and temporary memory. For our discussion, permanent memory will be the information that remains in the memory unit even after power is removed. Temporary memory will be that which is lost when power is discontinued. Mainly, this second type of memory unit is solid-

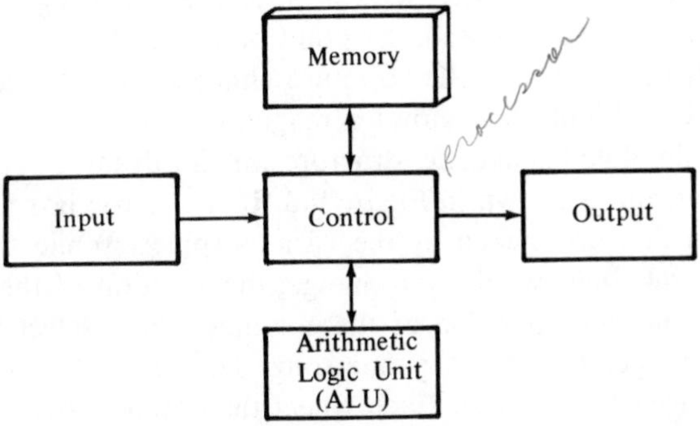

Memory

Input

Control

Output

Arithmetic Logic Unit (ALU)

Figure 9.1 Digital Computer Block Diagram

state memory. A flip-flop is an example of this type of memory device. Once the flip-flop is set, it can be said to have a logic 1 (or a 0) stored in it. This logic state may be lost when the power is discontinued to the flip-flop.

9.2 MEMORY DEVICES

Information can be stored in a variety of ways. There are magnetic devices that use magnetic fields to store 1's and 0's. Punch cards are another means of storing binary information. Here, holes are punched in cards in a pattern recognizable by a computer for retrieving this information. In some systems, information is formed on thin film by holograms. There are new methods using capacitive charge devices in conjunction with solid-state systems.

9.3 MAGNETIC DRUM

In a magnetic drum system, a paramagnetic cylinder is coated with channels of a magnetic substance and rotated at high speeds. Located along the length of the drum are magnetic heads that are used as read-write units. There are several types of magnetic drums and many types of magnetic head arrangements. Some systems use single-purpose heads, one each for read and write, whereas others use a combination read-write head. Placing information on the drum is called the *write cycle,* and retrieving information from the drum is known as the *read cycle.* As the drum is rotated, the write head is pulsed when a logic 1 is to be stored. When the head is pulsed, the point on the drum directly opposite the head will have its magnetic structure changed. This pattern will be called a logic 1. Figure 9.2 shows a magnetic drum with a read-write head and a simulated magnetic structure on the drum. An analogy to the magnetic drum is shown in Figure 9.3. It is a music box spine wheel and the tines that are picked by the various spines. While the plate of tines for several songs would not change, the position of the spines on the wheel would. The position of these spines would select the proper notes at the proper time for any one song. To change a song, only the spine wheel need be changed. Each wheel then "remembers" one song. Magnetic drum memory has a very large capacity. The cost per bit is also usually very low. The access time for these drums is very slow.

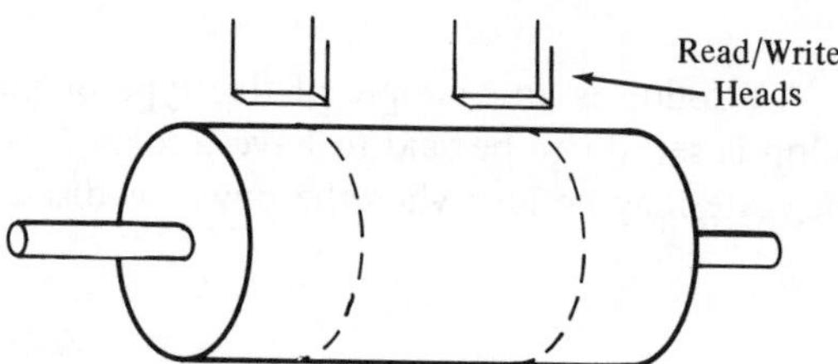

Figure 9.2 Magnetic Drum

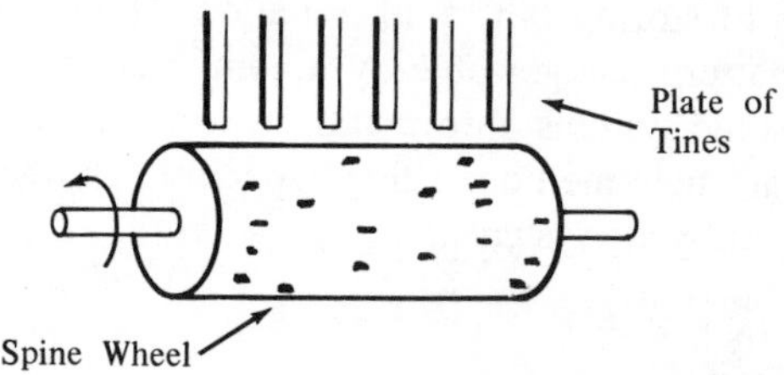

Figure 9.3 Music Box Spine Wheel and Tines

Magnetic Disks

In a magnetic disk system, disks that resemble phonograph records are coated on both sides with a magnetic substance. These disks are then stacked and spaced on a shaft that rotates at a constant high speed. See figure 9.4.

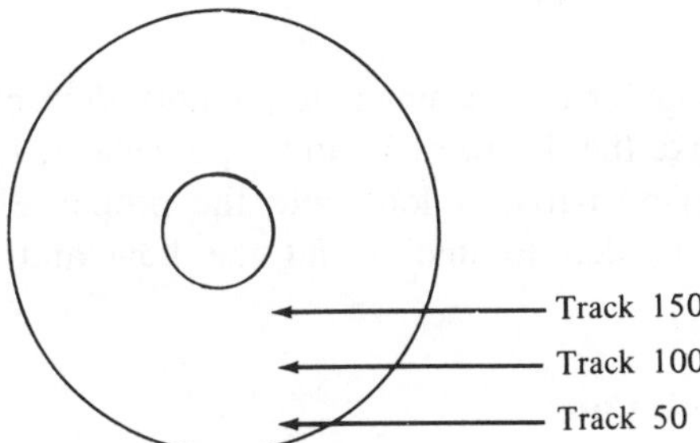

Figure 9.4 Magnetic Disk

Read and write heads are mounted on access arms that straddle each disk and can address both sides of the disk at the same time. Compared to core accessing, the magnetic disk accessing is much slower. Disks are generally used as additional storage devices to increase capacity.

9.4 MAGNETIC TAPES

A magnetic tape system uses a thin ribbon of plastic tape that is put on reels. A uniform coating of a magnetic oxide substance that will accept and hold magnetism permanently is placed on one side. The operation is similar to a home tape recorder in that the tape is moved past the read and write heads. See figure 9.5.

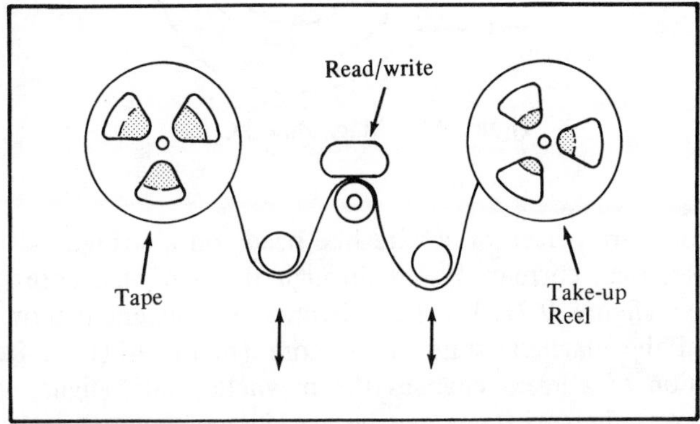

Figure 9.5 Magnetic Tape and Drive

9.5 FERRITE-CORE MEMORY

One of the most popular of the magnetic memory devices is the ferrite core memory. Unlike the drum, disk, and tape devices, ferrite core has no mechanical moving parts. A look into the properties of magnetic-core memories is needed to understand the how and why of these devices.

Properties of Magnetic Cores

If a wire is wound about the core and a current passed through the wire, a magnetic field surrounds the wire and core. A similar condition results when the wire is passed straight through the core. Part of this field will pass through the core which, being of low reluctance (offering little opposition to magnetic lines of force), takes a circular path within the core, as shown in figure 9.6. This field presents a magneto-motive force which, if strong enough, magnetizes the core to one of two possible states. The core is designed so that it retains most of the magnetic energy contained in the field after the current is removed. *Remanence* is the term used to define the amount of magnetic energy retained from the applied field; ideal cores have high retentivity, that is, their remanence is more than that of less efficient cores under similar conditions.

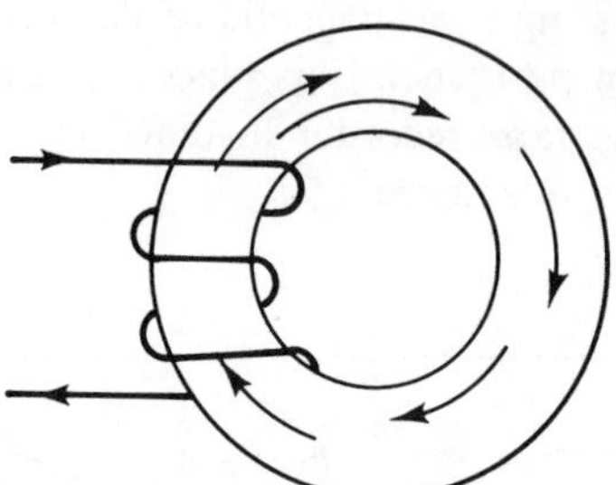

Figure 9.6 Magnetized Core

If cores are placed on a wire like beads on a string, and a strong enough electrical current is sent through the wire, the cores become magnetized (figure 9.7(a)). The direction of current determines the polarity of the magnetic state of the core (figure 9.7(b)). Reversing the direction of current changes the magnetic state (figure 9.7(c)). Consequently, the two states can be used to represent 0 or 1, plus or minus, yes or no, or on off conditions. For computer purposes, this is the basis of the binary system of storing information. More than a million cores are used in large storage units. Because any specified location of storage must be instantly accessible, the cores are arranged so that any combination of 1's and 0's representing a character can be written into or read out of memory when needed.

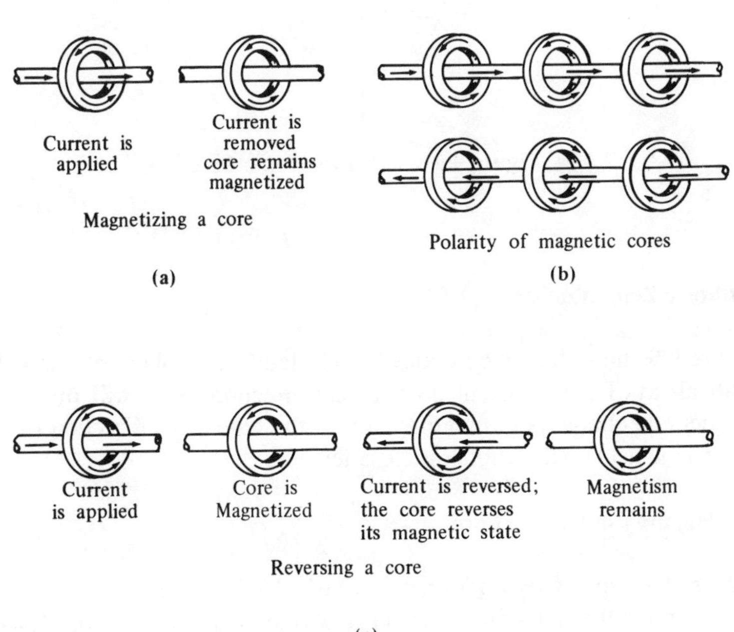

Figure 9.7 Magnetic Polarity of Core

Core Selection

Since each core in the memory represents one bit of information, it is necessary to be able to select any one of the cores independent of all the others. Figure 9.8 shows how this can be accomplished. Since we know it takes a certain amount of current through a core to magnetize that core, or change its direction of magnetization, a method for "addressing" a particular core has been devised. Here two wires (X and Y) are passed through a core. Each wire supplies half the current needed to magnetize the core. Now it takes current through each wire to magnetize a core. Figure 9.9(a) shows a plane containing 81 cores. Each core has an X and a Y wire passing through it. By selecting the proper X and Y wires, a particular core can be selected, as shown in figure 9.9(b).

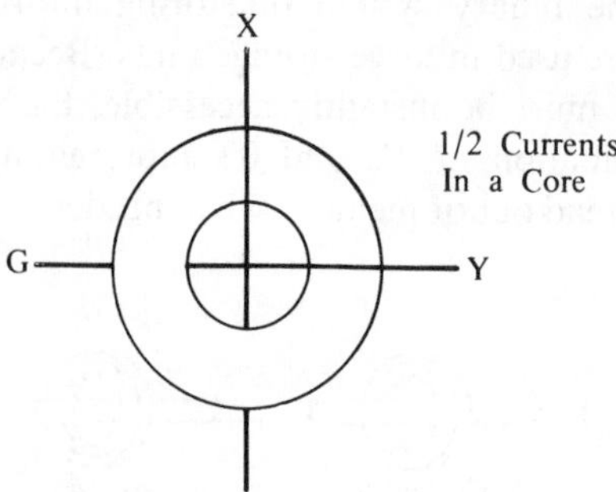

Figure 9.8 ½ Currents in a Core

Writing a Zero into Core

Figure 9.9 shows how a core may be selected from a plane of cores. We shall always let the current flow in one direction. This will produce a magnetic field always in one direction in the core. This direction of flux (magnetic field) will denote the logic level 0.

Reading the Contents of Core

Figure 9.10 shows two additional wires added to the ferrite core. We will examine the function of the sense wire first. Let us suppose for our discussion that a particular core is magnetized such that it has a logic 0 stored in it. Let us further set the conditions such that a negative pulse to the X and Y wires produces that magnetic field. To read the contents of a particular core, a pair of positive pulses are applied to the X and Y wires. These pulses will cause the magnetic field about the core to change direction. This change of flux field will cause a current to be induced in the sense wire. If a logic 1 were stored in the core, these positive pulses will not create any change of direction of flux and therefore no current will be induced in the sense wire. It can be seen, then, that monitoring the sense wire while applying positive pulse to the X and Y wires will tell us the condition of core. We will be able to *read* the contents of the core. However, once this is done, should there have been a logic 0 stored in core, it would have changed state. Figure 9.11 shows a method of pulsing the core to assure that the state of the core will not change during the read cycle. Figure 9.11(a) shows the waveshapes used to read (or write) a 1. First a pair of positive pulses are

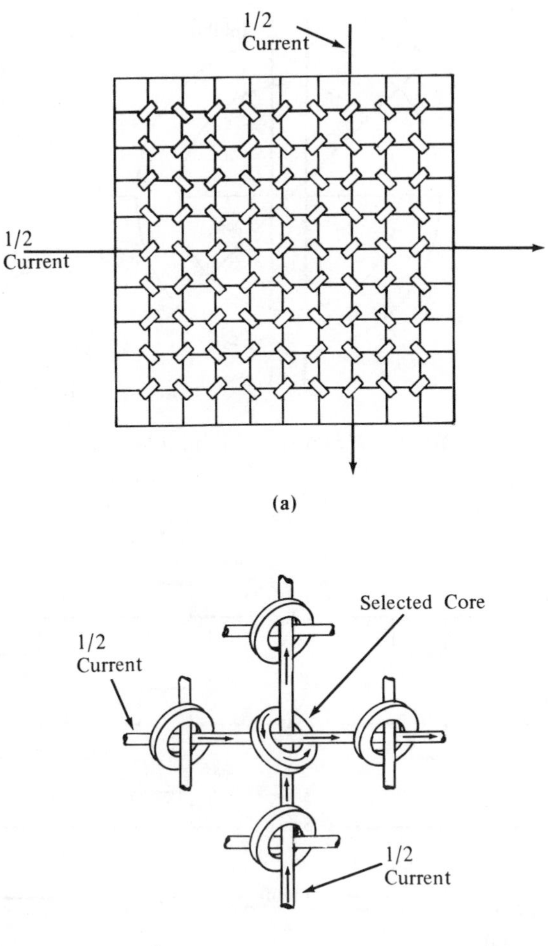

Figure 9.9 **(a)** Core Plane. **(b)** Close-Up of Plane.

fed to the X and Y wires. If a 1 were stored in the core, there would be no change of flux and no pulse of current on the sense line. No current on the sense line would signal the inhibit to create a positive pulse in time with the two negative pulses being fed to the X and Y wires. This positive pulse holds the total current in core windings below that needed to create a flux change, therefore the core remains a logic 1. Should there have been a 0 stored in the core, a look at figure 9.11(b) shows

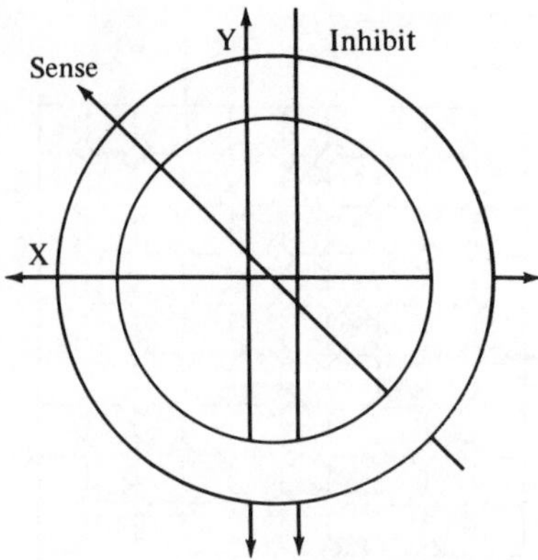

Figure 9.10 Sense and Inhibit Wires

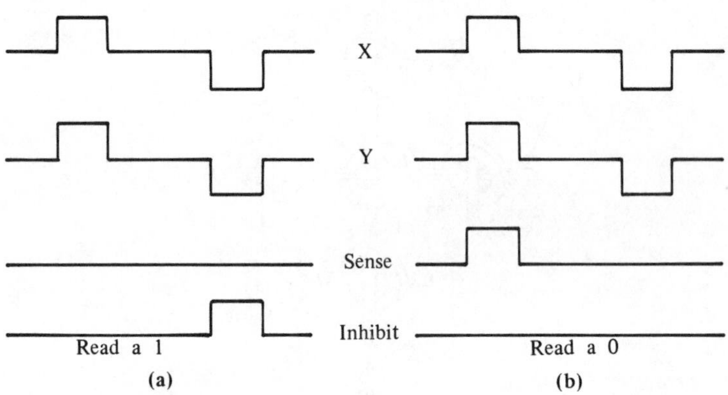

Figure 9.11 Waveshapes for Reading Core

what happens. The original two positive pulses to the X and Y wires will create a change of flux in the core. This change will produce a pulse on the sense wire. A pulse on the sense wire causes no pulse on the inhibit wire. In this case, the two negative pulses that appear now at X and Y will cause the core to again change state (back to 0). Using this pulsing arrangement enables the information in core to be read and restored. This method is known as *nondestructive read*.

Figure 9.12 shows the wiring diagram for a nine-core plane. Fig-

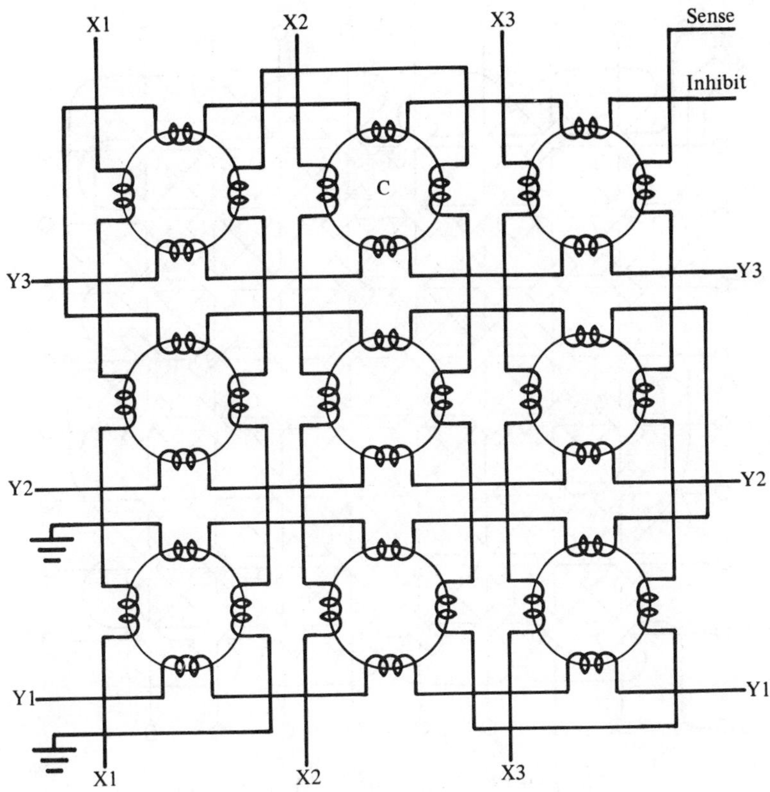

Figure 9.12 Wiring of a Nine-Core Plane

ure 9.13 shows a sixty-four-bit memory plane. Notice that the ferrite cores are actually standing on edge. They are held to a memory board by a touch of glue and the various wires threaded through each ferrite core. By selection of the proper X and Y lines, any one of sixty-four bits may be selected.

9.6 BINARY WORD SELECTION

Although it is important to be able to select any particular bit from memory, it is usually necessary to select a complete binary (or octal) word from memory at one time. Figures 9.14 and 9.15 illustrate how

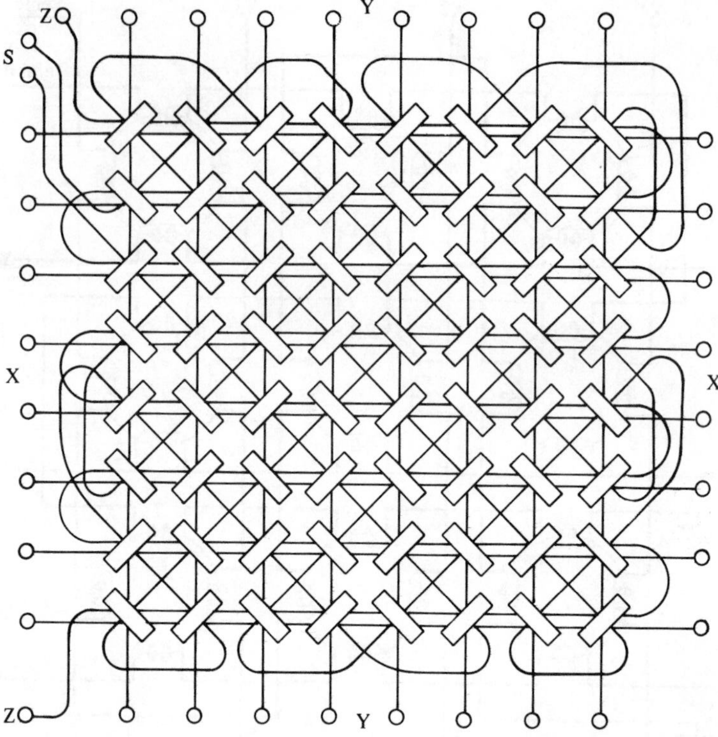

Figure 9.13 Sixty-Four-Bit Memory Plane

this is accomplished. Figure 9.14 shows 2 four-bit planes stacked. Each plane in the stack has identical bits. Therefore, let us say that the X_1, Y_1 bit of plane 1 has been selected. The same bit of plane 2 is selected at the same time. This enables memory to respond with a two-bit binary word. Figure 9.15 shows a $4 \times 4 \times 4$ core array. Using this configuration, any one of 16 four-bit binary words may be selected. The size of such an array is theoretically unlimited.

9.7 CORE MEMORY SYSTEM

The block diagram of a typical core memory system shows the circuitry required to implement memory according to the principles previously discussed (see figure 9.16).

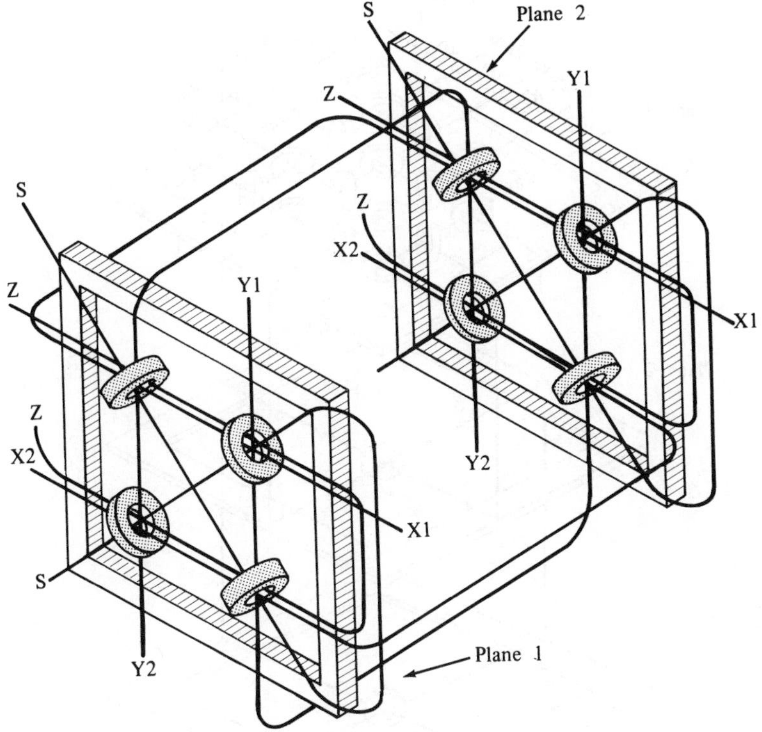

Figure 9.14 Memory Plane Stacking

The drive currents necessary for operation of the stack are produced by the circuits within the system. The timing pulses are required for maintaining accurate relationships between the read, write, inhibit, and other current pulses.

During a read or clear cycle, a drive-current pulse is set through the core stack in a direction that sets all the cores of the selected word to the 0 state. This read current is used for interrogating the cores; if any of the cores had been in the 1 state, they would be switched, inducing a pulse on the associated sense wires. The sense amplifiers detect the induced pulse during the read cycle and are gated to control the data register. As soon as the data register flip-flops are set, the data are available to control equipment associated with the memory system.

A clear cycle is used when it is desired to set the cores of a selected word to 0 prior to a write cycle. For this operation the sense amplifiers are not gated.

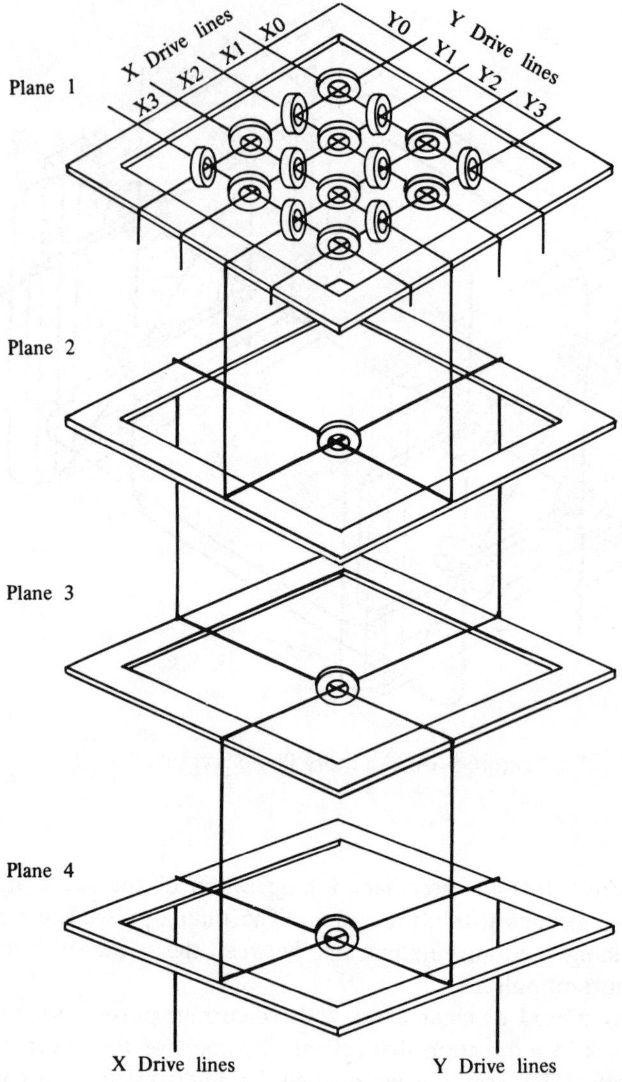

Figure 9.15 A 4 × 4 × 4 Core Array

During the write or restore cycle, the write current pulse is set through the core stack in the direction that sets the selected cores to the 1 state. This current is equal in magnitude to the read current, but in the opposite direction.

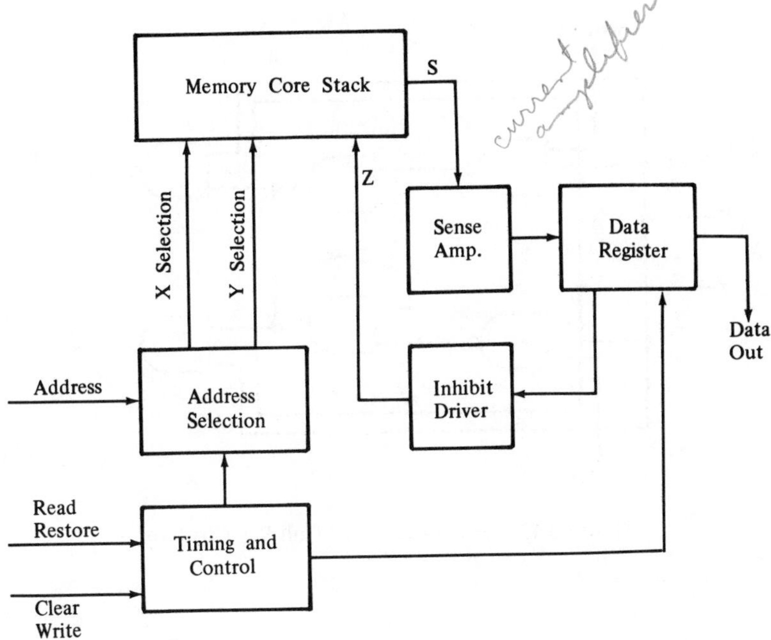

Figure 9.16 Block Diagram Core Memory System

The same cores that had been set to 0 by the read current now receive write current. Where certain of the cores are to be left in the 0 state, the inhibit current for these word bits is turned on, preventing switching by the write current. The inhibit current is controlled by the inhibit timing input.

9.8 SOLID-STATE MEMORIES

Solid-state memory devices have been available for some time but core memories have been used more extensively because of their comparatively low cost. Now, however, due to greater development in this area, the cost of solid-state memories has been greatly reduced. Some advantages of the solid-state memories over core memories include greater speed, smaller size, less power required for operation, and design flexibility. Figure 9.17 is a typical single-memory-cell R-S flip flop.

A 4 × 4 memory matrix comprised of sixteen flip-flops, eight select drivers, and two sense/write circuits is shown in figure 9.18.

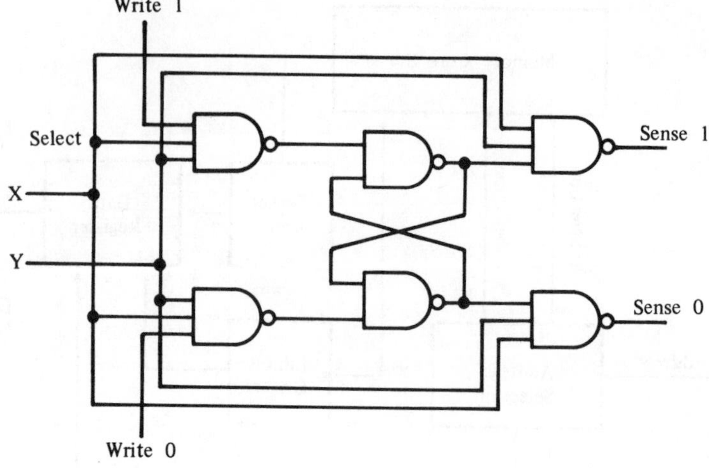

Figure 9.17 Single-Memory-Cell R-S Flip-Flop

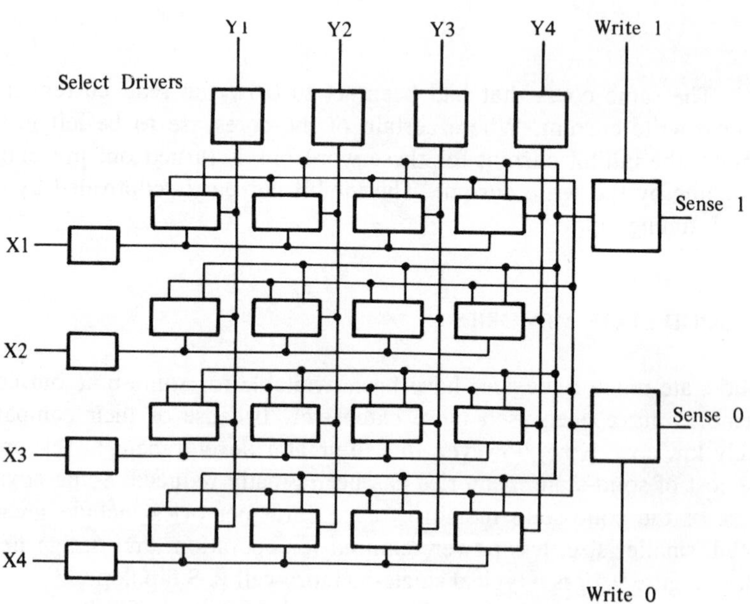

Figure 9.18 Sixteen Flip-Flops (4 × 4 Matrix)

A 1 on an X and Y select driver input selects the desired flip-flop in the matrix. If a 1 is to be written, then a 1 is entered into the write 1 line while the select driver inputs are enabled. If a 0 is to be written, a 1 must be entered into the write 0 line while the select driver inputs are enabled.

To read or sense the output of either complementary output (1 or 0) again the appropriate select lines must select the desired flip-flop. The state or condition of the flip-flop may then be read without affecting the output condition. This is called *nondestructive readout* or NDRO.

Solid-State Memory Devices

In general, solid-state memory devices are simply a number of transistor (or gate) flip-flops connected in matrix fashion to act as storage devices. One of these devices is the 7489 shown in figure 9.19. The 7489 is a sixteen-pin integrated circuit. It is capable of holding 16 four-bit binary words. Through proper coding of the four select inputs, one of the sixteen rows can be selected. Information may be written into or read from the memory through proper coding of pins 2 and 3. These pins are the memory and write enable lines to the device. The function table of figure 9.20 gives the logic levels needed to operate these inputs. Referring to figure 9.19, to place a binary word into memory location of row 16-7, follow these steps:

1. Pin 2 to logic 0
2. Pin 3 to logic 0
3. Binary word to be stored place on pins 4, 6, 10, and 12
4. Pins 1, 15, 14, 13 to logic binary 7 (1, 1, 1, 0)
5. The complement of the binary word present at pins 4, 6, 10, 12 has now been stored at location 16-7.

To retrieve the word from memory, use the following steps:

1. Pin 2 to logic 0
2. Pin 3 to logic 1
3. Pins 1, 15, 14, 13 to logic binary 7 (1, 1, 1, 0)
4. The complement of the word stored in location 16-7 will now appear at pins 5, 7, 9 and 11.

Since the complement of the binary word present at the input was stored in memory, when the stored word is complemented the original word appears at the output pins.

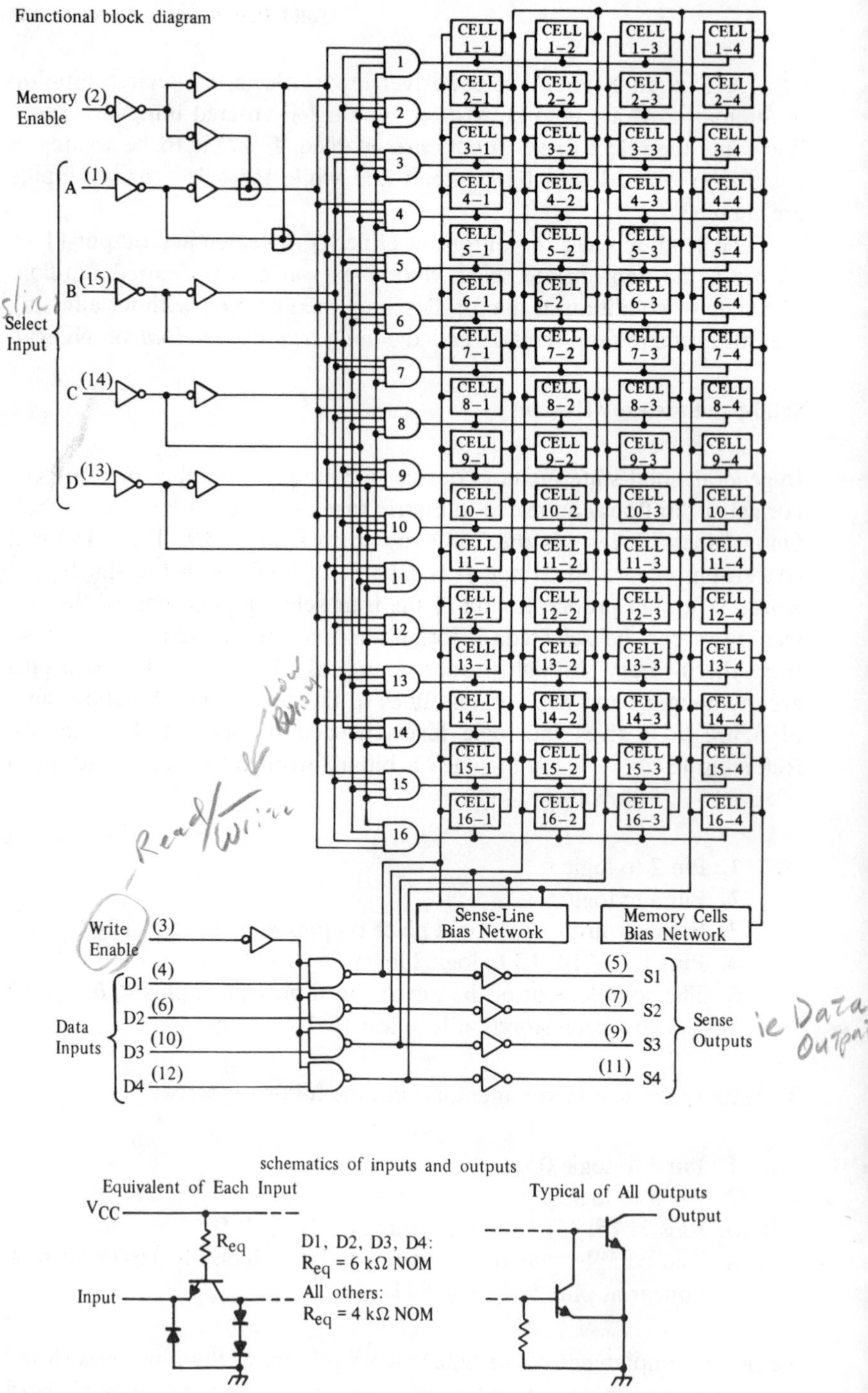

Functional block diagram

Memory Enable (2)

Select Input: A (1), B (15), C (14), D (13)

Write Enable (3)

Data Inputs: D1 (4), D2 (6), D3 (10), D4 (12)

Sense-Line Bias Network

Memory Cells Bias Network

Sense Outputs: (5) S1, (7) S2, (9) S3, (11) S4

schematics of inputs and outputs

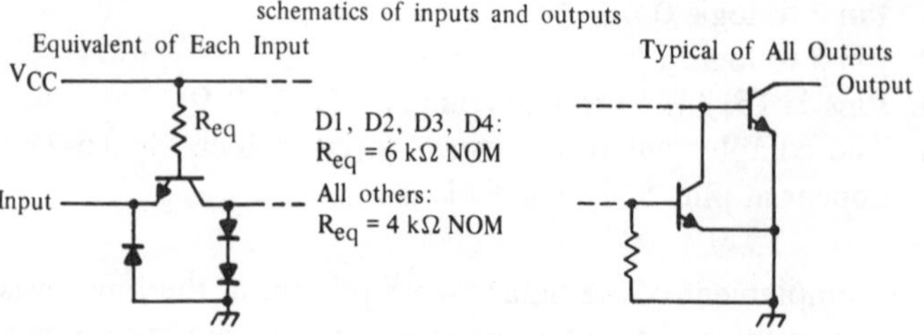

Figure 9.19 Type SN7489 Sixty-Four-Bit Read/Write Memory

For Application as a "Scratch Pad" Memory with Nondestructive Read-Out

Fully Decoded Memory Organized as 16 Words of Four Bits Each Words

Fast Access Time . . . 33 ns Typical

Diode-Clamped, Buffered Inputs

Open-Collector Outputs Provide Wire-AND Capability

Typical Power Dissipation . . . 375 mW

Compatible with Most TTL and DTL Circuits

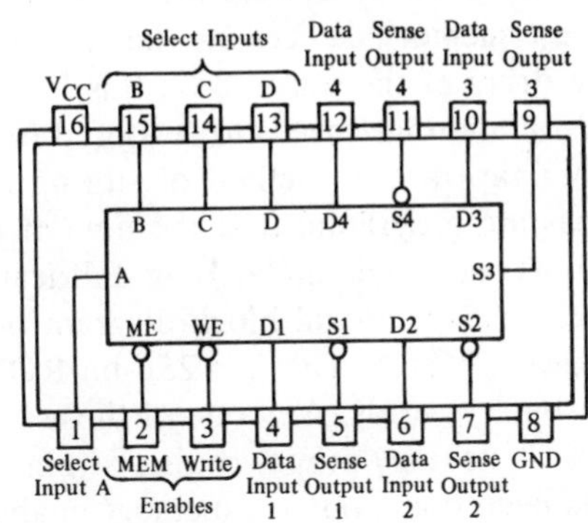

positive logic: see description

†Pin assignments for these circuits are the same for all packages.

Function Table

ME	WE	Operation	Condition of Outputs
L	L	Write	Complement of Data Inputs
L	H	Read	Complement of Selected Word
H	L	Inhibit Storage	Complement of Data Inputs
H	H	Do Nothing	High

write operation

Information present at the data inputs is written into the memory by addressing the desired word and holding both the memory enable and write enable low. Since the internal output of the data input gate is common to the input of the sense amplifier, the sense output will assume the opposite state of the information at the data inputs when the write enable is low.

read operation

The complement of the information which has been written into the memory is nondestructively read out at the four sense outputs. This is accomplished by holding the memory enable low, the write enable high, and selecting the desired address.

Figure 9.20 SN7489 Sixty-Four-Bit Read/Write Memory

9.9 RAMS/ROMS

The devices we have seen to this point are known as *random-access memories* (RAMS). This is so because any cell (or bit) on the device may be accessed randomly, as opposed to sequentially. As a further point of classification of this device, the RAM may also have information read into it as well as read out. The information stored in memory can be changed. These are also called *read-write memories.*

ROMS

Not all memory devices are capable of having their contents changed. These memory devices are called *read-only memories* (ROMS). Recall the drum of the music box; this is a type of read-only memory. When information just needs to be stored for future reference, read-only memory is a low-cost method of storing that information. Read-only memories are programmed at the time of their manufacture (the music box drum) or while undergoing fabrication (solid-state ROMS). Figure 9.21 is a functional block diagram of one type of solid-state ROM. It shows a 7488A. This is a 256-bit ROM. The ROM is actually 32 eight-bit binary words. Any one of those thirty-two words may be addressed. This is done by proper coding of pins 10, 11, 12, 13, and 14. Once this has been done, pin 15, memory enable, is put low and the word stored at the address selected will appear at the output pins. To inhibit a word from appearing at the output, pin 15 is simply held high (logic 1). Figure 9.22 gives the pin layout and functional description of the operation of the device.

ROM Programming

The ROM described in the preceding paragraph must be programmed during fabrication. This is accomplished by connecting (or not connecting) each of the 256 programmable links, as shown in figure 9.21(c).

There are also available at higher cost devices called PROMS (programmable read-only memory). These devices can be programmed (and erased) by the user.

ROM Application

One use for a ROM is as a look-up table. Here information such as multiplication or trigonometry tables can be stored in ROMS and recalled as needed. Another use for a ROM would be to store a series of binary-coded instructions for a computer (or calculator). These instructions could then be recalled in sequence when needed by a circuit such as figure 9.23, which shows a counter with its output connected to the select inputs of a ROM. As the count increments, the row selected from the ROM is also incremented.

This type of ROM is called the *masked ROM* because during manufacture the IC chip is made in two steps. These steps help to keep the overall cost down and enable many users to specify their own requirements. Figure 9.24 is a pictorial representation of a masked ROM.

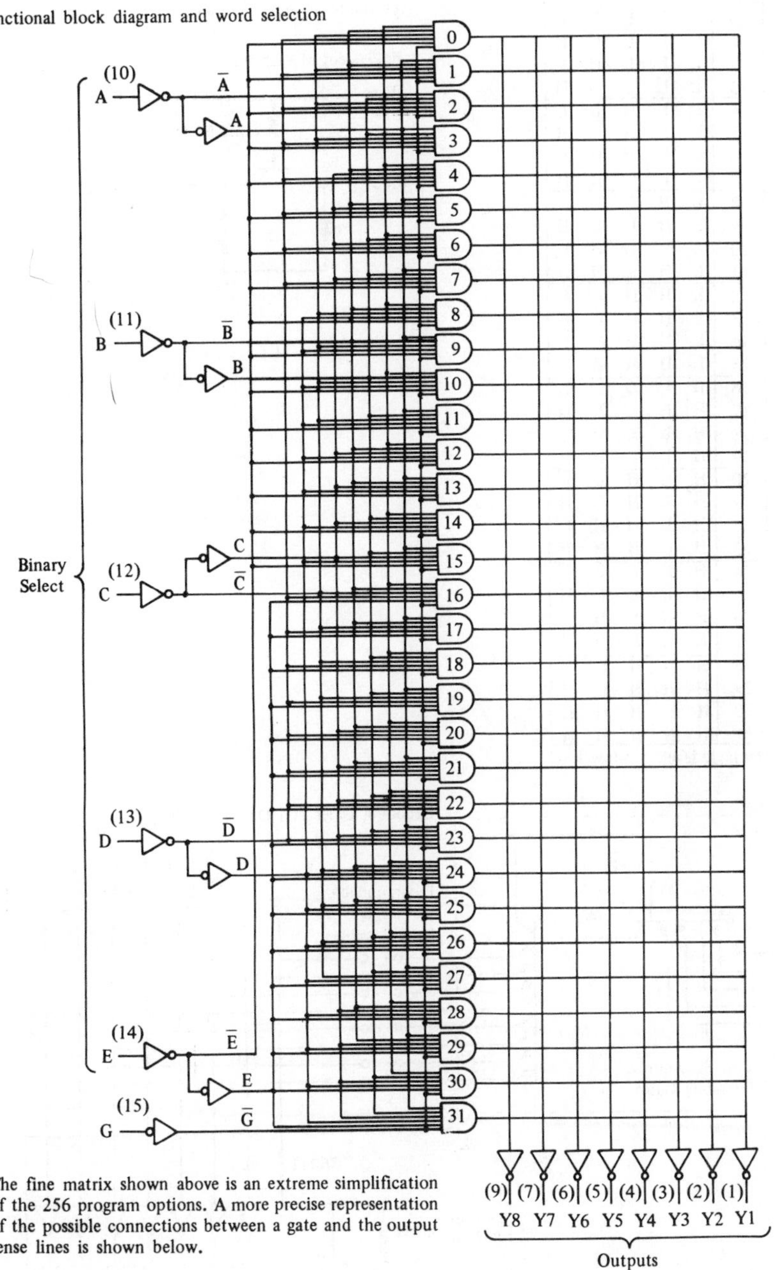

The fine matrix shown above is an extreme simplification of the 256 program options. A more precise representation of the possible connections between a gate and the output sense lines is shown below.

(a)

Figure 9.21 Type SN7488A 256-Bit Read-Only Memories

Word-Select Table

Word	Inputs				
	E	D	C	B	A
0	L	L	L	L	L
1	L	L	L	L	H
2	L	L	L	H	L
3	L	L	L	H	H
4	L	L	H	L	L
5	L	L	H	L	H
6	L	L	H	H	L
7	L	L	H	H	H
8	L	H	L	L	L
9	L	H	L	L	H
10	L	H	L	H	L
11	L	H	L	H	H
12	L	H	H	L	L
13	L	H	H	L	H
14	L	H	H	H	L
15	L	H	H	H	H
16	H	L	L	L	L
17	H	L	L	L	H
18	H	L	L	H	L
19	H	L	L	H	H
20	H	L	H	L	L
21	H	L	H	L	H
22	H	L	H	H	L
23	H	L	H	H	H
24	H	H	L	L	L
25	H	H	L	L	H
26	H	H	L	H	L
27	H	H	L	H	H
28	H	H	H	L	L
29	H	H	H	L	H
30	H	H	H	H	L
31	H	H	H	H	H

H = high level, L = low level

schematics of inputs and outputs

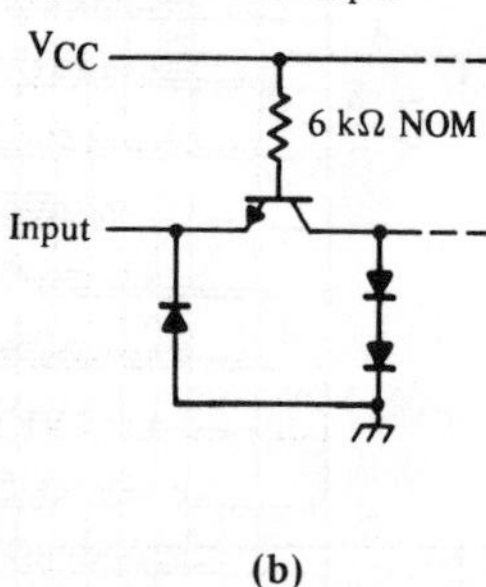

(b)

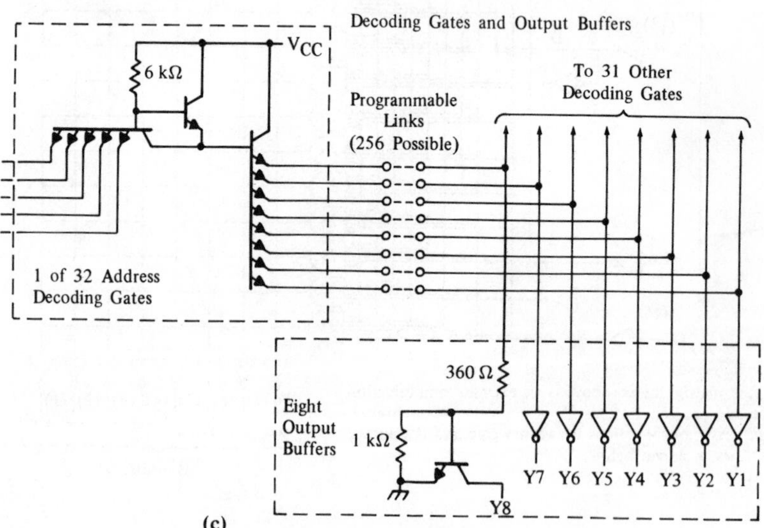

(c)

Figure 9.21 *continued*

Applications in Computer Subroutines

Useful in Display Systems and Readouts

Memory Organized as 32 Words of 8 Bits Each

Input Clamping Diodes Simplify System Design

Open-Collector Outputs Permit Wire-AND Capability

Typical Access Time: 25 nanoseconds

Typical Power Dissipation: 285 milliwatts

Fully Compatible with Most TTL and DTL Circuits

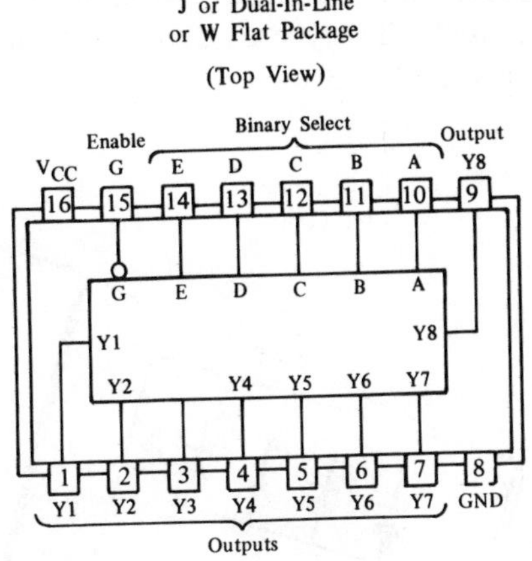

Figure 9.22 SN7488 256-Bit Read-Only Memory

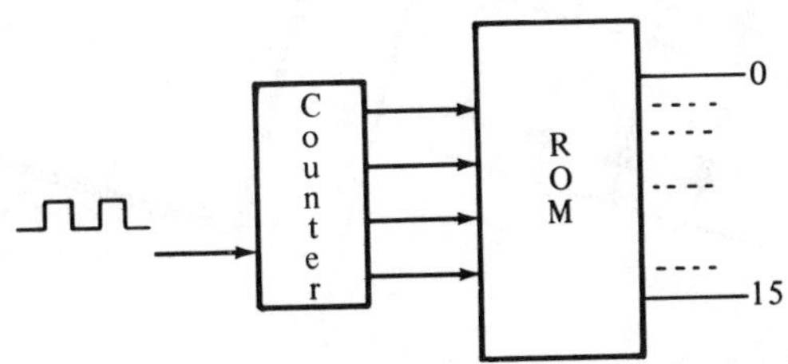

Figure 9.23 ROM Addressing via Counter

On the silicon wafer, a number of IC diodes are fabricated. The anodes of all diodes in a particular row are connected together. The diagram shown indicates a 4 × 4 sixteen-bit read-only memory configuration. There are four rows of four diodes each. Figure 9.25 gives the circuit diagram for this configuration. If a diode is to be connected, a logic 1 at that location, the mask layer of the IC creates a connection at that diode's cathode. This is shown in figure 9.24. Notice that figure 9.25 shows all the diodes. Those with connected cathodes, via masking, are shown with a dot, and those without the masking are shown simply not connected. Figure 9.26 shows a simplified diagram of the 4 × 4, sixteen-bit ROM.

Even though this type of structure has enabled manufacturers to keep costs down, masked ROMs are still expensive and can only be justified when a large number of the same masked devices are going to be used. This type of device is available using diodes, bipolar or MOS transistors.

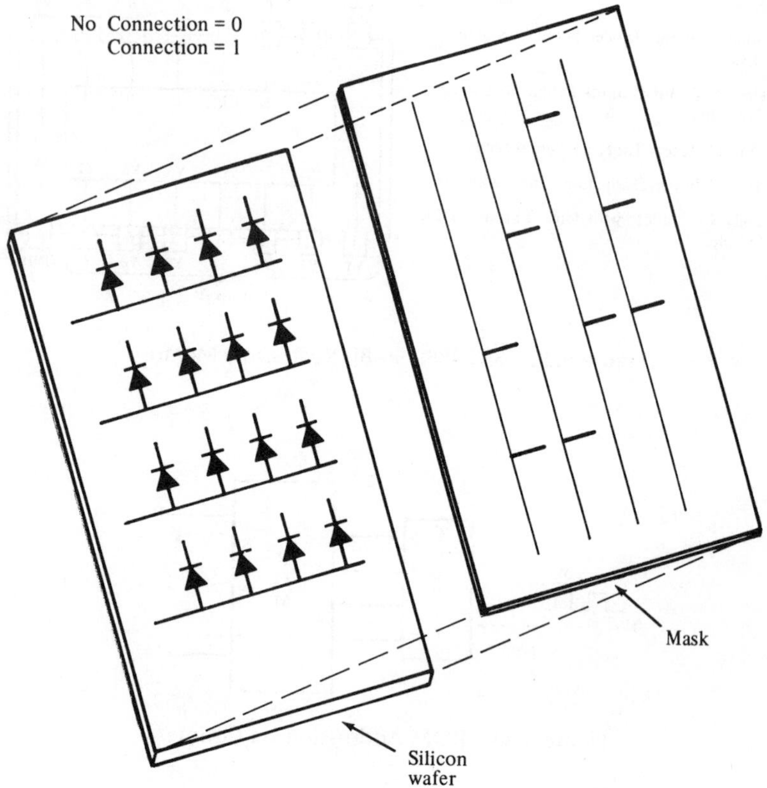

Figure 9.24 Masking Process for Diode Matrix

PROMS

PROMS are field-programmable or user-programmable memory devices; that is to say they came from the factory with all bits set to either a 1 or a 0 and the end user may change this pattern. These programmable read-only memory devices are excellent when the user needs to develop a stored program or perhaps store some newly calculated data into a circuit. The user simply "burns" the program into the PROM and it is ready to use. These PROMS are not alterable. Once programmed, they are set with the bit pattern permanently.

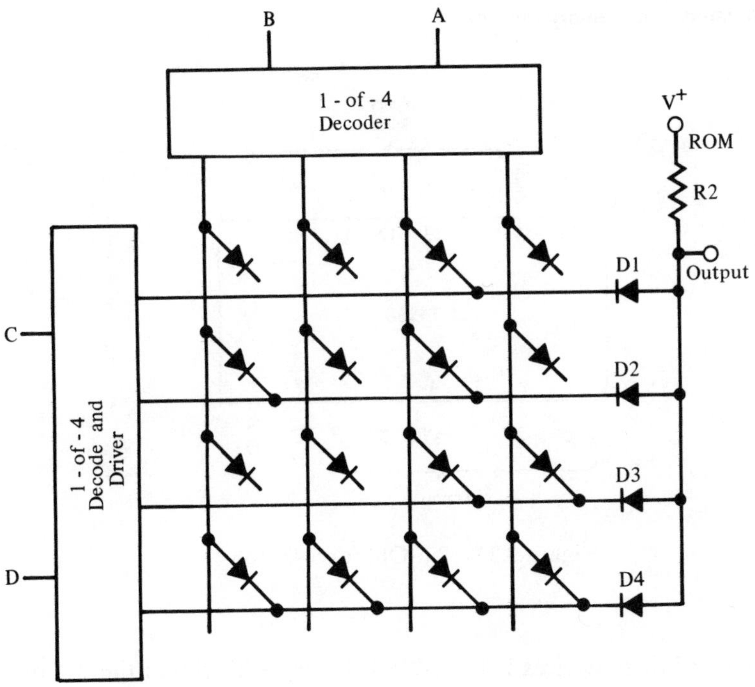

Figure 9.25 Sixteen-Bit ROM Using Diode Matrix

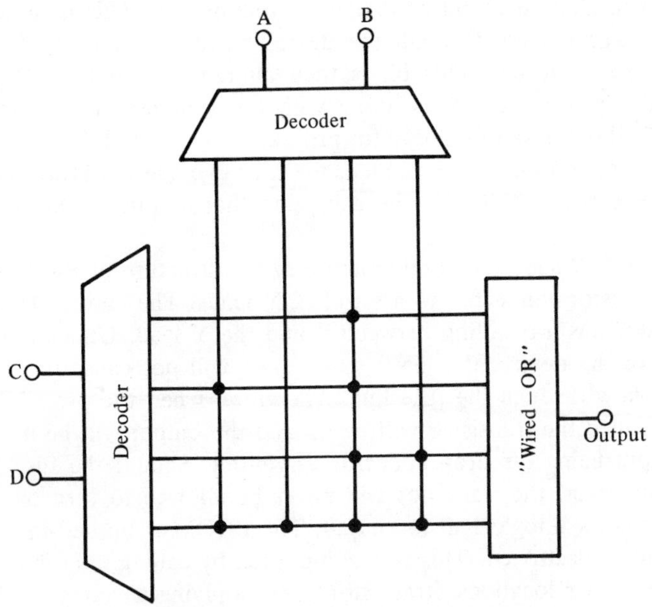

Figure 9.26 Simplified Version of Figure 9.25

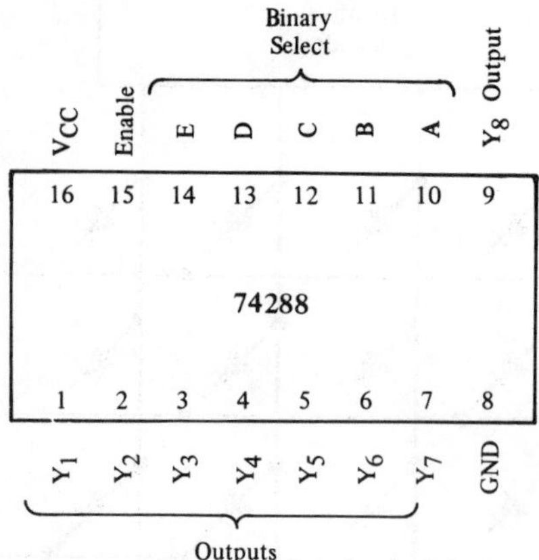

Figure 9.27 Pin-Out for SN74288

One of these devices is the 74288. Figure 9.27 shows the pin-out of one of these devices. This particular device has tri-state output. It is arranged as a 32 by 8 array. It contains 32 eight-bit words. When the device is to be programmed, the user simply addresses the proper location and then alters each bit of that word one-by-one. This is done by passing a large current through the device when only one bit in the array has been selected. This bit is then altered from a logic 0 to a logic 1. The procedure can be done with the simplest of equipment. Figure 9.28 shows a simple circuit for programming the 74288.

If tri-state output is not needed, the user can use a 74188, which is identical to the 74288 with the exception that its outputs are open-collector.

Figure 9.29 shows a simple picture of the structure of PROM device. A transistor connected to a set of X-Y leads. The emitter lead of the transistor has a fuse link between it and the Y lead. Under normal operation of the device, the +5.0 volt (V_{CC}) will not cause excessive current to flow through the fuse link. Therefore, when the proper X-Y pattern is called the transistor will be on and the output will be a logic 0, the output being the drop collector to emitter. Should the fuse link be open, however, the transistor will never be allowed to turn on and will always show a logic 1 at its output. The PROM is burned-in when the fuse link is destroyed. This is accomplished by calling (via the X-Y leads) the proper locations (transistor) and applying a voltage at V_{CC} greater than 5.0 volts. This voltage level produces between 20 to 30 milliamperes of emitter current and the fuse link opens. It should be

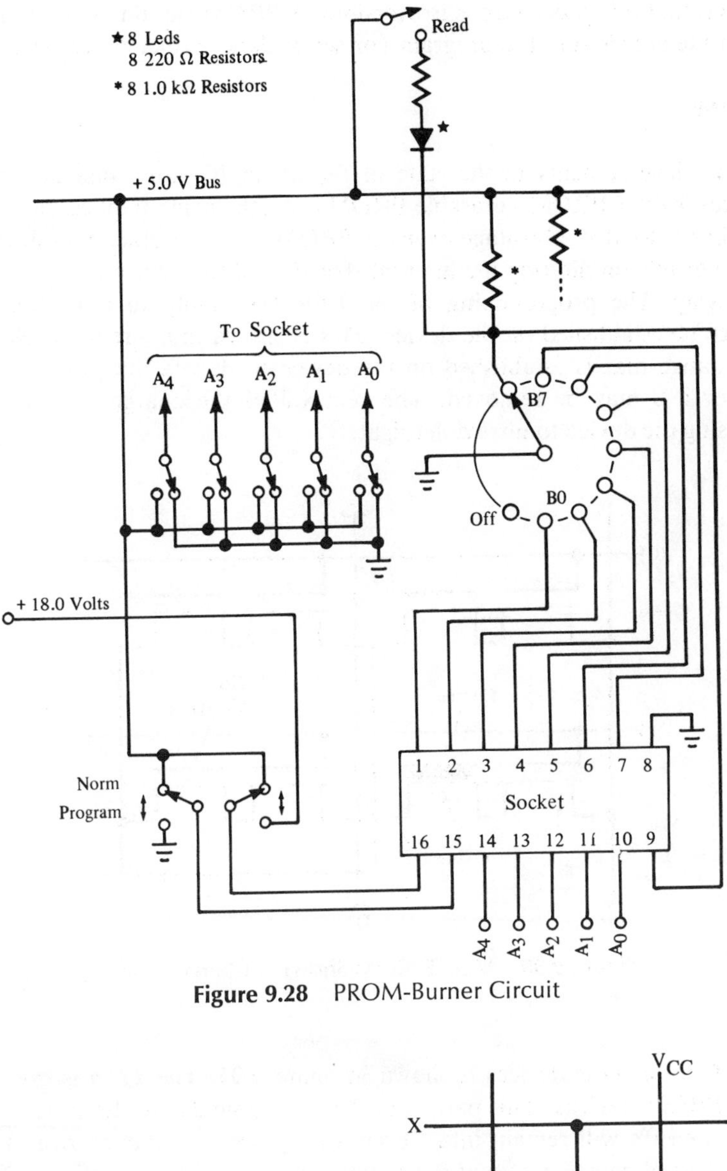

Figure 9.28 PROM-Burner Circuit

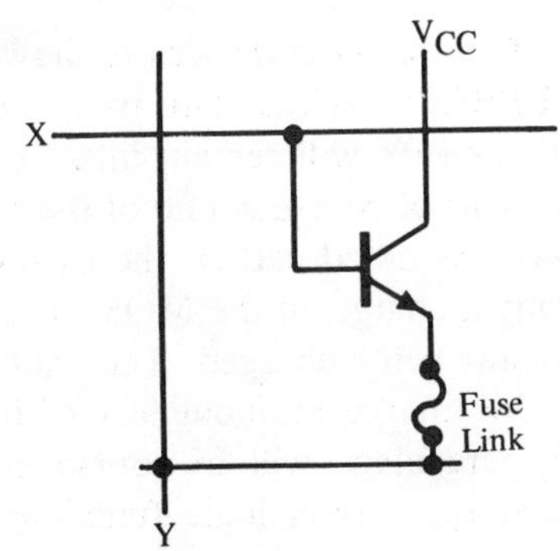

Figure 9.29 Typical PROM Fuse Link

evident that this procedure is irreversible. A PROM should be used only when the user is sure the program (or set of data) is exactly as wanted.

EPROMS

Recent developments in the state of the art in IC solid-state memory devices is the EPROM (erasable PROM). As the name implies, this device has a decided advantage over the PROM, it is alterable. This device does not rely on altering the internal structure of the device in a permanent way. The programming of the EPROM simply requires that a charge be established on the device. This is shown in figure 9.30. Here, the change that is established on the device to denote the presence of a 1 or a 0 may be removed. The removal of the charge is done by exposing the device to ultraviolet light.

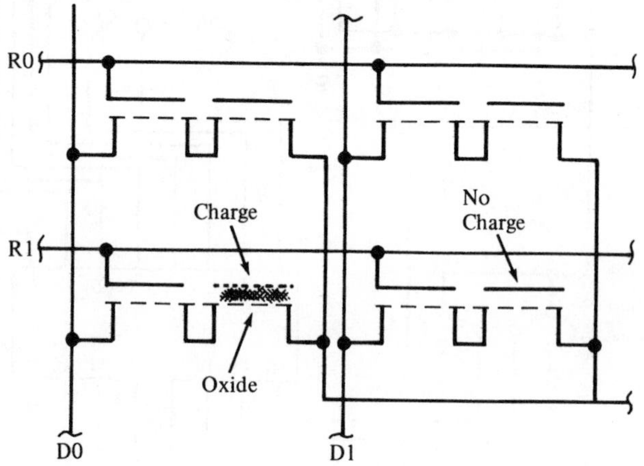

Figure 9.30 MOS EPROM Showing Charge Field

One of these devices is shown in figure 9.31. The 2716 is one of the EPROM devices. This particular device is nonvolatile. Its stored set of 1's and 0's will remain intact even after power is removed from the chip. This of course is one of the distinct advantages of the entire ROM family. As noted earlier, the storage of memory in the chip is done by placing a charge on the MOS transistor, which acts much the same as a capacitor being charged. The MOS transistor characteristics will maintain this charge at about 90° of its initial value for close to 10 years. This percentage will of course be enough to maintain the 1-0 level needed for correct logic functions. As can be seen in the diagram, a

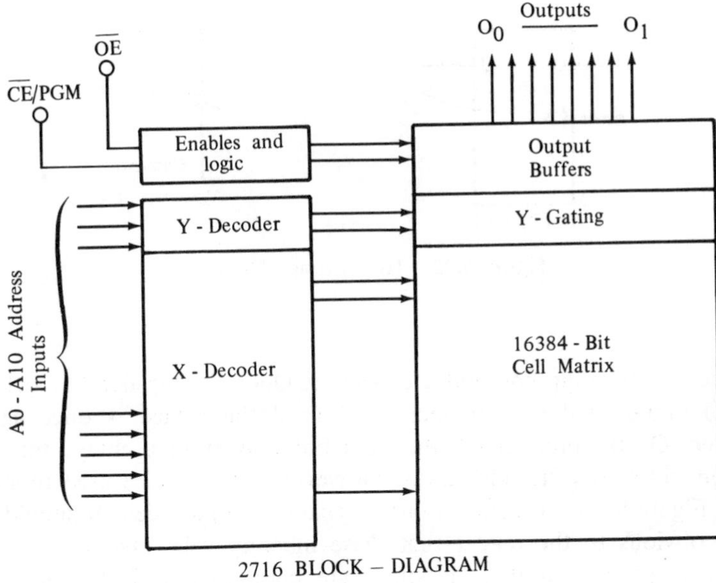

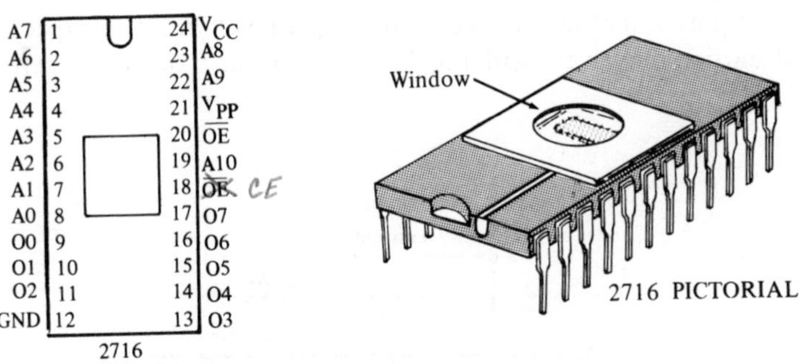

Figure 9.31 2716 Block Diagram and Pictorial

small window is cut into the top of the chip case. When an ultraviolet light is shown in this window, the charge on the MOS transistor is altered and the bit pattern is erased.

9.10 STATIC AND DYNAMIC MEMORY

There are two basic ways data may be stored in memory, statically or dynamically. Figure 9.32 shows these two methods schematically. Here

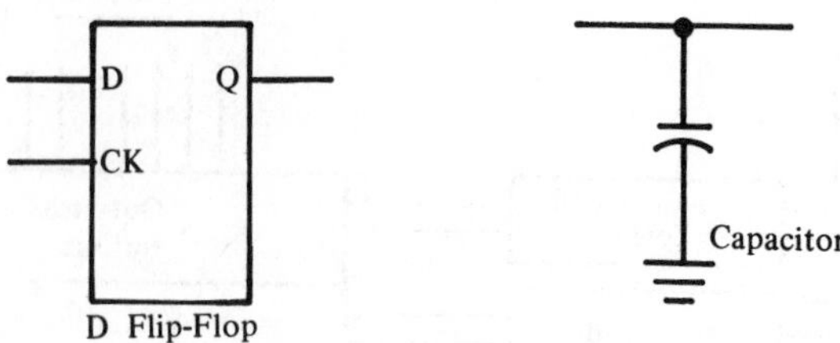

Figure 9.32 Two Storage Devices

we see a D type flip-flop and a capacitor. Once the flip-flop is clocked, the Q output will remain unchanged until the device is once again clocked. On the other hand, the capacitor may store a charge but the charge will slowly "trickle" away. However, both may be used to store data. Figure 9.33 represents a basic dynamic memory cell. It should be quite obvious to the reader that these memory cells take up far less physical space than their bipolar-state counterparts. Refer back to figure 9.11. There can be about four dynamic cells placed in the same physical space as one static cell. This means less cost, less operating power required, and faster access time for a given bit. The capacitor, as stated earlier, however, will not hold a charge; it must be refreshed.

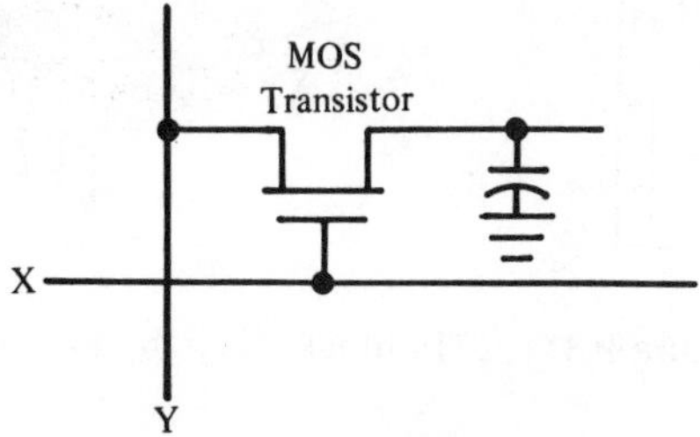

Figure 9.33 Dynamic Cell

Thus it requires a separate circuit to continually refresh each bit in memory, that is, it must constantly restore 1's and 0's at each memory location. This requires extensive circuitry to support the chip. This may be a small price to pay if larger amounts of memory in a small space at low power consumption is needed.

SUMMARY

This chapter has taken a look at computer memory. In short, memory is the storing of information in a location from where it can be retrieved when needed. Think of memory as a storage box as shown in figure 9.34, which shows a box with twelve compartments. Compartment A, for example, is in row 1, column X. Compartment H is given by row 3, column Y. These various storage locations can be found electronically in a number of devices. Several of those were shown in this chapter. Memory may be permanent, such as tape, or may need to be refreshed such as the RAMS. A ROM, on the other hand, is also a permanent memory device that may be field-programmed.

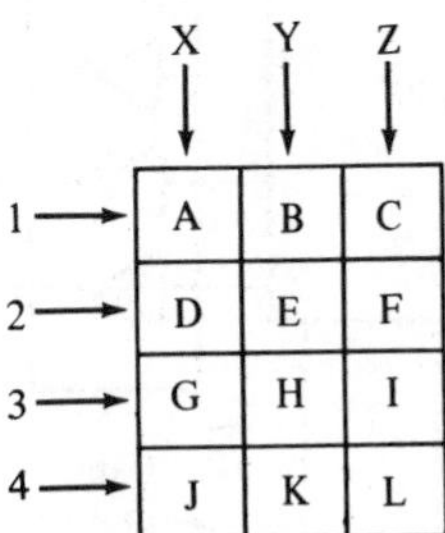

Figure 9.34 Twelve-Address Memory

EXERCISES

1. Build the sixty-four-bit read/write memory system shown in figure 9.35. The 7493 binary counter is used for generating the address inputs to the 4064 read/write memory. It may be pulsed from a push-button clock.
2. Write in a sequence of 16 four-bit words by applying the word to the data input lines as its address is set up by the address generator. After all words have been entered into the memory, read the data back out by stepping the counter back through all sixteen addresses.

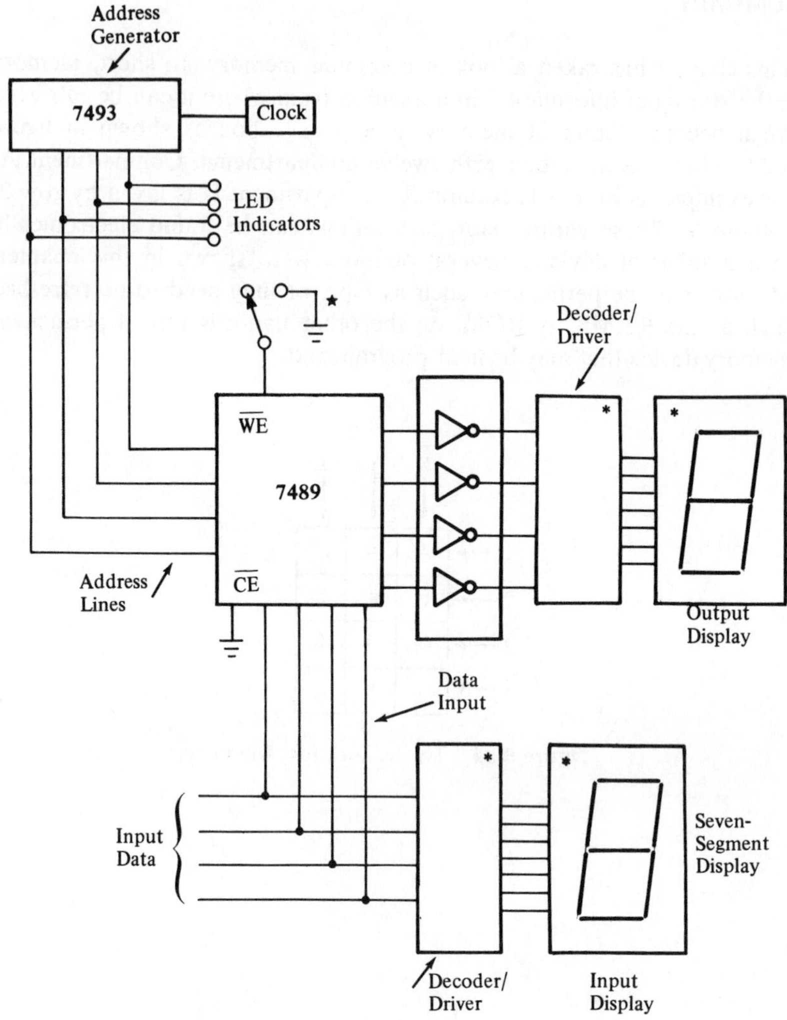

* Decoder/driver and Display must be compatible.

★ WE	OPERATION
LOW	Write to memory
HI	Read from memory

Figure 9.35 Circuit for Student Exercise

QUESTIONS AND PROBLEMS

9.1 Match the following names to the proper diagrams in the unit.

(a) Memory stack ______
(b) Solid-state cell ______
(c) Magnetic drum ______
(d) Memory plane ______
(e) Memory system ______
(f) 4×4 memory matrix ______
(g) Magnetic disk ______
(h) Single toroidal core memory cell ______

9.2 Match the following names to the proper advantage:

1. Magnetic Cores ______
2. Magnetic Disks ______
3. Solid State Memory ______
4. Magnetic Tapes ______
5. Magnetic Drums ______

(a) Due to recent developments the cost is now about the same
(b) When compared to core memories they have greater capacity and less cost per bit but greater access time
(c) Greater speed of any device
(d) Fastest access time of any magnetic device
(e) Smaller size than any other memory
(f) Generally used as additional storage devices to increase capacity
(g) Less power required than any other memory
(h) A thin ribbon of plastic tape with a coating of magnetic oxide on one side
(i) Greater design flexibility than any other memory
(j) A magnetic device with the best stability

9.3 Define *memory*.

9.4 Name the two major types of memory.

9.5 Where would core memory be used today?

9.6 What are the advantages of solid-state memory?

9.7 What is the basic solid-state memory unit?

9.8 Name the distinctive features of:

(a) RAM
(b) ROM
(c) PROM
(d) EPROM

9.9 Where would each of the above probably be used in a digital system?

9.10 Explain what is meant by a masked ROM?

9.11 What is the major difference between a programmable and a field-programmable ROM?

9.12 Explain the major difference between static and dynamic memory.

9.13 Which, static or dynamic, is least expensive to use in a system? Why?

9.14 What extra precautions must be taken when using EPROMS?

9.15 Give a complete pin-by-pin wiring diagram for figure 9.25.

9.16 Of what practical value is a circuit such as the one shown in figure 9.25?

9.17 Design a circuit that will constantly refresh the dynamic memory shown in figure 9.27.

Computer Arithmetic Circuits

10

OBJECTIVE: To introduce the concepts and circuits required to perform computer arithmetic functions. These circuits will be shown in simple gate format before being shown as complete circuits contained in a single IC package.

Introduction: The idea that a machine could perform arithmetic functions has been around for a long time. In fact, in 1642 Blaise Pascal designed and built the first mechanical adding machine. The basic principles behind Pascal's calculator are still in use today.

Not long after Pascal, Gottfried Wilhelm von Leibniz also built a calculator. His could not only add, but it could subtract, multiply, and divide. Leibniz also introduced the binary number system in 1671.

Today the size and circuitry may be different, but the basic ideas are the same as they were in 1671. In this unit, we shall take a look at how binary arithmetic handles its various operations. We shall see the power, ease and limitations of the system. We shall see how simple circuits may be used to perform complex mathematical operations with the speed of light.

Figure 10.1 is a block diagram of a small digital computer. The circuitry we shall focus on in this chapter will be that contained in the *arithmetic* unit. Note that this unit interfaces directly with the memory unit. Recall from the chapter on memory that all information stored in the memory unit is in binary form. This means that our arithmetic unit must operate in the binary system.

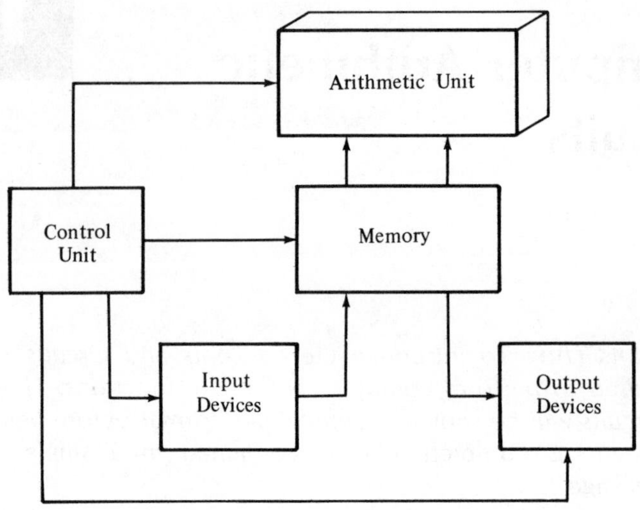

Figure 10.1 Block Diagram of a Small Digital Computer

10.1 BINARY ARITHMETIC

This section is intended as a review of the basics of binary addition and subtraction. As you remember from your study of number systems, the binary number system consists of only two digits, 0 and 1.

EXERCISES

1. Write the decimal equivalent of the following binary numbers.

Binary	*Decimal*
100	
010	
1011	
1111	
0001	
0101	

2. Write the binary equivalent of the following decimal numbers.

Decimal	*Binary*
10	
7	
13	
8	
3	
12	

The answers to the first part are 4, 2, 11, 15, 1, and 5. The answers to the second part are 1010, 0111, 1101, 1000, 0011, and 1100.

3. For the following decimal numbers, select the correct binary number.

(a) 8 in decimal equals

1011 1000 1111 0001

The correct answer is 1000.

(b) 16 in decimal equals

10000 11111 10010 10001

The correct answer is 10000.

4. For the following binary numbers, select the correct decimal number.

(a) 11001

21 25 16 8

The correct answer is 25.

(b) 1001

5 6 7 9

The correct answer is 9.

Addition

The addition of numbers in the binary system may appear puzzling at first, but by remembering the rules of addition we can arrive at our answers quickly and accurately. These rules are:

1. 1 + 0 equals 1 and a carry of 0.
2. 0 + 1 equals 1 and a carry of 0.
3. 1 + 1 equals 0 and a carry of 1.
4. 0 + 0 equals 0 and a carry of 0.

Let's look at an example of the addition of two numbers in the binary system.

Example Add the binary number 101 to the binary 011.

Solution

Step 1. Place the two numbers in columns as shown.

101
011

Step 2. Add the two digits to the far right first. (These are the least-significant digits.) Adding the 1 to the 1 according to our rules gives us a sum of 0 and a carry of 1.

This gives us the following:

1
1 0 1
0 1 1
0
carry
sum
1 + 1 = sum 0, carry 1

Step 3. Add the two digits in the middle column next. This gives us the following: A 1 with a carry of 0.

0 1 carry
1 0 1
1
0 1 1
0

or

Step 4. Next add this 1 to the 1 from the carry of the first column. This gives us

1 carry
0 1 carry
1 0
1
0 1
0 0
sum

The 1 and 1 give a 0 to the sum and a carry of 1 to the far left column. The far left column now has four digits in it.

Step 5. Add the digits two at a time. This yields

```
  ⎧1
1 ⎨
  ⎩0 1
  ⎧1 0 1
1 ⎨
  ⎩0 1 1
 ────────
     0 0
```

Step 6. Add the sum of Step 5, which gives:

```
  carry ⎧1   1
  ←     ⎪    0 1
 |  |   ⎪    1 0 1
 ↓  |   ⎩1   0 1 1
    └──────────────
    sum  →   0 0 0
```

Step 7. This final carry of one is brought down giving a final answer of

```
    1 0 1
    0 1 1
  ───────
  1 0 0 0
```

To check your answer simply write the decimal equivalent of the three binary numbers and see if the decimal arithmetic is correct.

```
    1 0 1      5
    0 1 1      3
  ───────     ──
  1 0 0 0      8
```

Since the answer checks in decimal form, we have correctly performed our binary addition.

EXERCISES

1. Add the following pairs of binary numbers.

(a) 110
101

(b) 111
101

(c) 111
110
———

(d) 111
111
———

(e) 1101
0101
———

You should have the following answers: 1011, 1100, 1101, 1110, 10010.

2. Select the correct answer for the following problem.

(a) 101 added to 110 equals

1001 1010 1011 1100

The correct answer is 1011.

(b) 111 added to 101 equals

1001 1010 1011 1100

The correct answer is 1100.

(c) 1011 added to 0111 equals what in decimal?

15 16 17 18

The correct answer is 18.

(d) 1011 added to 1011 equals what in decimal?

20 21 22 23

The correct answer is 22.

For the addition of numbers in base 8, it is important to remember that the numeral 7 is the highest allowed in that base. The numbers 8 and 9 do not appear in base 8. So, to add the numbers 7 and 8 to the number 1 base 8, we simply say

1
7
——
carry 10

or 7 plus 1 equals 0 carry the 1.

To add 7 base 8 to 2 base 8 we would have

1
2
7
——
11

or

$$7 + 2 = 7 + \underbrace{1 + 1}_{2}$$

then

$$\underbrace{7 + 1}_{} = 0 \text{ carry } 1$$
$$\underbrace{0 + 1}_{} = 1$$
$$1 + \text{carry } 1 = 11$$

To add 7 base 8 to 3 base 8, we have

$$\underbrace{7 + 1}_{} + 2 \quad 7 + 1 = 0 \text{ carry } 1$$
$$0 + 2 = 2$$
$$2 + \text{carry } 1 = 12$$

Remember, keep carry digit to left.

EXERCISE

1. Add the following numbers in base 8.

$$\text{(a)}\ \begin{array}{r} 7 \\ 4 \\ \hline \end{array} \qquad \text{(b)}\ \begin{array}{r} 3 \\ 3 \\ \hline \end{array} \qquad \text{(c)}\ \begin{array}{r} 4 \\ 5 \\ \hline \end{array}$$

The answers are 13, 6, 11. Remember, the digits 8 and 9 cannot be used for base 8 numbers.

Subtraction

Subtraction of numbers in base 2 can be accomplished by two methods, either directly or by complementary subtraction. We will look at complementary subtraction in this review.

Example Subtract binary 101 from binary 111.

Solution

Step 1. Write the two numbers in columns as shown.

$$\begin{array}{r} 111 \\ 101 \\ \hline \end{array}$$

Step 2. Complement the lower number giving:

111
010
———

Step 3. Add the two numbers.

111
010
———

Step 4. Add the least digit on the left to the least-significant digit.

 111
 010
———
1001
└→1
———
 010

Step 5. This gives the final answer.

Step 6. To check, convert both digits to decimal and subtract.

111	7
101	5
010	2

EXERCISE

1. Subtract the following binary numbers using the complement method.

(a) 101	**(b)** 110	**(c)** 110	**(d)** 111
001	010	101	011

The answers are

(a) 100 $(5 - 1 = 4)$
(b) 100 $(6 - 2 = 4)$
(c) 001 $(6 - 5 = 1)$
(d) 100 $(7 - 3 = 4)$

10.2 THE HALF-ADDER

If we let the binary 1 equal a high logic level and the binary 0 equal a low logic level, figure 10.2 will add two binary numbers. The circuit of figure 10.2 is made up of an exclusive-OR and an AND gate. The output of the exclusive-OR gate will always be the sum of the two binary numbers (A and B) at its input. The output of the AND gate will only be high, a logic 1, when both of its inputs are high. This gate gives us the carry digit for binary addition. If A were 1 and B were 1, the sum (output of the exclusive-OR) would be a logic 0. At the same time the output of the AND gate would be a logic 1. Looking at the output of figure 10.2, then, one would see

Carry 1 0 Sum

or a binary number 2. So, a binary 1 was added to a binary 1 and the answer was a binary 2. This simple circuit has accurately performed binary arithmetic.

Since all computer arithmetic must be done with binary numbers, it can be easily seen that the simple half-adder will be the backbone of our arithmetic circuitry.

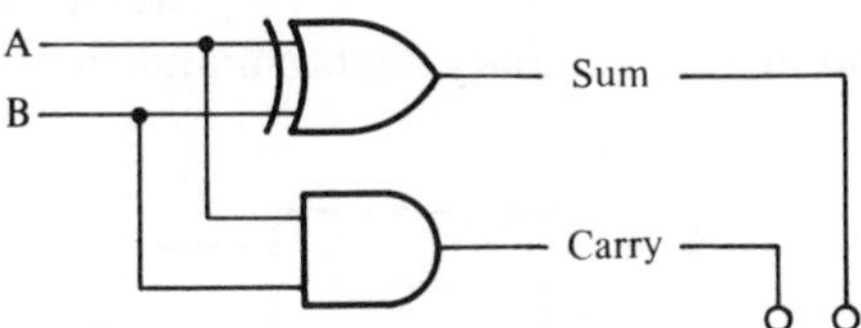

Figure 10.2 The Half-Adder Circuit

10.3 THE FULL-ADDER

The term *full-adder* is often misleading. As shown in section 10.2, the circuit of figure 10.2 was indeed able to add two binary numbers. Yet this circuit was designated as a half-adder. If the circuit could satisfactorily add two binary numbers, why was it called a half-adder? The following example will show the reason.

Example Suppose we wish to add the number 3 to the number 3. In binary this would be

$A{-}11_2$ 3_{10}
$B{-}11_2$ 3_{10}

the half-adder circuit would have no trouble adding the first set of digits in the problem, that is, the least-significant set of A and B.

Addition of these two digits by the half-adder will yield the results as shown:

	1	carry
A	1	1
B	1	1
		0 sum

The addition of binary 1 to binary 1 yields a sum of 0 and a carry of 1. The carry digit now must be added to the next set of A and B digits of the original problem. The half-adder circuit of figure 10.2 is not capable of doing this, however. The half-adder circuit has provision for only two inputs. The major difference then between a full-adder and a half-adder is this: The full-adder has provision for addition of two binary digits plus a carry digit from a preceding addition. Figure 10.3 shows this in block diagram form. Figure 10.4 shows the logic circuit diagram of the full-adder. Note that the full-adder is simply two half-adders connected to perform the correct logic operations. The circuit of figure 10.4 has the capability of adding three binary digits simultaneously.

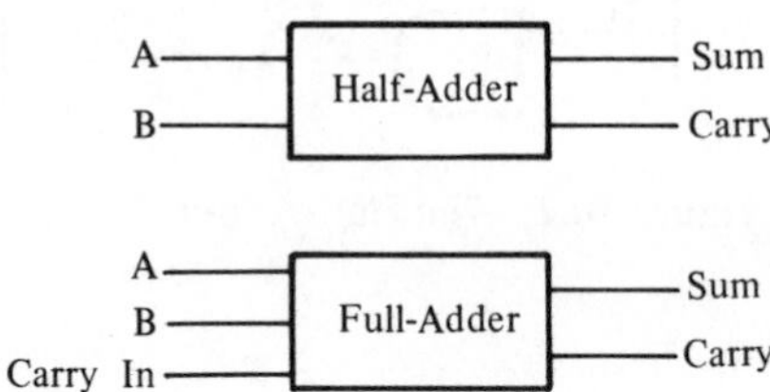

Figure 10.3 Block Diagram of Half- and Full-Adders

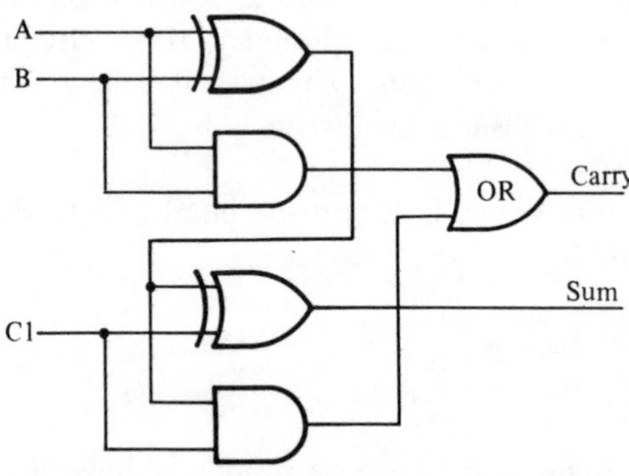

Figure 10.4 Full-Adder Logic Diagram

10.4 SERIAL ADDITION

To use the adder units to perform binary addition requires that the digits to be added be presented to the adder in proper sequence. One of the simplest forms of this type of addition is called *serial addition* and is accomplished by a circuit similar to that shown in figure 10.5. For clarity, the timing circuitry has been omitted. Recall from the chapters on shift registers, that the information stored in the registers will shift one location for each clock pulse. This means that the registers will present the two numbers to be added, one set at a time for each clock pulse. The adder will add these two numbers and create a sum and carry bit for the two. The sum bit will be present at the input-to-shift-register 3, while the carry bit will be at one side of a toggle flip-flop. The next clock pulse will present two new binary bits from registers 1 and 2 to the adder. This clock pulse will also toggle the flip-flop to present the carry from the previous addition to the input of the full-adder. This insures that the carry bit arrives at the input to the adder in the proper time for correct addition. During this clock pulse, the sum out of the adder from the first addition process is entered into shift register 3. Five clock pulses will be required to correctly add the four bits contained in register 1 to the four bits contained in register 2.

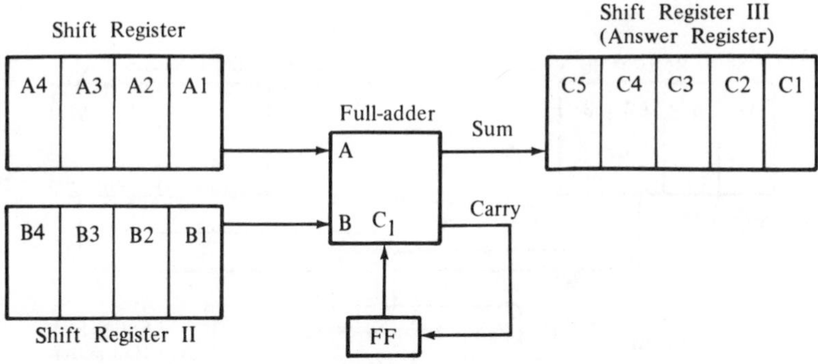

Figure 10.5 Circuit to Perform Serial Addition

Since the binary digits are presented to the adder in a series, this type of addition is called serial addition. In serial addition, one full-adder is needed to perform the addition. The registers may be of any length. The number of clock pulses needed to complete the addition process will be N + 1, where N is number of bits in register 1 or register 2 (whichever is larger). Serial addition is very slow. The circuitry needed for serial addition is quite simple, however.

In what year was the first mechanical adding machine built?

What is the major difference between a full-adder and a half-adder?

What is the purpose of the flip-flop in figure 10.5?

10.5 PARALLEL ADDITION

For those applications where serial addition is too slow, a method called *parallel addition* may be used. Although parallel addition is faster, the circuitry is more complex and more costly. The technician will be required to make a decision in cost versus speed. Figure 10.6 shows a block diagram representation of the circuitry required for parallel addition of two binary numbers. Notice that in the parallel circuitry the registers have the same number of bits; that is, the answer register has provision for one bit more than the largest of the other two registers. In parallel addition, one half-adder and one or more full-adders are required. Each set of bits (A, B, etc.) is sent to its own adder circuit. The sum out of each adder is sent to one of the bit inputs of the answer register. The carry out bit of each adder is sent to the next adder (last carry goes to last bit of answer register). In this way, each set of bits is

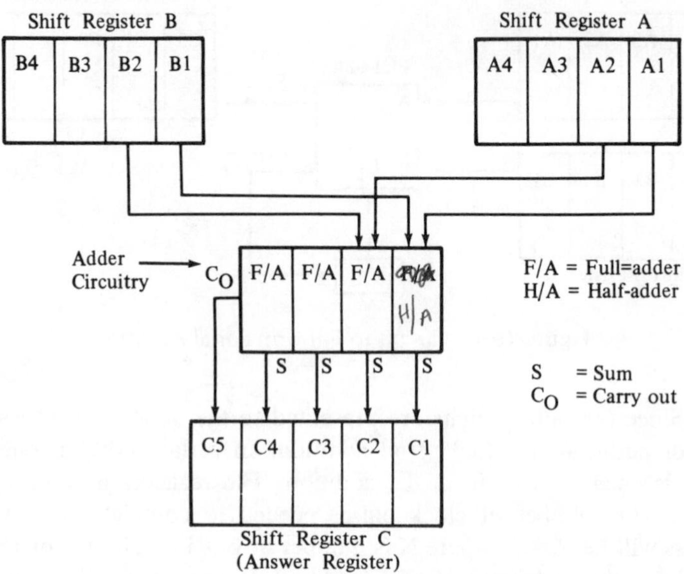

Figure 10.6 Circuitry to Perform Parallel Addition

added at the same instant as each of the other sets. The speed of a parallel adder is very fast compared to the serial adder. The cost is increased by the use of additional full-adder circuits.

> Why is the first adder in the adder circuitry of figure 10.6 a half-adder?
>
> How many clock pulses are required to add the contents of registers I and II of figure 10.5?

10.6 ADDER CHIPS

The state of the art in integrated circuitry is moving very fast. The devices shown in this text may already be obsolete. However, the ideas behind the operation of these circuits are very important and their study will lead to a better understanding of newer and more complex chips.

10.7 THE TYPE SN7482 TWO-BIT BINARY FULL-ADDER

Figure 10.7 is the pin configuration for the type SN7482. This circuit has the capability to add 2 two-bit binary numbers A1A2 and B1B2. It also has the capability of accepting a carry in from a preceding circuit (C0). The output (binary-coded) result of this addition is found at pin 1, ($\Sigma 1$) and pin 12, ($\Sigma 2$) with a carry out present at pin 10, (C2). The arithmetic is performed as shown below:

```
       C0
     A2A1
   + B2B1
   ------
  C2 2 1
```

A function table for the operation of this circuit is shown in figure 10.8.

Example To add the two numbers 3_{10} and 1_{10} with no carry from a preceding stage we would first convert the decimal numbers to their binary equivalents. This produces

```
                 C0
  11           A2A1
  01           B2B1
  --         ------
             C2 2 1
```

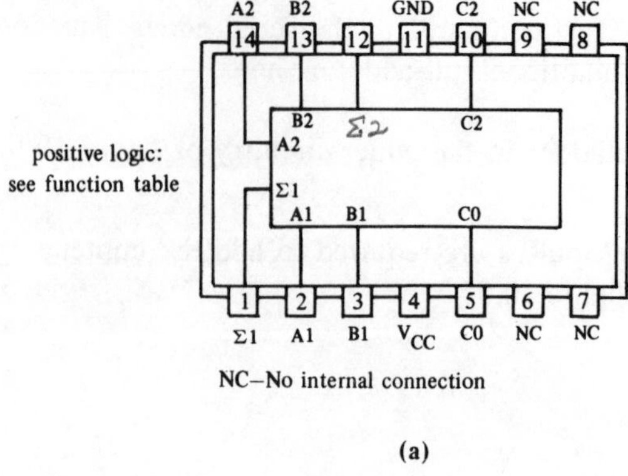

(a)

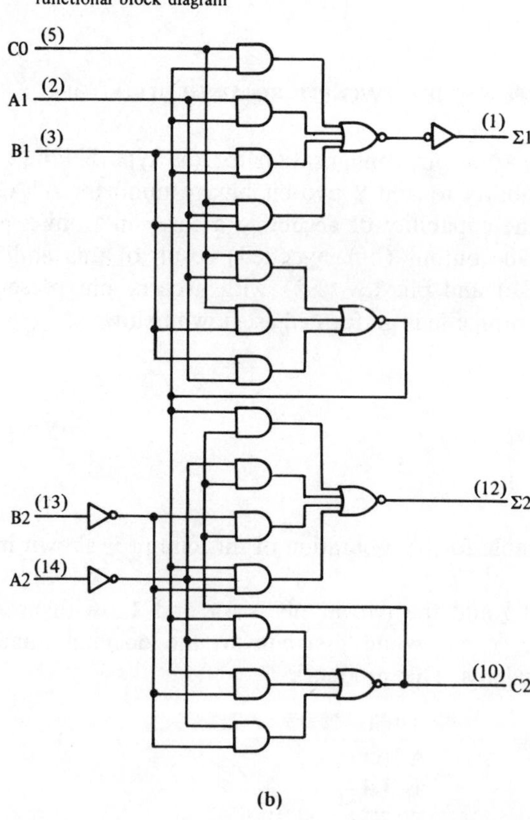

(b)

Figure 10.7 Type SN7482 Gated Full-Adder

Function Table

Inputs				Outputs					
				When CO = L			When CO = H		
A1	B1	A2	B2	Σ1	Σ2	C2	Σ1	Σ2	C2
L	L	L	L	L	L	L	H	L	L
H	L	L	L	H	L	L	L	H	L
L	H	L	L	H	L	L	L	H	L
H	H	L	L	L	H	L	H	H	L
L	L	H	L	L	H	L	H	H	L
H	L	H	L	H	H	L	L	L	H
L	H	H	L	H	H	L	L	L	H
H	H	H	L	L	L	H	H	L	H
L	L	L	H	L	H	L	H	H	L
H	L	L	H	H	H	L	L	L	H
L	H	L	H	H	H	L	L	L	H
H	H	L	H	L	L	H	H	L	H
L	L	H	H	L	L	H	H	L	H
H	L	H	H	H	L	H	L	H	H
L	H	H	H	H	L	H	L	H	H
H	H	H	H	L	H	H	H	H	H

H = high level, L = low level

Figure 10.8 Function Table for SN7482 Gated Full-Adder

Since for this example, both A1 and A2 are logic 1's, they are both high (H). The B1 bit is also high, but the B2 bit is a 0 (L) bit. This set of conditions is met on line 8 of the function table. To find the output levels we could expect for these inputs, we need go to the output column marked "When CO = L". This is true since our carry in is a 0 (L). The output levels then are:

C2 2 1
H L L

This corresponds to a 100_2 or decimal 4. Since we added a decimal 3 to a decimal 1 this answer is correct.

10.8 THE TYPE 7483 FOUR-BIT BINARY FULL-ADDER

Figure 10.9 shows a four-bit binary full-adder. This circuit is very similar to the SN7482 with the obvious increase in its capability. The
~~43 PULSE & DIGITAL CIRCUITS & SYSTEMS (Castellucis)~~
SN7483 has the capability to add two sets of four-bit binary numbers of the form:

A4A3A2A1
B4B3B2B1

C4 4 3 2 1

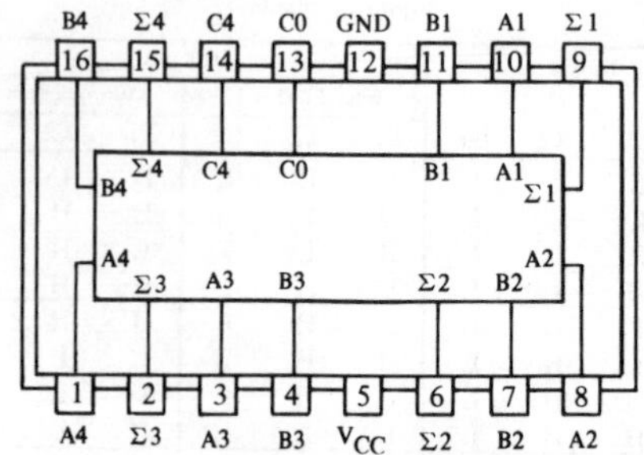

Figure 10.9 Four-Bit Binary Full-Adder

As did the SN7482, this chip has the capability to accept a carry bit from a preceding stage. It also generates a carry out bit from the fourth bit.

Figure 10.10 is a function table for the SN7483. The reading of this table is slightly more complex than the function table for the SN7482.

Example If we wish to add the number 7_{10} to the number 5_{10} wth a C0 = 0(L), the following procedure is used.

Step 1. Convert the decimal number to its binary equivalent

7_{10}	0111_2	A4A3A2A1
5_{10}	0101_2	B4B3B2B1

Step 2. Equate each of the A and B bits with its binary equivalent.

A4 = 0	B4 = 0	C0 = 0
A3 = 1	B3 = 1	
A2 = 1	B2 = 0	
A1 = 1	B1 = 1	

Step 3. Go to the function table. Locate A1, A2, B1, B2 and C0 that match the values of Step 2. This occurs on line 8.

When C0 = L

A1/A3	B1/B3	A2/A4	B2/B4	Σ1/Σ3	Σ0/Σ4	C2/C4
Line 8						
H	H	H	L	H	H	L

FUNCTION TABLE

INPUT				OUTPUT					
				WHEN C0 = L / WHEN C2 = L			WHEN C0 = H / WHEN C2 = H		
A1 / A3	B1 / B3	A2 / A4	B2 / B4	$\Sigma1$ / $\Sigma3$	$\Sigma2$ / $\Sigma4$	C2 / C4	$\Sigma1$ / $\Sigma3$	$\Sigma2$ / $\Sigma4$	C2 / C4
L	L	L	L	L	L	L	H	L	L
H	L	L	L	H	L	L	L	H	L
L	H	L	L	H	L	L	L	H	L
H	H	L	L	L	H	L	H	H	L
L	L	H	L	L	H	L	H	H	L
H	L	H	L	H	H	L	L	L	H
L	H	H	L	H	H	L	L	L	H
H	H	H	L	L	L	H	H	L	H
L	L	L	H	L	H	L	H	H	L
H	L	L	H	H	H	L	L	L	H
L	H	L	H	H	H	L	L	L	H
H	H	L	H	L	L	H	H	L	H
L	L	H	H	L	L	H	H	L	H
H	L	H	H	H	L	H	L	H	H
L	H	H	H	H	L	H	L	H	H
H	H	H	H	L	H	H	H	H	H

H = high level, L = low level

NOTE: Input conditions at A1, B1, A2, B2, and C0 are used to determine outputs $\Sigma1$ and $\Sigma2$ and the value of the internal carry C2. The values at C2, A3, B3, A4, and B4 are then used to determine outputs $\Sigma3$, $\Sigma4$, and C4.

Figure 10.10 Function Table for Binary Full-Adder

This condition determines the output levels of $\Sigma1$, $\Sigma2$ and C2 (C2 is an internal carry). Thus far then the solution to our addition problems looks like this:

C2 C0	(Hi)
A4 A3 A2 A1	A4 A3 A2 A1
B4 B3 B2 B1	B4 B3 B2 B1
C4 4 3 2 1	(Low) (Low)

Step 4. Next, return to the function table and locate the values of C2, A3, A4, B3 and B4 that match the values of Step 2.

When C0 = H/C2 = H

A1/A3	B1/B3	A2/A4	B2/B4	$\Sigma1/\Sigma3$	$\Sigma2/\Sigma4$	$\Sigma2/\Sigma4$
Line 4						
H	H	L	L	H	H	L

This condition occurs on line 4 of the function table. From line 4, we can now determine the output levels of Σ3, Σ4 and C4. These are shown below:

C4	Σ4	Σ3	Σ2	Σ1
L	H	H	(L)	(L)

The final solution to the problem is now shown to be

01100_2 12_{10}

Using the function table of figure 10.10, which lines are needed to produce the answer to the addition of decimal number 4 and decimal number 4?

SUMMARY

This chapter has progressed from simple binary arithmetic to the use of MS1 technology for the solution of complex addition problems. The gating circuitry used in a type SN7483 chip is shown in figure 10.11. Modern minicomputers and microprocessors may use larger capacity circuits, but much of the addition is carried on in the manner as shown in this unit.

EXERCISES

1. Construct a four-bit serial adder to illustrate the application of some of the basic devices discussed earlier.

Serial Adder

In binary addition the number to be increased by having another number added to it is called the *augend.* The number added to the augend is called the *addend.*

1001	Augend
+0101	Addend
1110	Sum

A simple four-bit serial adder is shown in Figure 10.12. It uses several

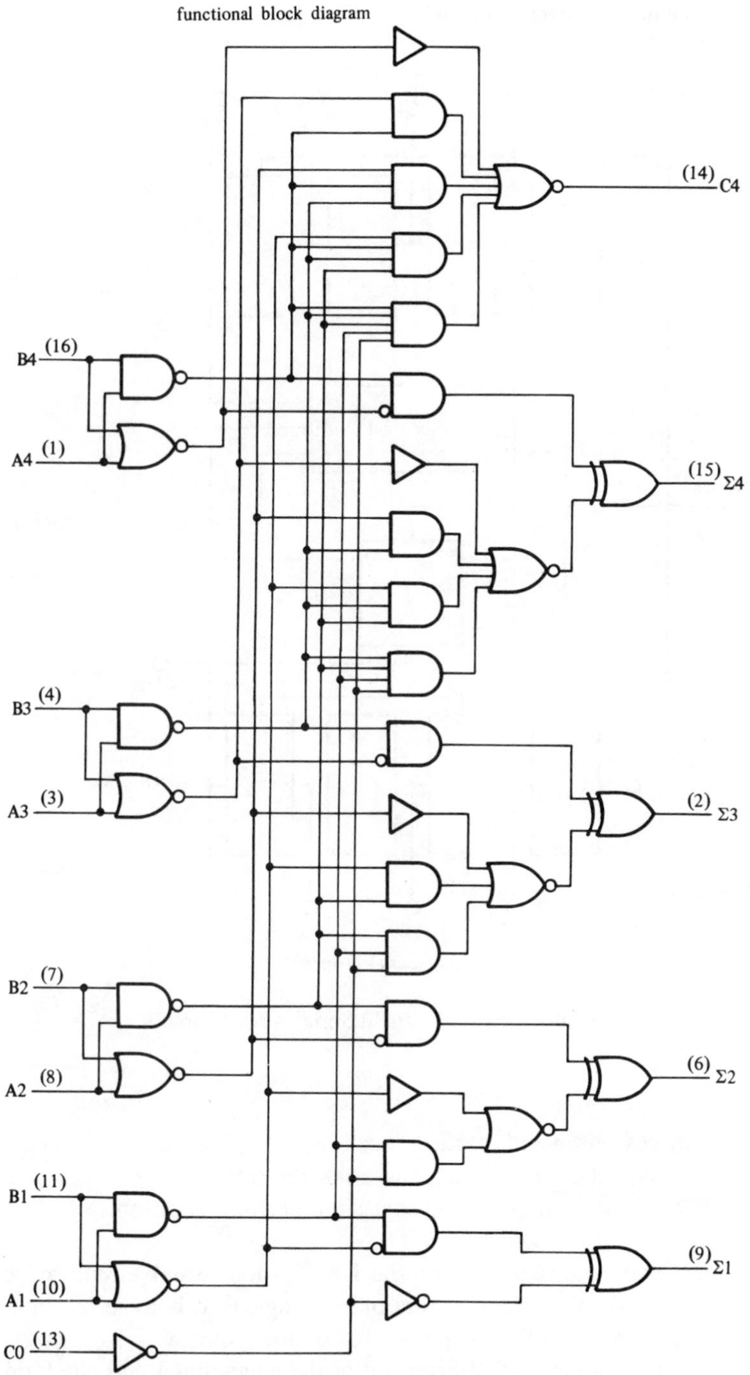

Figure 10.11 7483 Functional Block Diagram

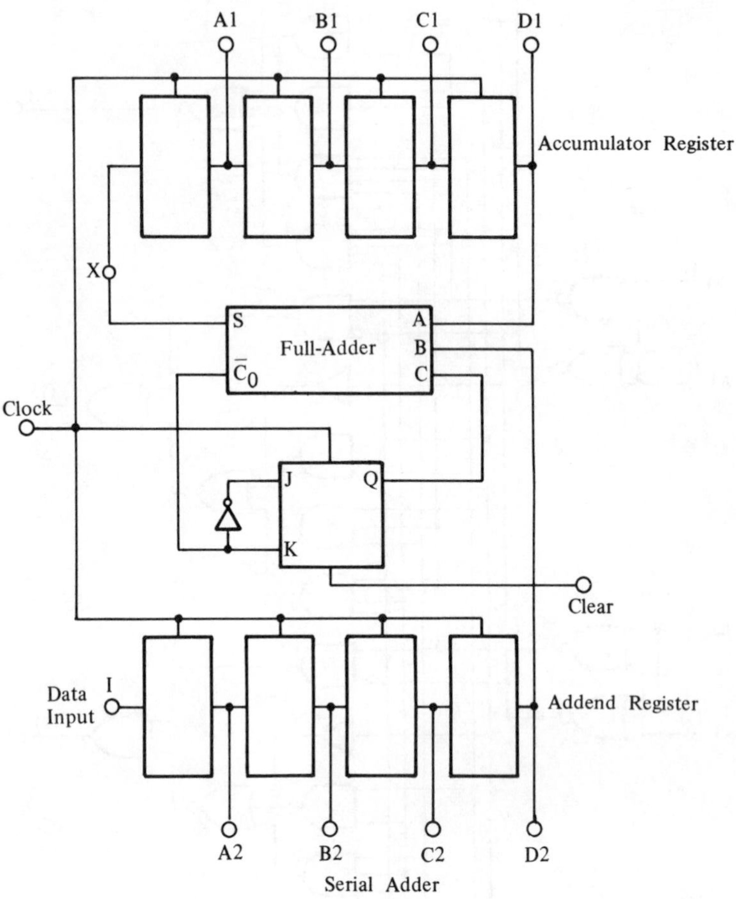

Figure 10.12 Four-Bit Serial Adder Circuit

basic devices discussed earlier, such as a full-adder, a bounceless switch, 2 four-bit shift registers, and a J-K flip-flop.

The following steps are those used to perform addition.

1. Clear both registers and the J-K flip-flop. Break circuit at point X. (To clear the registers place a logic 0 at both serial data inputs and clock four times.) Reconnect point X.
2. Enter the least-significant bit of the augend at I and clock once. Enter the next three bits in order clocking each time.
3. Enter the addend using the procedure of step 2.
4. Set I to 0 and clock four more times to cycle the numbers

through the full adder. The sum appears in the accumulator register. If the sum is more than four bits the carry is indicated by Q.

The augend enters the augend register through the addend register. After the augend is entered in the addend register, the addend is entered. As the addend enters, the augend is "pushed" out of the addend register through the full adder into the augend register. This register serves a dual purpose in that it is both the augend and the accumulator register. We now have the augend and the addend in their respective registers. The J-K flip-flop is used as a delay for the carry-out signal, so that it arrives at the carry-in input at the proper time, since it must arrive with the next significant bit. As the two numbers are added in the full adder the sum is entered in the accumulator register. The sum enters one end of the register as the augend exits the other end.

Construct the four-bit serial adder shown in Figure 10.12. A SN7476, two SN7495s, a SN7480, and a SN7400 will be required to construct the circuit. Connect the SN7495s for serial shift-right operation. The necessary information for connection of the SN7480, the full-adder, is given in Figure 10.13 on page 226. Perform the following additions and fill in Tables 10.1, 10.2, and 10.3.

(a) 1010
+0101

TABLE 10.1 Truth Table

Clock Pulse	*I*	*A2*	*B2*	*C2*	*D2*	*A1*	*B1*	*C1*	*D1*	*Q*
0	0	0	0	0	0	0	0	0	0	0
1										X
2										X
3										X
4										X
5										X
6										X
7										X
8										X
9										X
10										X
11										X
12										X

X = Don't care

C_i	A*	B*	S	$\overline{C_o}$
0	0	0	0	0
0	0	1	1	1
0	1	0	1	1
0	1	1	0	0
1	0	0	1	1
1	0	1	0	0
1	1	0	0	0
1	1	1	1	0

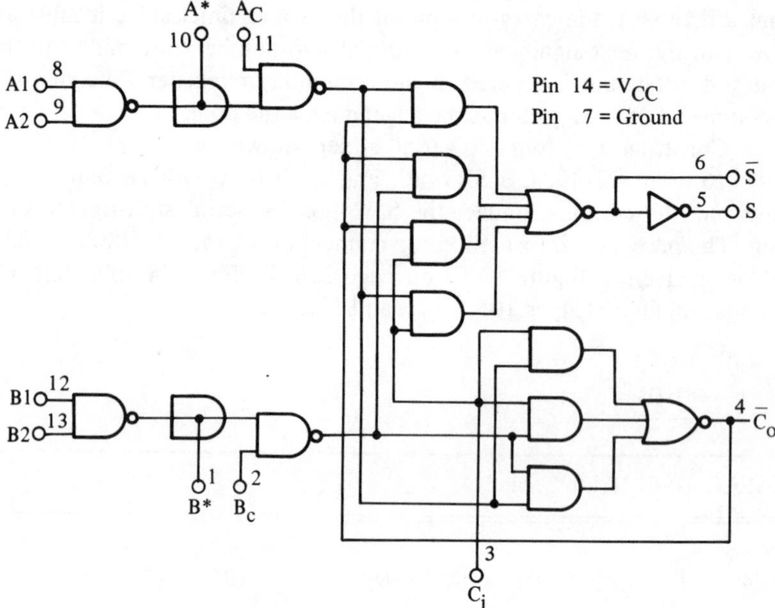

Figure 10.13 Logic Diagram

The SN7480 gated full-adder has capabilities beyond our requirements; however, it is easily adapted to our needs. Use A* and B* for the inputs. Inputs A1, A2, Ac, B1, B2, and Bc must be grounded. Since only a not-carry output is provided, it must be fed into the K input of the J-K flip-flop.

The 7483 Adder

2. Using a type 7483, verify the function table for the addition of the same numbers as used in Exercise 1. Use a circuit similar to the one

(b) 0111
+0010

TABLE 10.2 Truth Table

Clock Pulse	*I*	*A*2	*B*2	*C*2	*D*2	*A*1	*B*1	*C*1	*D*1	*Q*
0	0	0	0	0	0	0	0	0	0	0
1										X
2										X
3										X
4										X
5										X
6										X
7										X
8										X
9										X
10										X
11										X
12										X

X = Don't care

(c) 0101
+1110

TABLE 10.3 Truth Table

Clock Pulse	*I*	*A*2	*B*2	*C*2	*D*2	*A*1	*B*1	*C*1	*D*1	*Q*
0	0	0	0	0	0	0	0	0	0	0
1										X
2										X
3										X
4										X
5										X
6										X
7										X
8										X
9										X
10										X
11										X
12										X

X = Don't care

shown in figure 10.6. Figure 10.14 shows a device diagram to perform this exercise.

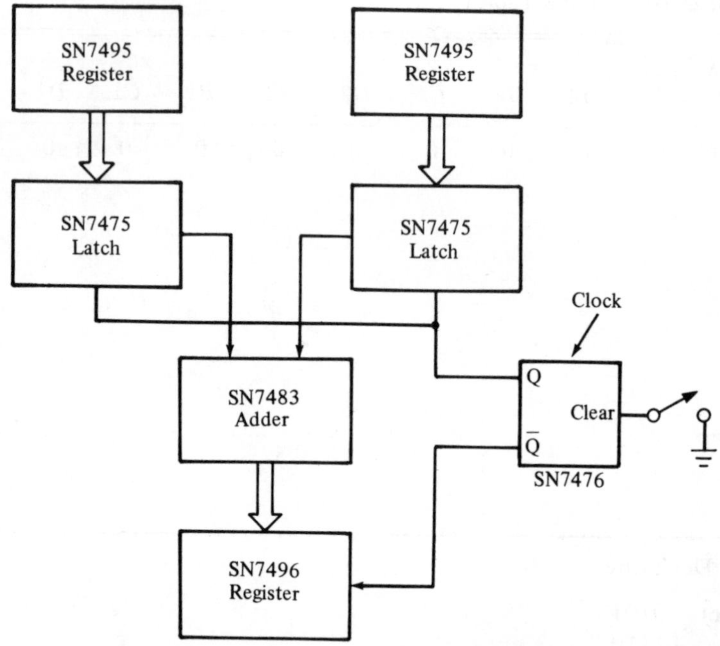

Figure 10.14 Diagram for Exercise 2

QUESTIONS AND PROBLEMS

10.1 Add the following sets of binary numbers, and select the correct answer.

101	(a) 1011
110	(b) 1100
	(c) 1001
	(d) 1111

110	(a) 1011
110	(b) 1100
	(c) 1001
	(d) 1111

1001	(a) 10101
1101	(b) 10110
	(c) 11001
	(d) 10001

10.2 Subtract the following sets of binary numbers and select the correct answers.

1001	(a) 0001
1000	(b) 0101
	(c) 0110
	(d) 1000

1101	(a) 3
1001	(b) 4
	(c) 5
	(d) 6

10.3 From your study so far, which gates are capable of performing arithmetic operations in binary?

10.4 How might binary subtractions be performed?

10.5 Briefly explain the major differences between a half-adder and a full-adder.

10.6 When using half-adders and full-adders to add 2 four-bit numbers, what is the number of each that must be used when adding in parallel?

10.7 What is the difference in circuitry, speed accuracy, between a parallel adder and a serial adder?

10.8 Show how you would wire several 7482 devices to perform the addition of 2 six-bit binary numbers.

10.9 Design the same system as problem 10.8 but use 7482^s, 7483^s or both.

10.10 For the circuit shown in figure 10.12 how many additional full-adders would be needed if 2 eight-bit numbers were to be added?

10.11 For the circuit shown in figure 10.12 what changes would need to be made so that 2 eight-bit numbers could be added?

Section III

11

The Operational Amplifier

OBJECTIVE: To introduce the operational amplifier as a device for use in the digital field. Characteristics and applications of the device will be studied.

Introduction: To this point, we have studied digital signals and digital quantities. These were defined as quantities having discrete states or levels. A signal that does not have discrete levels or states, but which varies continuously is called an *analog signal*. In our work in digital controls, circuits and systems are found signals that contain both digital and analog quantities. A circuit or system that contains both types of signals is called a *hybrid*. Often in hybrid systems one of the quantities (digital or analog) needs to be converted to the other type of signal. When an analog signal needs to be converted to a digital signal, this conversion is called *analog-to-digital* or *A/D conversion*. In a like manner the conversion of a digital signal to an analog signal is called *digital-to-analog* or *D/A conversion*. An example of such a system would be in a digital voltmeter. The analog signal (the voltage to be measured) is converted to a digital signal by a D/A converter. This digital signal can then be processed by digital logic circuits and a direct readout of the voltage can be displayed.

One of the basic building blocks for the A/D and D/A converter is a device called the *operational amplifier*. This chapter discusses the operational amplifier's characteristics, parameters, and circuit operation.

11.1 THE IDEAL OPERATIONAL AMPLIFIER

The operational amplifier (op-amp) is basically a very high-gain amplifier that also has provision for external feedback. The output of such an amplifier can be made to depend primarily on externally connected passive elements. To insure that this is the case, the amplifier device has three major requirements.

1. The amplifier open loop gain (A_{VOL}) must be very high, usually above 80 decibels.
2. If it is a multiple-stage amplifier, the stages must be direct-coupled.
3. It must have provision for obtaining negative feedback via a single external resistor connected from the output to the input terminal of the amplifier.

The operational amplifier's most common form is shown in figure 11.1. This symbolizes an amplifier that has a differential input and a single-ended output. The ideal properties of such an amplifier are as follows:

1. Infinite gain
2. Infinite bandwidth
3. Infinite impedance between the input terminals
4. Infinite impedance from each input terminal to ground
5. Infinite output-current capability
6. Zero output impedance
7. Zero input voltage offset
8. Zero input current
9. Infinite common-mode rejection as a differential amplifier
10. Each of the above must be true at all temperatures.

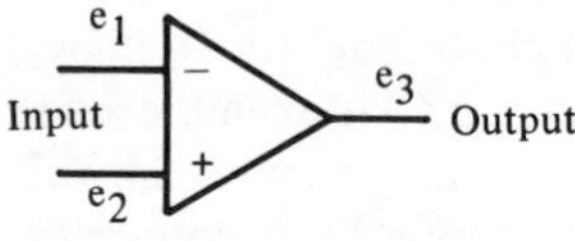

Figure 11.1 The Operational Amplifier

Of course, such a device cannot be manufactured or constructed. Amplifiers can, however, be constructed which, for many calculations, the above may be considered true. Op-amps may also be designed so that for a large number of applications, the actual values of the above properties will not cause a significant change in the performance of the device over what might be expected under ideal conditions.

11.2 THE OP-AMP AS A SINGLE-PACKAGE DEVICE

In modern usage, the operational amplifier is a self-contained complex circuit, manufactured and sold in a single package. Figure 11.2 shows a schematic diagram of a common type of operational amplifier. This circuit is constructed on a single integrated circuit (IC) chip and is packaged such that only the eight external connections are available to the user.

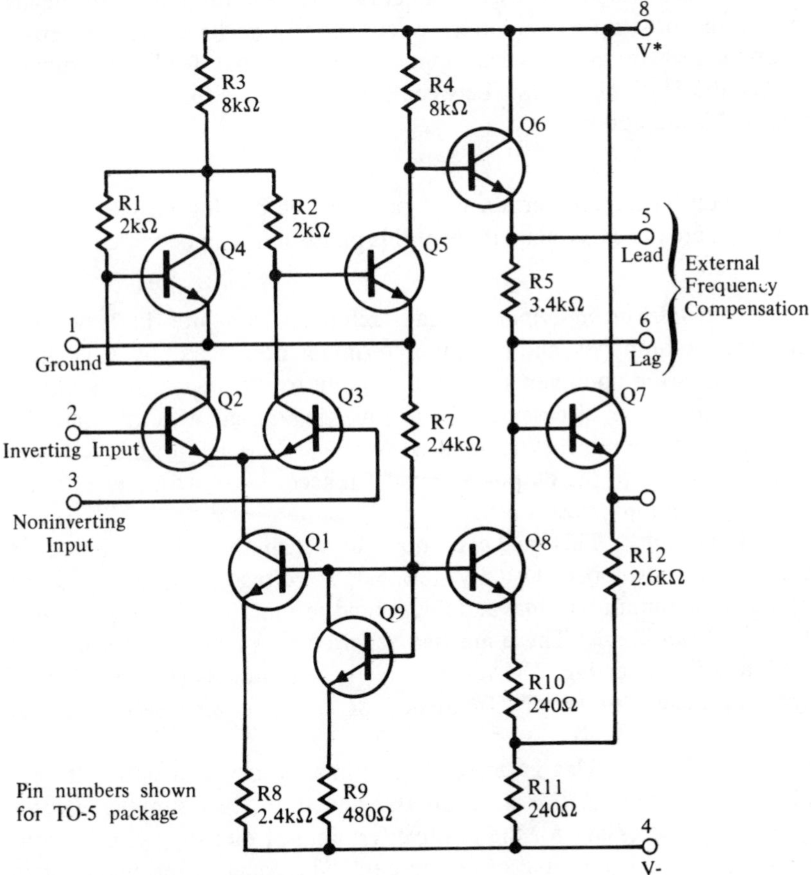

Figure 11.2 Equivalent Circuit of IC Op-Amp (Courtesy of Fairchild)

11.3 FUNCTIONS OF THE EXTERNAL CONNECTIONS

1. *Ground.* This terminal, as for most electronic circuitry, is the common reference point for both input and output signals. It is also the reference point and connection for one of the terminals of an external dc supply that is required to power the device.

2. *Inverting input.* This is one of the two signal input terminals for the device. Refer to figure 11.1. Here in simplest block diagram form, the op-amp is shown as a device having a differential input pair. One of the inputs is labeled with a negative sign (−) and the other with a positive sign (+). The terminal with the negative sign is called the *inverting input* because any signal applied to this terminal will undergo a 180° phase shift between this terminal and the output terminal. This concept must hold true for any configuration of the op-amp. This means that if the op-amp is operating as an open-loop device (no external components) or as a closed-loop device (external feedback components) the 180° phase shift between the inverting input and the output terminal is maintained.

Note: There are certain configurations employing signals to both inputs where the exact magnitude of this phase shift can be varied.

3. *Noninverting input.* Again referring to figure 11.1, we see that the op-amp has another input terminal designated by a positive sign. This is the noninverting amplifier input to the device. It should be obvious from the discussion of the inverting input, exactly what the function of the noninverting input is. Any signal applied to this terminal will be reflected to the output terminal (in accordance with any external circuitry) with no phase shift.

4. *Output.* This is the terminal at which the output signals of the op-amp developed. Usually, this output voltage (V_o) is developed between the output terminal and the ground terminal.

5. *Lead/Lag.* These are two separate terminals. One is marked lead, and the other lag. The function of these terminals is to provide an external means to supply frequency or phase compensation to the op-amp.

6. *+Volts.* This is one of two terminals to which power is supplied to the op-amp. For most op-amp functions, two supplies are required. It is necessary for the positive terminal of one supply to be connected to the +V terminal of the op-amp. The negative terminal of this supply must then go to the ground terminal of the op-amp circuit (op-amp does not have ground).

7. *−Volts.* The −volts terminal is the other terminal to which power is supplied to the operational amplifier. It is necessary that the negative terminal of the second supply be connected to the −V terminal of the op-amp, and that the positive terminal of this supply be connected to the ground terminal of the op-amp circuit.

Figure 11.3 shows a simple connection diagram for a typical op-amp. As noted earlier, the triangle is the standard block diagram symbol for the op-amp. The figure also illustrates a typical operational connection for an op-amp. Here are shown two supplies, each properly connected, an input signal (to the inverting input) and a load connected between the output and ground. No feedback elements (external passive elements) are shown in this diagram nor are connections to the lead or lag terminals. These items will be discussed later.

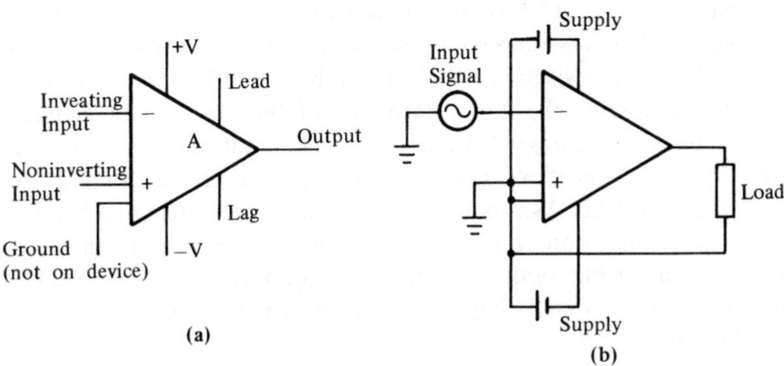

Figure 11.3 Lead Designations of the Typical Operational Amplifier

11.4 SOME OP-AMP TERMINOLOGY

Before we can actually look at some of the circuits and functions of the operational amplifier, we must become familiar with some of the terminology associated with the op-amp. The op-amp, although it may be considered just a high-gain amplifier that must possess the characteristics previously discussed, does have some terminology concerning various signals that are unique to this type of a device.

Figure 11.4 shows a typical block diagram arrangement for an operational amplifier. In most IC op-amps the circuitry consists of a cascade of four stages. Amplifiers A1 and A2 of figure 11.4 are differential amplifiers. In addition, A1 has a double-ended output. A2 is a differential amplifier with a single-ended output. A3 is an emitter fol-

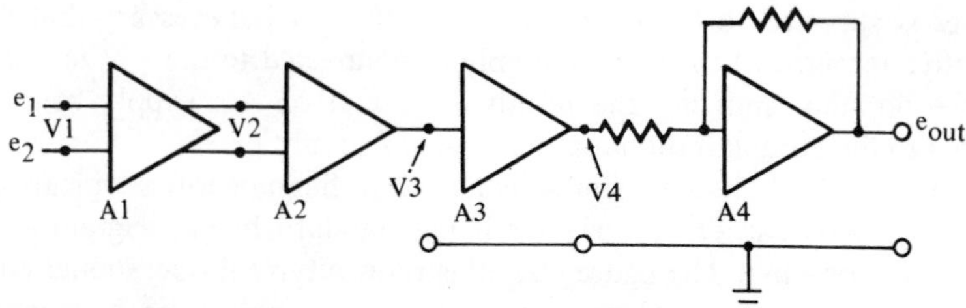

Figure 11.4 Block Diagram of Operational Amplifier

lower. This stage of course has a value of A_v approximately equal to 1. The last stage, A4 is a dc-level translator and the output driver for the entire circuit. Figure 11.5 shows the schematic diagram of the circuits.

Figure 11.6 is a simplified equivalent model of the op-amp. Notice that there is a resistance (R_{in}) shown between the two input terminals. This represents the input resistance of the op-amp. In actuality, referring back to figures 11.4 and 11.5, the input resistance is determined by the connections made to the input terminals. Figure 11.7 shows this to be true. As can be seen from the figure, there are in reality three input connections made to the op-amp. There is a connection made to the inverting input (e_1); there is a connection made to the noninverting input (e_2) and there needs to be a connection made to the ground terminal.

Difference Inputs

Suppose that an input voltage V_1 is to be applied between the inverting terminal of the op-amp and ground and that a second input voltage V_2 is to be applied between the noninverting terminal and ground. Further assume that these two voltages are exactly equal. From our earlier discussion, we know that a voltage placed on the inverting input will undergo a 180° phase, while a voltage placed across the noninverting terminal to ground will not undergo any such shift.

Figure 11.8 represents the input and output voltages; it can be seen that two output voltages are produced that are of equal amplitude but which are 180° out of phase with each other. This will produce an output of zero volts. This should be as expected. Further, in this condition the voltage across R_1 of figure 11.7 will be $e_1 - e_2$ (in this case $e_1 = e_2$ so $e_{R1} = 0$). If e_{R1} is 0, and the output is 0, it should be evident that in this circuit configuration the output of the amplifier should be

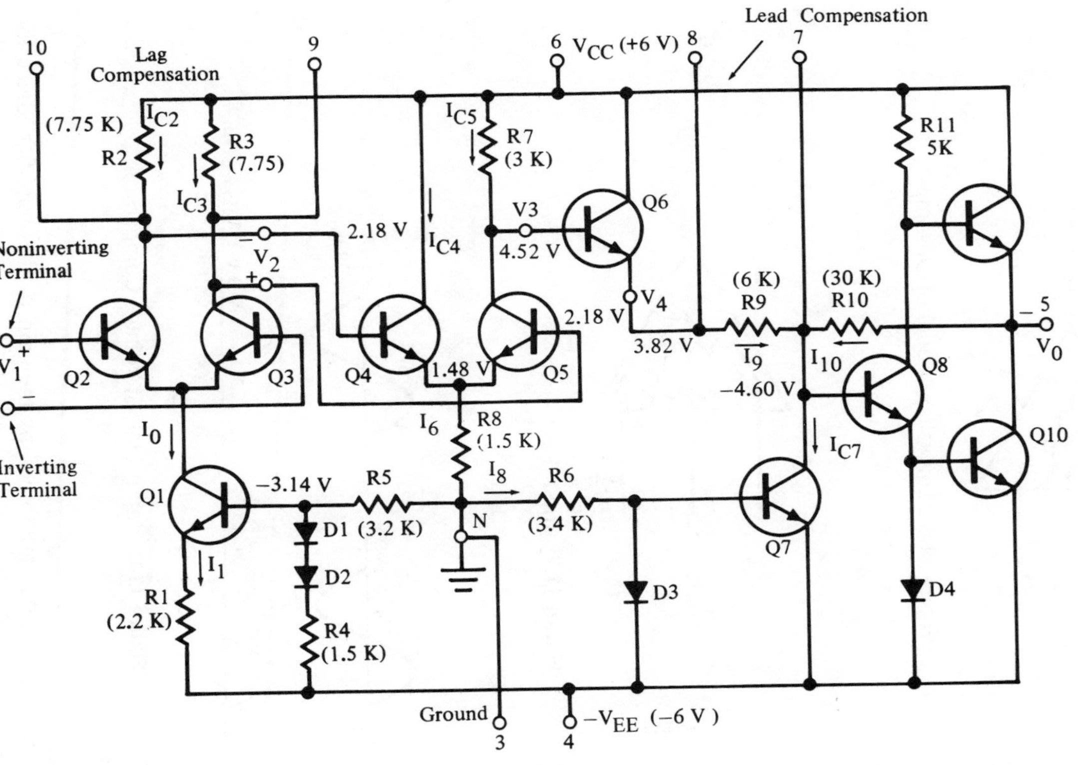

Figure 11.5 The Motorola MC1530 Operational Amplifier (Courtesy of Motorola)

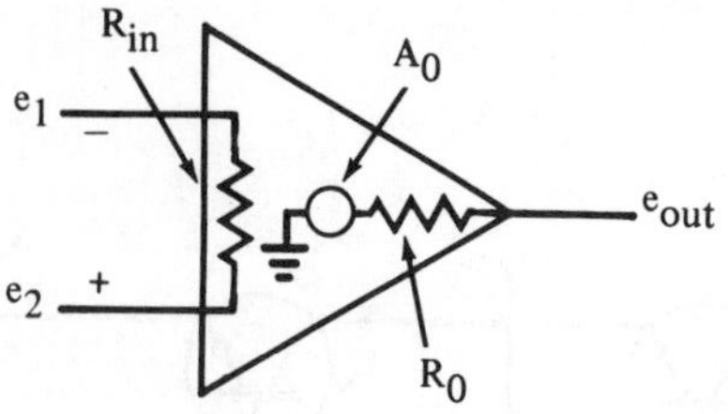

Figure 11.6 Equivalent Model of Op-Amp

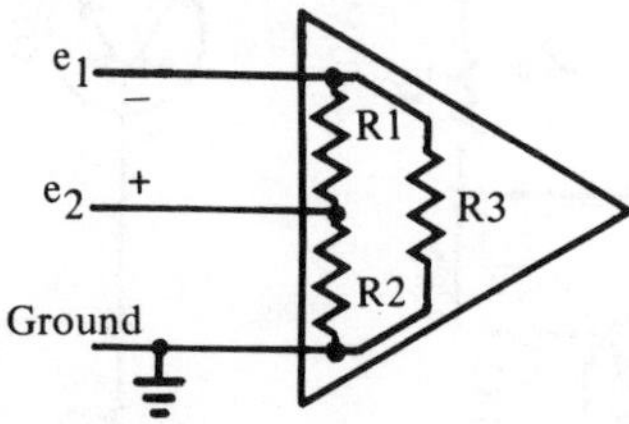

Figure 11.7 Input Resistance Model of Op-Amp

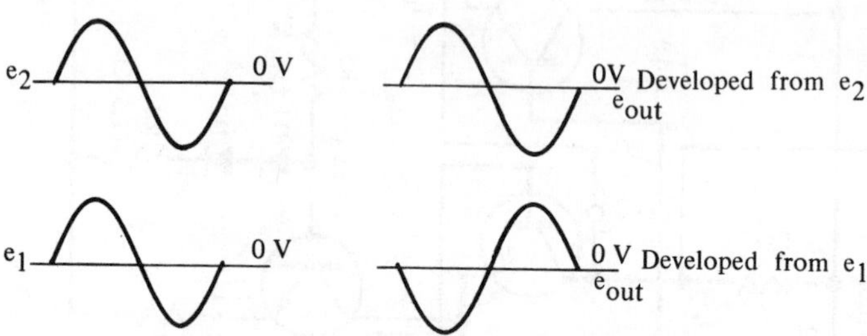

Figure 11.8 Input-Output Voltage Comparisons

the voltage across R1 multiplied by the gain of the amplifier. Since the gain of the amplifier is very large, the difference in the voltages applied at the input terminals will be greatly amplified. The amplifier when connected as a difference amplifier is very good at detecting small differences between two voltages.

In the difference amplifier mode, the gain of the device can be written as

$$A = \frac{e_{out}}{e_{diff}}$$

where e_{out} is the measured output voltage and where e_{diff} is the differential input signal. The differential input signal is the difference between the voltages applied at the input terminals.

Common-Mode Gain

The operational amplifier when connected as shown in figure 11.9 exhibits a common-mode gain characteristic. In this configuration, both inputs have the same signal and are tied together. We may think of the input in this circuit to be equal to:

$$\frac{e_1 + e_2}{2}$$

This input is called the *common-mode input* or e_{cm}. The common-mode gain then may be thought of as simply

$$\frac{e_{cm}}{\cdot\, e_{out}}$$

To measure this value, a signal is applied at the input terminals of figure 11.9 and this value is divided by the value of the output signal. A measure of how well the amplifier operates is found by calculating a ratio of the two gains of the amplifier. This ratio is called the *common-mode rejection ratio* (CMRR), and is simply:

$$CMRR = \frac{Ad}{A_{cm}} \qquad \begin{aligned} Ad &= \text{differential gain} \\ A_{cm} &= \text{common mode gain} \end{aligned}$$

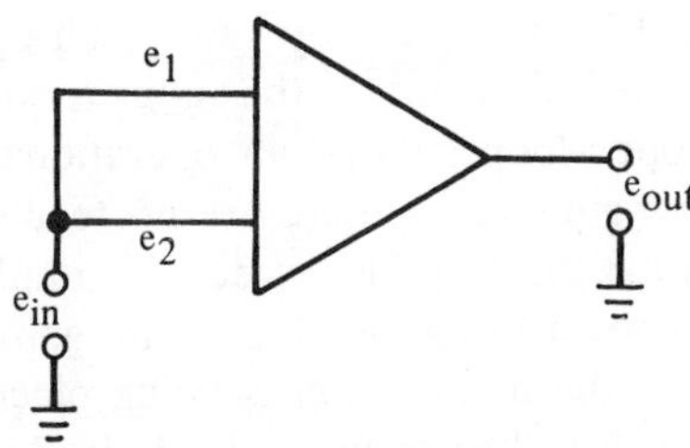

Figure 11.9 Common-Mode Input of Op-Amp

A good rule of thumb is that this ratio should be somewhere around 80 decibels. What this tells us is that we want the difference signals to be amplified greatly and any common signals to receive relatively no amplification. A practical example of what this means can be shown in the circuit of figure 11.10. Here we see an amplifier with two inputs. The signal e_d might be thought of as the signal from some low-level transducer. This is the signal that we wish to have amplified. There is usually some dc offset voltage or perhaps some form of noise that is also being picked up by the circuit and that we do not want amplified (e_{cm}).

An amplifier with a very high value of CMRR would allow us to amplify the difference signals (e_d) to adequate levels and would effectively shut out the common-mode signal. These two would be separated by about 80 decibels at the output. This separation is the same as if the common signal input did not exist.

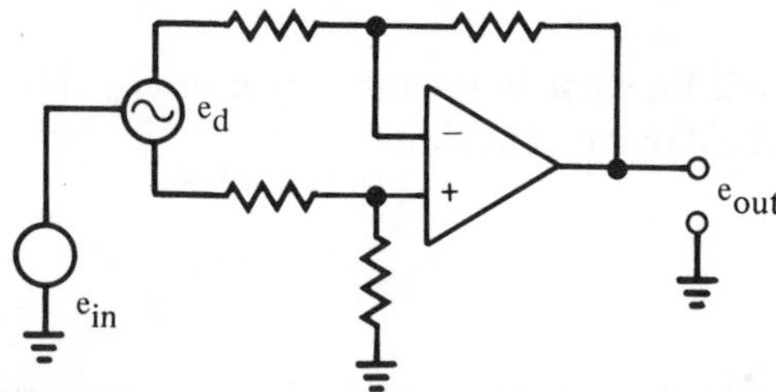

Figure 11.10 Circuit Showing Two Input Signals to Op-Amp

11.5 OPEN-LOOP GAIN

Since the operational amplifier is a device that must exhibit certain characteristics, we need to have some understanding of those characteristics. From section 11.1 we learned that the operational amplifier must have a very high open-loop gain figure (A_{VOL}) of around 80 decibels. Exactly what is the open-loop gain of an op-amp? It is simply the gain of a signal entered at one of the input terminals. Figure 11.11 illustrates how the open-loop gain of an operational amplifier is measured. Notice that in figure 11.11 there are no feedback components in the circuit. The gain of the amplifier circuit is simply the total gain of the device itself, as shown in figure 11.11. The gain would actually be a negative value since the input signal is being placed on the inverting (−) input to the amplifier. The open-loop gain is always extremely high for operational amplifiers.

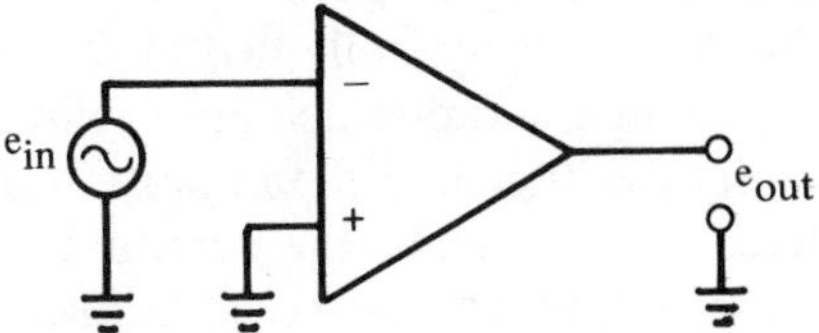

Figure 11.11 Open-Loop Gain

11.6 CLOSED-LOOP GAIN (THE FEEDBACK AMPLIFIER)

Figure 11.12 will be used to demonstrate several important concepts of the operational amplifier circuit. From our earlier discussions, we saw that the input impedance to an op-amp was considered to be infinite.

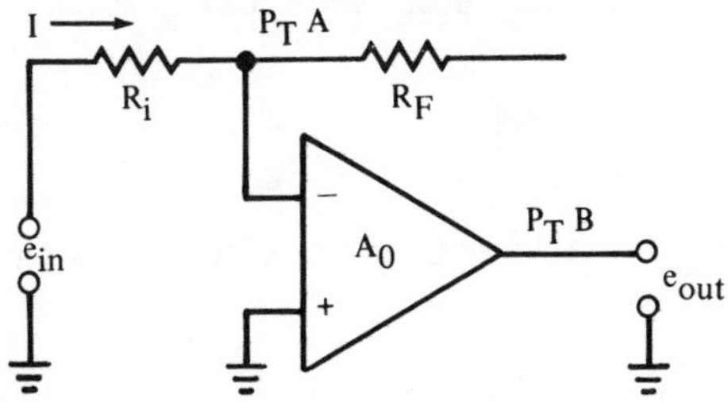

Figure 11.12 Feedback Circuit (Closed-Loop Gain)

This means that the current I (produced by e_{in}/R_i) as it reaches point A seeing two paths, one through R_F the other into the negative terminal, will choose to go through R_F. The voltage developed at the output, e_{out} will be:

$$e_{out} = A_o(V \text{ pt } A)$$

or, the voltage at point A will be:

$$\frac{e_{out}}{A_o}$$

since we know that A_o is a very large number (remember A_o is the open-loop gain of the operational amplifier) then the voltage at point A will be very small, almost 0. When this voltage is so small that it is 0 for all practical purposes, we say that point A is at *virtual ground*. This says that while point A is not physically tied to the ground, it is essentially at the ground potential and that no current flows between the two terminals. If no current can flow into the negative terminal and no current can flow between the terminals, the current I (e_{in}/R_i) must flow through R_F. Since the output voltage (e_{out}) is then from point B to ground, it should be evident that this voltage is the voltage drop on R_F. Remember one end of R_F is at ground potential (virtual ground). Therefore, e_{out} may be written as:

$$e_{out} = I \cdot R_F$$

or

$$e_{out} = \frac{e_{in}}{R_i} \cdot R_F$$

more simply, we can rewrite the equation for e_{out} as simply:

$$e_{out} = e_{in} \left(\frac{R_F}{R_i}\right).$$

The gain of the operational amplifier circuit as shown in figure 11.13 may be written as:

$$A_F = \frac{e_{out}}{e_{in}} = \frac{R_F}{R_i}$$

where A_F is the gain of the circuit of figure 11.12, or the gain of an operational amplifier with feedback. R_F is the feedback element and the feedback component is voltage.

The important concept to see in this discussion is that when the operational amplifier is in the feedback configuration, the circuit gain is dependent upon external components and not on the amplifier device. While the amplifier is acting as a passive element, it must be clearly understood that without the device, the circuit of figure 11.12 would not work. Consider figure 11.13. Here we see a simple voltage divider. We know that the voltage e_2 is equal to:

$$e_2 = e_{in} \left(\frac{R_2}{R_1 + R_2}\right)$$

Mathematically, we may rearrange this expression and solve for e_{in} if we have a value of e_2. This would give us

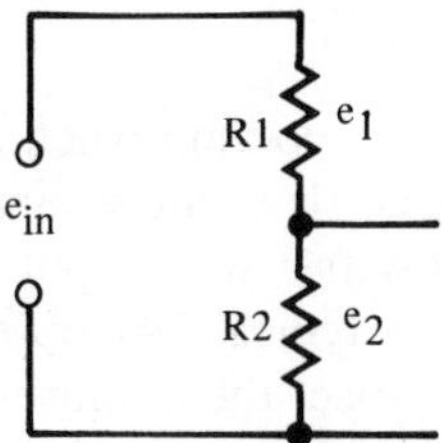

Figure 11.13 Voltage Divider Network

$$e_{in} = \frac{(R_2/R_1 + R_2)}{e_2}$$

A nice trick if we could do it. But alas, the math and the practical don't agree. However, by the use of the apparently passive operational amplifier, we have managed to do what the above expression implies can be done.

Example For the circuit of figure 11.12 calculate the gain of the circuit when $R_i = 3.0\ k\Omega$; $R_F = 6.0\ k\Omega$; $e_{in} = 2.1$ millivolts, also compute the values of e_{out} and I.

Solution The gain is simply given by:

$$A_F = \frac{R_F}{R_i} = \frac{6.0\ k\Omega}{3.0\ k\Omega} = 2$$

However, remember that we are using the negative terminal. Therefore, the gain needs to have a negative sign preceding it to show that there will be a 180° phase shift between input and output. So A_F equals -2.

To calculate the value of the current I we write Ohm's law:

$$I = \frac{E}{R} = \frac{e_{in}}{R_i} = \frac{2.1 \times 10^{-3} \text{ volts}}{3 \times 10^{3}\ k\Omega}$$

$$I = 0.7 \times 10^{-6} \text{ amperes}$$

The value of e_{out} is given by

$$\begin{aligned} e_{out} &= (e_{in})\ (A_F) \\ &= (2.1 \times 10^{-3} \text{ volts})\ (-2) \\ &= 4.2 \times 10^{-3} \text{ volts} \end{aligned}$$

11.7 SLEW RATE

One important feature of the op-amp when used in digital or pulse circuits is called *slew rate*. The slew rate is simply a parameter of the device that tells the user how fast the output voltage can change. As an example, the slew rate of a typical 741 op-amp is given as 0.5v/μsec. This tells us that we can expect the output voltage to have a maximum change of 0.5 volts in 1 microsecond. The slew rate of the device can be altered by the addition of external capacitors. It is important to know the slew rate of the device since we wish to have our square waves remain square when passing through the op-amp. Suppose for instance that we have a 741 (slew rate = 0.5v/μsec) and we are using a 5.0-volt square wave. Instead of an instantaneous change for 0.0 volts to 5.0 volts or from 5.0 volts to 0 volts we will have a change that takes 10.0 μseconds. This may be calculated from:

$$\text{Slew rate} = \frac{\text{output voltage change}}{\text{time}}$$

$$0.5\ \text{volts}/\mu\text{sec} = 5.0\ \text{volts}/\text{time}$$

$$\text{time} = \frac{5.0\ \text{volts}/\mu\text{sec}}{0.5\ \text{volts}} = 10\ \mu\text{seconds}$$

Slew rate also limits the maximum frequency that may be passed through the op-amp. This is shown by

$$F_{max} = \frac{\text{slew rate}}{6.28 \times \text{maximum output volts (undistorted)}}$$

It is extremely important to keep in mind the slew rate of the device as well as the frequency of the digital signal being processed when designing any type of digital circuit using op-amps.

11.8 SOME OP-AMP CIRCUITS

One of the advantages of using the operational amplifier as a circuit device was shown in the preceding section. The op-amp acts as a passive element and the circuit function is implemented by the external components. The ability of a device to perform in such a manner leaves the user with a variety of circuits that can be constructed with accurate and predictable results. The remainder of this chapter will be devoted to several of the more common circuits using op-amps in connection with digital circuits.

The Op-Amp Comparator

It is strange that as we talk about the use of the external components and their uses in op-amp circuits, the first circuit we will look at uses no external components. This circuit is called the op-amp comparator. The comparator is shown in figure 11.14. Notice the similarity between the comparator circuit and the basic op-amp as shown in figure 11.1.

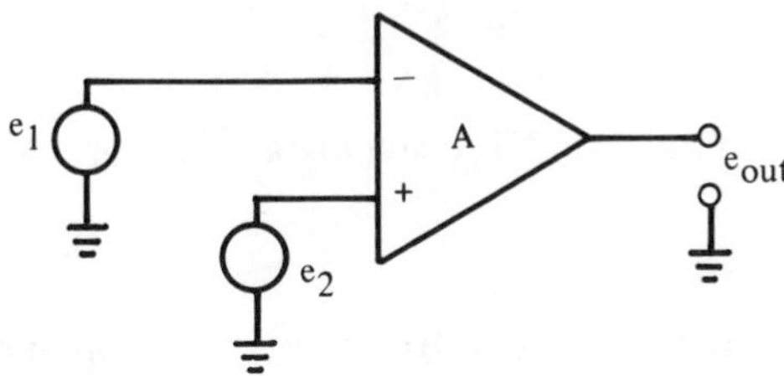

Figure 11.14 Op-Amp Comparator Circuit

The operation of the op-amp comparator is that the polarity of the output voltage (e_{out}) is a function of the input voltages (e_1 and e_2). Remember, the output voltage will be such that its polarity is 180° out of phase with the signal at the inverting terminal and its polarity will be the same as a signal at the noninverting input. If the input signals then are such that e_1 is greater than e_2 (both positive) the output voltage will be a negative. In a similar manner, if the signal at e_2 is greater than the signal at e_1 (both positive) the output voltage will have a positive sign. Further, should the input signals be such that $e_1 = e_2$ the output voltage will be 0. Table 11.1 shows several possible input levels and the resulting polarity of the output voltage. It is important to remember that the user of a voltage comparator is simply interested in which of the two inputs has the greater potential. The actual values of the signals

TABLE 11.1 e_{out} for Values of e_1 and e_2

e_1	e_2	e_{out}
+3.0 V	+3.0 V	0.0
+3.0 V	+2.0 V	negative
+3.0 V	+4.0 V	positive
−2.0 V	+4.0 V	positive
−2.0 V	−1.0 V	positive
+2.0 V	−1.0 V	negative

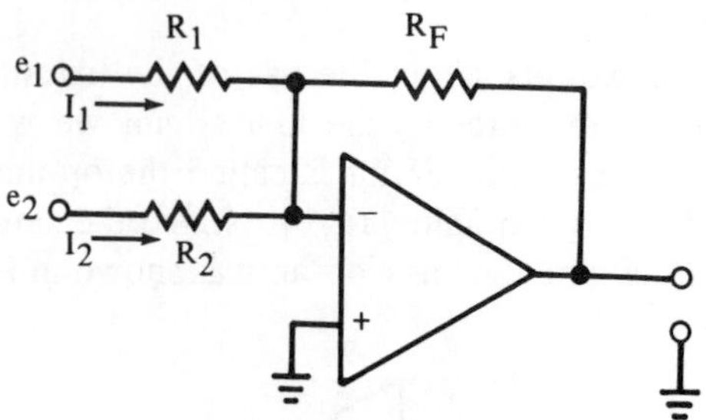

Figure 11.15 The Summing Amplifier

are of no importance. The use of a circuit such as the op-amp comparator should be quite evident in digital work. The comparator will produce an output that is the equivalent of a logic level (either positive or negative) depending on the levels of the input signals.

The Summing Amplifier

Refer to figure 9.12. Remember, we said that the input current, I, was also the current through the feedback resistor R_F. This, in turn, created the output voltage (e_{out}) as the product of I and R_F. In short, the current entering the junction of the input terminal and the feedback resistor is effectively shunted through the resistor due to the high input impedance of the op-amp. With this concept firmly in mind we should be able to calculate the current in the feedback resistor of figure 11.15.

Here we see two resistors, R_1 and R_2. They are effectively in parallel and in series with R_F. The equivalent circuit is shown in figure 11.16.

Here we have omitted the op-amp since it is effectively an open circuit. The current I_F now can be seen as the sum of the two current I_1 and I_2. This means that the voltage drop across R_F is given as:

$$U_{RF} = (I_1 + I_2)\ (R_F)$$

Since $I_1 = e_1/R_1$ and $I_2 = e_2/R_2$ we can write $V_{RF} = (e\ \cdot/R_1 + e_2/R_2)$ (R_F) or

$$V_{RF} = e_1 \frac{R_F}{R_1} + e_2 \frac{R_F}{R_2}$$

If R_1 and R_2 are equal, (R_x), we have the special case where:

$$V_{RF} = \frac{R_F}{R_x}(e_1 + e_2)$$

The voltage (V_{RF}) now can be shown to be the summation of the input voltages e_1 and e_2. The circuit of figure 11.16 is called a summing amplifier. An extended version of this circuit is shown in figure 11.17.

Figure 11.18 shows a summing amplifier which has four inputs.

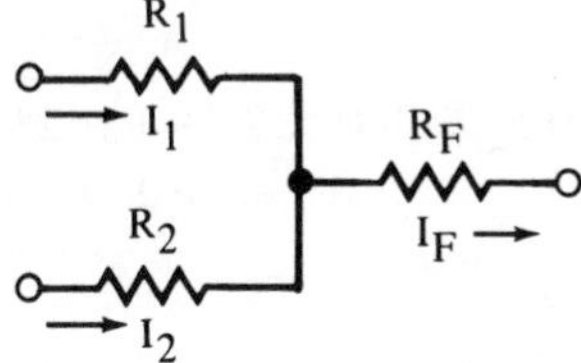

Figure 11.16 Equivalent Input to Summing Amplifier

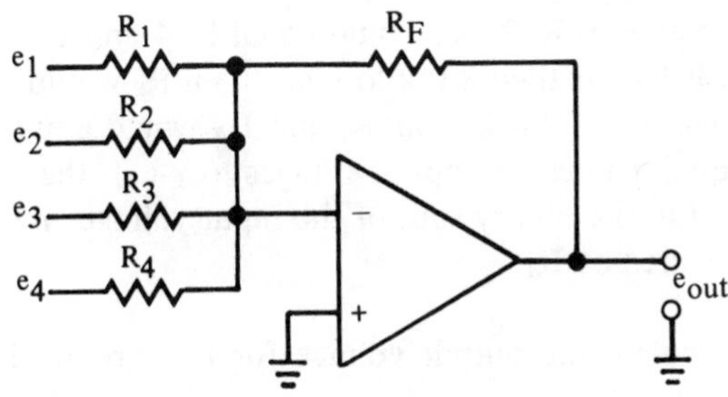

Figure 11.17 Four-Input Summing Amplifier

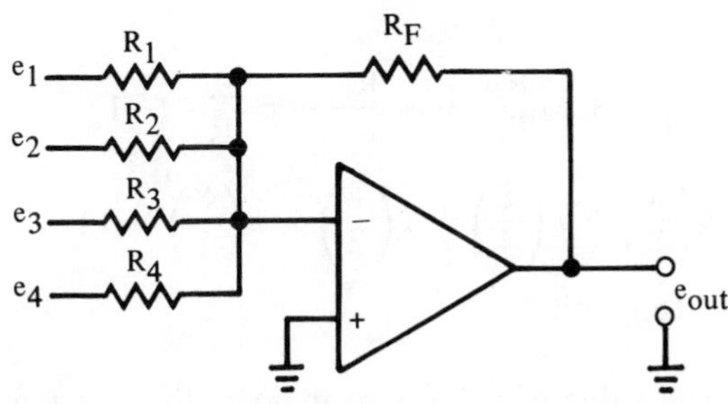

Figure 11.18 Summing Amplifier for Example 11.1

The output voltage then will be the following:

$$e_{out} = e_1 \frac{R_F}{R_1} + e_2 \frac{R_F}{R_2} + e_3 \frac{R_F}{R_3} + e_4 \frac{R_F}{R_4}$$

We can make good use of this fact in a very simple circuit that is of practical value in our digital work. Recall that to convert a binary number to its decimal equivalent, we form a weighted sum of 0's and 1's. For example, to convert the binary number 1001 to its decimal equivalent, we write:

$$\begin{aligned} 1001 &= 1 \times 2^3 + 0 \times 2^2 + 0 \times 2^1 + 1 \times 2^0 \\ &= 8 \qquad + 0 \qquad + 0 \qquad + 1 \\ &= 9 \end{aligned}$$

Since converting from the binary to decimal involves simply the addition of weighted binary digits, it should be apparent that we can use the summing amplifier of figure 11.17 to make our conversion. To accomplish this, we simply need to weight each of the series input resistors in binary fashion. This means that R_1 would have a value of R, while R_2 would have a value of R/2, R_3 would equal R/4 and R_4 would be R/8. For example, if R_1 equalled 8.0 kiloohms, then R_2 would equal 4.0 kiloohms, R_3 would equal 2.0 kiloohms, and R_4 would equal 1.0 kΩ. Now if we had equal values of input voltages (e_1–e_4) the output voltage would be equal to the binary sum of the input voltages multiplied by the value of R_F divided by R_1.

Example Calculate the output voltage for the circuit of figure 11.18.

$e_1 = e_2 = e_3 = e_4 = 1.0$ volt
$R_1 = 8.0\ \text{k}\Omega$, $R_2 = 4.0\ \text{k}\Omega$, $R_3 = 2.0\ \text{k}\Omega$, $R_4 = 1.0\ \text{k}\Omega$
$R_F = 8.0\ \text{k}\Omega$

Solution

$$\begin{aligned} e_{out} &= e_1 \frac{R_F}{R_1} + e_2 \frac{R_F}{R_2} + e_3 \frac{R_F}{R_3} + e_4 \frac{R_F}{R_4} (-1) \\ &= 1\left(\frac{8}{8}\right) + 1\left(\frac{8}{4}\right) + 1\left(\frac{8}{2}\right) + 1\left(\frac{8}{1}\right)(-1) \\ &= -15.0 \text{ volts} \end{aligned}$$

Note: The value of (−1) is to indicate the fact that we are using the inverting input.

Example Calculate the output voltage for the circuit of figure 11.18 for the following input voltages: $e_1 = 1.0$ volt, $e_2 = 0.0$ volt, $e_3 = 0.0$ volt, $e_4 = 1.0$ volt.

Solution

$$e_{out} = e_1\left(\frac{R_F}{R_1}\right) + e_2\left(\frac{R_F}{R_2}\right) + e_3\left(\frac{R_F}{R_3}\right) + e_4\left(\frac{R_F}{R_4}\right)(-1)$$

$$= 1\left(\frac{8}{8}\right) + 0\left(\frac{8}{4}\right) + 0\left(\frac{8}{2}\right) + 1\left(\frac{8}{1}\right)(-1)$$

$$= 1 \quad + 0 \quad + 0 \quad + 8 \quad (-1)$$

$$= -9.0 \text{ volts.}$$

(Does this value agree with the binary equivalent of the input voltages?)

EXERCISES

1. The 741 op-amp is a monolithic integrated circuit similar to other general-purpose operational amplifiers. It is an improved version of one of the earliest IC op-amps; the A709. The large voltage gain of 80 decibels for this amplifier can cause unstable and nonlinear operation when the unit is used as an open-loop amplifier. If the unit is operated with negative feedback, stable voltage gains of 60 decibels or less can be obtained.

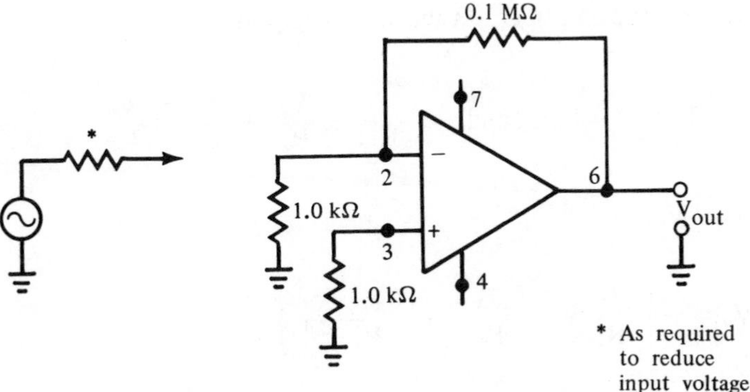

Figure 11.19 Test Circuit for Op-Amp

(a) Connect a 741 as shown in Figure 11.19.

(b) Operate the amplifier with a very small kilohertz input signal. Disconnect the end of the 0.1 megohm feedback resistor. Record your observations of amplifier linearity and stability. Reconnect the feedback resistor.

(c) Apply a small one-kilohertz signal to terminal 1. Record input and output waveshapes. Repeat this step with the input signal applied to terminal 2.

(d) Measure and record the maximum output signal voltage for a 1-kilohertz signal applied to terminal 3. Also record the value of input signal voltage used. Repeat this step for an input on terminal 2.

(e) Record enough data for a gain frequency curve with (a) $R_F = 0.1$ megohm, (b) $R_F = 10$ kiloohms and (c) $R_F = 1$ kiloohms. Use terminal 3 for the input signal in this step.

Results Is the output of this op-amp 0 without an input signal? Briefly summarize your observations of step 2.

2. The summing amplifier is a basic application of op-amps in many signal-processing systems. This circuit can do algebraic summing of ac or dc input voltages. The summed inputs can also be multiplied by a fixed constant determined by the feedback and input resistors. The object of this exercise is to verify the design and operation of a three-input summing and constant-multiplier amplifier circuit. Consider a situation where it is desired to average three dc levels. One method of implementing an analog averager would be to sum the three dc signals and multiply by one-third. In figure 11.20 the summing point, pin 2, can be considered at 0 potential and that 0 current enters the amplifier. At the summing point:

$$I_1 + I_2 + I_3 = I \qquad \text{and} \qquad \frac{V_1}{R_1} + \frac{V_2}{R_2} + \frac{V_3}{R_3} = \frac{V_o}{R_F}$$

solving for V_o;

$$V_o = -\left[\frac{R_F}{R_1} V_1 + \frac{R_F}{R_2} V_2 + \frac{R_F}{R_3} V_3\right]$$

If the ratio of the feedback resistor to each input resistor is made equal to one-third then:

$$-V_o = \frac{1}{3\ (V_1 + V_2 + V_3)}$$

V_o is the average of the three input voltages. The negative sign on V_o can be removed with an inverter stage if necessary.

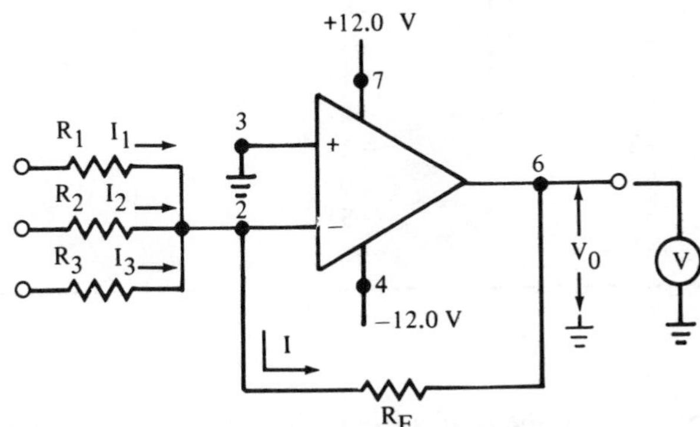

Figure 11.20 Three-Input Summing Amplifier

Construct the circuit shown in Figure 11.20. Use 100 kΩ resistors for R_1, R_2, R_3. Select the value of R_F to give a constant multiplier of one-third. A method to generate the input voltages is shown in the following figure.

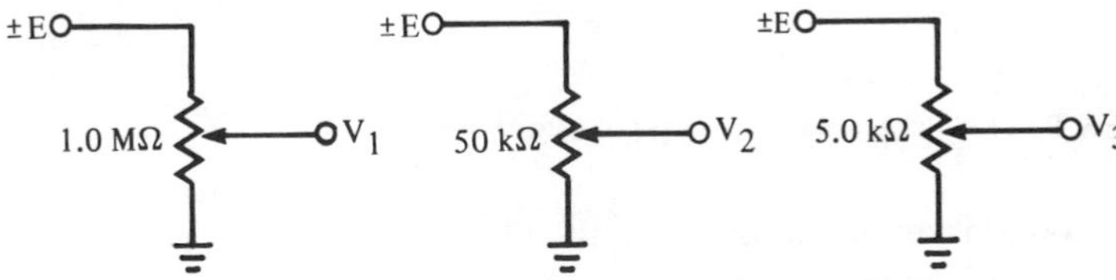

Complete table 11.2 on page 254 using your circuit and calculations.

TABLE 11.2

	Input (Volts)		*Output (Volts)*	
V_1	V_2	V_3	*Measured*	*Calculated*
0	0	0		
+1	+1	+1		
+4	+4	+4		
+4	−1	−3		
−2	+1	f1		
+1	+2	+3		
−2	−3	+1		
+1	+2	+3		
−2	−3	−1		
+3	−3	0		
+4	−4	−3		
−1	−3	+2		

Results

1. Summarize the results from table 11.2. Were there problems in setting the input voltages?
2. Draw a block diagram and label any important component values or ratios for a circuit which can perform the following analog computations. Combine the average of three daily quiz grades, a final exam grade and a tab grade in the proportions 70%, 10%, 20% to produce a course grade as the final result. You may use more than one amplifier if necessary.
3. Explain the operation of your circuit sketched in step 2.

QUESTIONS AND PROBLEMS

11.1 What is an operational amplifier?

11.2 What is meant by open-loop gain?

11.3 Some operational amplifier ICs do not have a ground lead. How is circuit ground, input ground, and output ground accomplished in this case?

11.4 Explain what is meant by double-ended output and single-ended output of an op-amp.

11.5 For most op-amps, what would be a typical value for the CMRR?

11.6 Define CMRR.

11.7 Define the term *virtual ground.*

11.8 In closed-loop operations, one advantage of the op-amp circuit is that the gain is now dependent only upon the external components. Explain how this is possible.

11.9 For the circuit of figure 11.12, calculate the gain when: R_i = 4.7 kΩ; R_F = 10.0 kΩ; e_{in} = 1.3 millivolts. Also, compute the values of e_{out} and I.

11.10 For the circuit of figure 11.12, using the values given in the problem, 11.9, what is the largest input voltage that will not cause the output to be clipped?

11.11 For the circuit shown in figure 11.14, what would be the value of output voltage when the inputs are:

a. $e_1 = -3.0$ volts $c_2 = -2.0$ volts
b. $e_1 = +3.0$ volts $c_2 = +2.0$ volts
c. $e_1 = +3.0$ volts $c_2 = -3.0$ volts
d. $e_1 = -3.0$ volts $c_2 = +3.0$ volts
e. $e_1 = -2.0$ volts $c_2 = +3.0$ volts

11.12 For the circuit shown in figure 11.18, calculate the output voltage if $c_1 = c_2 = c_3 = c_4 = 1.0$ volt; R_1 = 8.0 kΩ, R_2 = 4.0 kΩ, R_3 = 2.0 kΩ, R_4 = 2.0 kΩ. **Note:** Check your answer carefully with data sheets.

11.13 Design a circuit using the op-amp configuration of figure 11.18. Have the circuit perform some practical application.

11.14 Give an example where common-mode input connections to the op-amp are practical.

11.15 Give an example of an application using the op-amp comparator circuit.

11.16 Explain the function of the 1.0 kΩ resistors used in figure 11.19.

11.17 Redesign the circuit of figure 11.20 so that it may be used as a digital- (binary-coded) to-analog converter.

11.18 Show the signal wave-shape through the block diagram of figure 11.4. Use an inverting amplifier configuration.

11.19 Using slew rate, calculate the maximum frequency of a 741 op-amp.

11.20 How long will it take the output voltage of an op-amp comparator (using a 741) to change 13.0 volts?

11.21 Define slew rate.

11.22 How may the slew rate of a 741 op-amp be increased?

11.23 Explain the operation of the summing amplifier.

12

Digital-Analog–Analog-Digital Conversion Circuits

OBJECTIVE: To introduce the student to the use of and the need for digital-to-analog and analog-to-digital circuits and conversions. To show the various types of conversion circuits and techniques.

Introduction: Digital electronics deals with discrete quantities, 0's and 1's. The circuits of digital electronics deal mainly in these discrete signal levels. The quantity of the change in the level is not nearly as important as the fact that there has been a change. Most quantities to be measured or controlled appear as analog signals. This means that they are continuously changing quantities or signals. Take a temperature change, for example. As the temperature changes from 80° to 83°, the change does not occur as a discrete step, but as a continuous change from 80° to 83°. Since most of the signals we deal with are analog, we need devices that can convert from the analog to the digital state or vice versa. A good example of such a conversion is the digital voltmeter. Here, the analog signal (the voltage to be measured) is converted to a digital signal, then processed by digital circuitry and displayed on a digital readout. This chapter will look at several circuits and devices that perform the digital-to-analog or the analog-to-digital conversions.

12.1 REVIEW OF THE SUMMING AMPLIFIER

As discussed in a previous chapter, the summing amplifier can be thought of as a digital-to-analog conversion circuit. The summing amplifier is an operational amplifier configured so that its output current will be the sum of several input currents. (The output signal voltage will also be the sum of the various input voltages.) Figure 12.1 shows the basic operational amplifier circuit used to sum voltages.

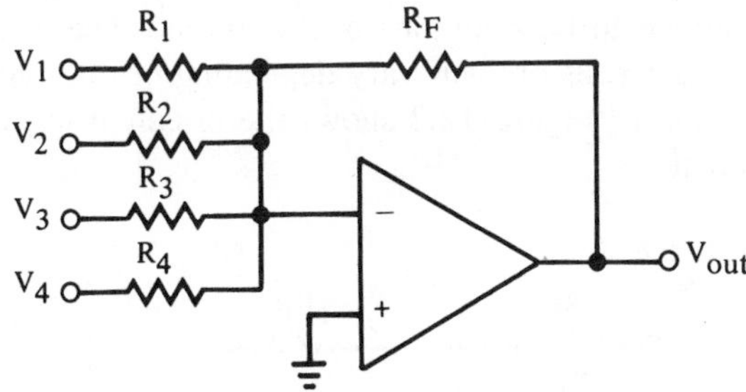

Figure 12.1 Basic Summing Amplifier

For the circuit of figure 12.1, the output voltage (V_{out}) may be written as:

$$V_{out} = -\left[\frac{R_F}{R_1}(V_1) + \frac{R_F}{R_2}(V_2) + \frac{R_F}{R_3}(V_3) + \frac{R_F}{R_4}(V_4)\right]$$

The output voltage then is the weighted sum of the individual input voltages. The weighted value assigned to each input voltage becomes the ratio of the circuit feedback resistor R_F, and the series input resistor of that particular voltage. For example, should all of the input resistors be of equal value, and they, in turn, equal to the feedback resistor such that:

$$R_1 = R_2 = R_3 = R_4 = R_F$$

then, V_{out} would be equal to:

$$V_{out} = -(V_1 + V_2 + V_3 + V_4) \qquad \textbf{(12.1)}$$

Remember that V_{out} will be the negative of the sum of the inputs since we are using the inverting input of the operational amplifier. This negative indicates a 180° phase inversion between the input and output voltages. Should the input resistors not be of equal value, then the output would be the sum of the weighted values of inputs. For example, should $R_1 = R_F$, $R_2 = R_F/2$, $R_3 = R_F/4$ and $R_4 = R_F/8$ the value of V_{out} would be:

$$V_{out} = -(V_1 + 2V_2 + 4V_3 + 8V_4). \tag{12.2}$$

This means that each input will be multiplied by its weighting factor before being summed to become part of the output. The exact values assigned to the input resistors will vary depending on the requirements of the summing circuit. Figure 12.2 shows the circuit using the weighting values as given above.

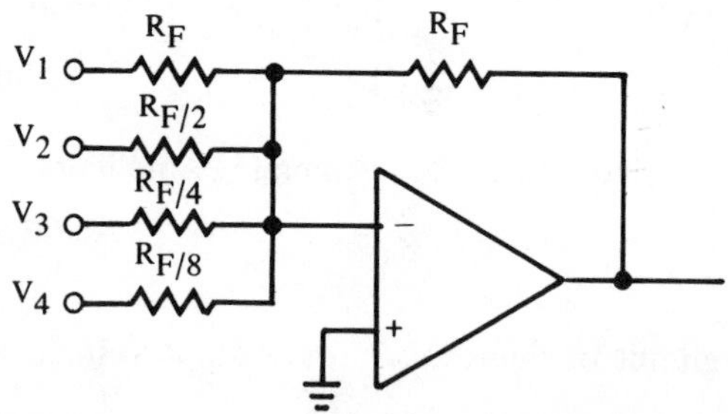

Figure 12.2 Binary-Weighted Summing Amplifier

Example Assume for figure 12.2 that the input voltages will be:

$V_1 = 1.0$ volt $\qquad V_3 = 0.0$ volt
$V_2 = 1.0$ volt $\qquad V_4 = 1.0$ volt

Calculate the output voltage, V_{out}.

Solution We know that the output voltage will be the weighted sum of the individual inputs. Substituting the input values into equation 12.2 yields

$$\begin{aligned} V_{out} &= -(1.0 + 2(1.0) + 4(0.0) + 8(1.0)) \\ &= -(1.0 + 2.0 + 0.0 + 8.0) \\ &= -11.0 \text{ volts} \end{aligned}$$

Suppose that we had assigned the inputs of the circuit of figure 12.2 binary number equivalents, we could have written the binary number

$$1011 = V_4\ V_3\ V_2\ V_1$$

This would have given V_1 the weighting of 2^0, V_2 the weighting of 2^1, V_3 would be 2^2, and V_4 would have a weighting of 2^3. This weighting would yield the same result as equation 12.2. Recall that to convert a binary number to its decimal equivalent, we simply formed the weighted sum of 0's and 1's. Exactly the same as for the example.

When configured as in figure 12.2 this summing amplifier is a form of digital-to-analog converter. The digital input in this case being the four input voltages, V_1 to V_4. Should a larger digital number be required as an input, the addition of more input stages would be used. Before we can use the op-amp as a digital-to-analog converter, one additional problem associated with the operational amplifier must be overcome. As you recall from the discussion of op-amps and op-amp characteristics, the op-amp output voltage swing is restricted (usually to a value slightly less than the supply voltages used to power the device). If this is the case, and say the amplifier of figure 12.2 is restricted to an output swing of 10 volts, we will have an error if the input were to go to a value of binary 1111. This is equal to decimal 15 and the op-amp summer will attempt to reach an output of −15.0 volts and become clipped at −10.0 volts. The summer would be in error and our digital-to-analog conversions inaccurate for values above 10. One way to overcome this problem is to decrease the gain of the op-amp circuit. This can be most easily accomplished by decreasing the value of the feedback resistor.

A solution to this problem is not as difficult as it may first appear. We have a binary 15 that must be represented by a decimal 10. Therefore, the output at maximum input must be equal to two-thirds of that input. To obtain an op-amp with this gain, we simply select R_F to equal $\frac{2}{3}R$. By keeping all other weighting values the same in equation 12.2 we now have:

$$V_{out} = -(R_F/R)\ (V_1 + 2V_2 + 4V_3 + 8V_4)$$

and since $R_F = \frac{2}{3}R$ we can write

$$V_{out} = -\tfrac{2}{3}\ (V_1 + 2V_2 + 4V_3 + 8V_4)$$

The circuits shown in figures 12.1 and 12.2 are indeed digital-to-analog converters. Although they are simple to construct, they do have several poor qualities. Most of the disadvantages of such simple circuits will be shown and discussed later in this chapter.

R2R Ladder D/A Converter

Figure 12.3 shows an improved variation over the simple D/A converters shown in figures 12.1 and 12.2. This circuit is shown here for a comparison of accuracy and simplicity. One of the advantages of the circuit of figure 12.3 is that the op-amp will see a constant input impedance regardless of the number of inputs that are on at any one time.

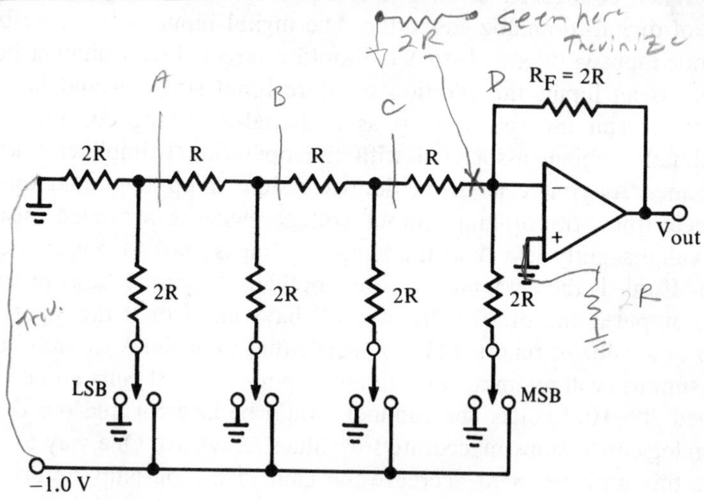

Figure 12.3 R2R Ladder D/A Converter

Another advantage of this circuit is that only two values of resistors are required (R and 2R). The maximum output voltage of this circuit (for a binary 15) is 1.875 volts. To change the maximum to any other value would require changing the value of R_F. The equation to do this is:

$$\frac{\text{max output}}{1.875} \times 2R = R_F \qquad \textbf{(12.3)}$$

Example To change the maximum output voltage of the circuit of figure 12.3 to 10 volts, what value of R_F should be used?

Solution Substitute 10.0 volts in equation 12.3.

$$\frac{10}{1.875} \times 2R = 10.67\ R$$

So we would pick a value of $R_F = 10.67\ R$. If, for example, the value of R were 1.0 kΩ then we would use $R_F = 10.67$ kΩ and $2R = 2.0$ kΩ.

Example Calculate the values of R_F and 2R required if R is 4.7 kΩ, and we wish V_{out} max to be 7.5 volts.

Solution

$$2R = (R \text{ time } 2) = (4.7\ \text{k}\Omega \times 2) = 9.4\ \text{k}\Omega$$

$$R_F = \left(\frac{7.5}{1.875}\right) \times 9.4 = 37.6\ \text{k}\Omega$$

Example For a circuit using the values as found in the previous example, what would the output voltage be if the input were a binary 6 (0110)?

Solution Use table 12.1.

$$V_{out} = V_{out}\ (\text{from table 12.1}) \times \frac{V_{max}}{1.875}$$

Therefore

$$V_{out} = 0.750 \times \frac{7.5}{1.875}$$

$$= 3.0 \text{ volts.}$$

Note: Use table 12.1 on page 262 to find the output volts for the original circuit of figure 12.3.

12.2 DATA CONVERSION SYSTEMS: ANALOG-DIGITAL-ANALOG

Before we continue our discussion of digital-analog or analog-digital conversion, a brief overview of data conversion systems will be presented. As mentioned at the beginning of this chapter, most industrial systems are not completely analog or completely digital. This requires the use of conversion techniques. There are times when input and out-

TABLE 12.1

Decimal	*Decimal Input* a_8	a_4	a_2	a_1	*Output (Volts)*
0	0	0	0	0	0.000
1	0	0	0	1	0.125
2	0	0	1	0	0.250
3	0	0	1	1	0.375
4	0	1	0	0	0.500
5	0	1	0	1	0.625
6	0	1	1	0	0.750
7	0	1	1	1	0.875
8	1	0	0	0	1.000
9	1	0	0	1	1.125
10	1	0	1	0	1.250
11	1	0	1	1	1.375
12	1	1	0	0	1.500
13	1	1	0	1	1.625
14	1	1	1	0	1.750
15	1	1	1	1	1.875

put data for a system need be in an analog quantity, but the handling and manipulating of that data are most easily performed in digital quantities. Figure 12.4 represents a typical data conversion system.

For a system such as the one shown in figure 12.4, both analog-to-digital and digital-to-analog conversion are used. Block I is the analog input circuit. The function of block I is to capture the physical quantity to be measured or handled (in this case the sun's heat) and convert this physical quantity (heat energy) to an electrical signal. The circuit of block I is an energy transducer. The various types of energy transducers will be discussed later in this chapter. This electrical energy, in the form of an electrical analog signal, is sent to a comparator circuit. The comparator is a form of op-amp circuit that compares two input signals and produces an output signal whose polarity is dependent upon the two inputs. Block III is the analog-to-digital comparator. Here the analog signal is converted to a digital quantity. This digital quantity is then read on a digital readout, block V. The digital signal is also sent to a digital-to-analog converter (block IV), where the analog is sent back to block II to be compared with the original quantity. This provides a feedback control for the A-D converter of block III. Each of these blocks will be broken down and discussed in later sections of this chapter.

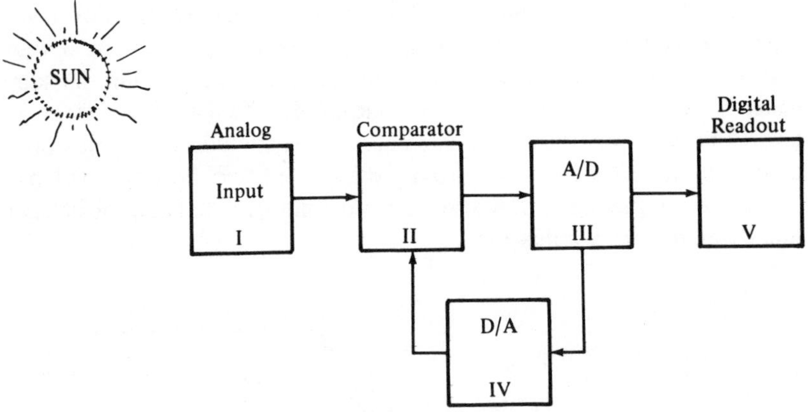

Figure 12.4 Typical Data Conversion System

12.3 QUALITY MEASUREMENTS OF D/A CONVERTERS

As we get into our discussion of D/A converters, there are some terms that are used to describe how well the converter is performing, which should become familiar to the reader. These terms are given here and should be referred to as the reader proceeds through the other sections of this unit.

Accuracy: Accuracy is a general term used to relate how close the actual output is to an expected output. The accuracy is usually stated as a percentage of full-scale output. If, for example, a D/A converter has an accuracy of 0.1 percent full scale, and full-scale reading is 10.0 volts, then the output for a digital input should not vary more than 10.0 millivolts from the expected output for any digital input.

Resolution: Resolution is determined by the number of binary bits of input to the converter. Commonly, D/A converters are found to have inputs of eight, ten, and twelve bits. A converter with eight-bit input for example would have a resolution of one part in 2^8 (256). This is expressed as a resolution of 1/256 or 0.0039. To express this as a percentage of full scale, we could say it has a resolution of 0.39 percent of full scale.

Linearity: Linearity, as with accuracy, tells the user the maximum deviation from an expected output. Figure 12.5 illustrates linearity.

In figure 12.5, we have plotted the input versus the output. For 1-volt input, we would expect 1-volt output, for 2-volts input, 2-volts output, and so on. This expected curve is plotted as a straight line. On the same chart is plotted the actual recorded outputs for each of the voltage inputs. This is the nonlinear line (actual). Linearity tells us the maximum deviation the actual line takes from the expected line. So with a 0.1 percent linearity, the curve of figure 10.5 must not deviate from the straight line by more than 5 millivolts.

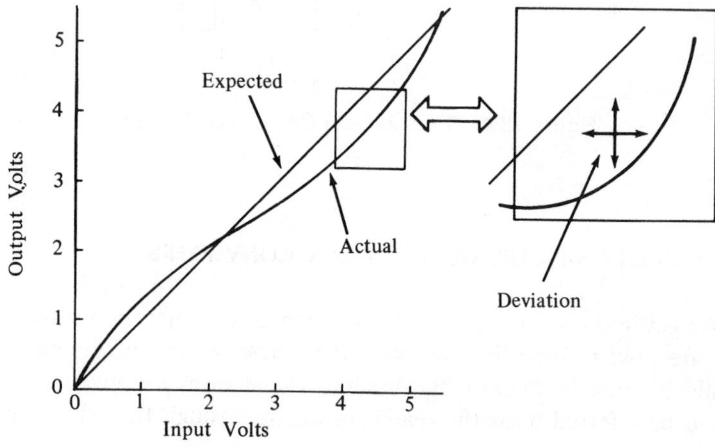

Figure 12.5 Linearity Curve

12.4 THE ANALOG SIGNAL

Analog signals for purposes of our discussion are those quantities that are continuous in nature. The heat from the sun is an analog quantity. We can measure the various analog quantities with instruments that give us continuous readings. A simple thermometer, using a column of mercury and scaled in degrees (Fahrenheit or Celsius) is a very accurate and simple device for measuring the temperature. The light from the sun or a bulb is another type of analog signal. Pressure, the flow of water through pipes, air pressure, etc. are all physical quantities that are analog in nature. Why then the demand for the conversion of these quantities to digital quantities? The advent of the digital computer for one. It is easier to store and manipulate digital quantities than analog quantities. In most cases, digital readouts are simpler to use than the analog counterpart. It is easier, for example, to look at a display that reads 20.013 than to find that point on a continuous dial with a pointer. The analog quantities are all around us, and so are the digital systems for handling, measuring, and controlling them. It is our job then to fit the two worlds together and to use the best of both to our advantage.

12.5 ANALOG-TO-DIGITAL CONVERSION

Analog-to-digital conversion is the conversion of a continuous signal to a signal composed of discrete voltage or current levels. One reason for such a conversion is given in the following example. Suppose we have a digital computer (or microprocessor) controlling the processing of a paper mill. In the process control of the paper mill, we have a number of analog quantities such as temperature, humidity, liquid flow, liquid level, etc., which must be monitored and controlled. These quantities are sensed and converted to electrical analog signals via various transducers. Temperature, for example, is represented by a voltage output of a thermocouple. To enable our process to communicate with the controller (digital computer) these analog signals must be converted to a digital form.

Figure 12.6 is a basic op-amp comparator circuit. The op-amp comparator develops an output whose polarity is dependent on the level of the two input voltages. The circuit of figure 12.6 "compares" the input voltages and produces an output voltage based on that comparison. If the two input voltages (A and B) are equal in phase and magnitude, the output of the comparator will be 0. (Review the material on op-amps.) If the voltage at input A is greater than the voltage at input B, the output at C will be a negative voltage. If the voltage at input B is greater than the voltage at input A, the output voltage at C will be positive. Therefore, we have two analog signals at A and B creating a digital voltage at point C. The comparator is a very simple form of A/D converter.

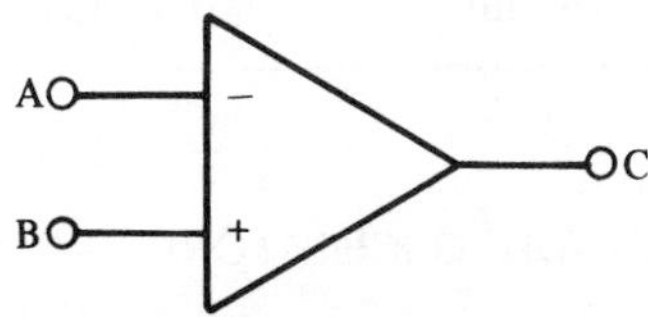

Figure 12.6 Basic Op-Amp Comparator Circuit

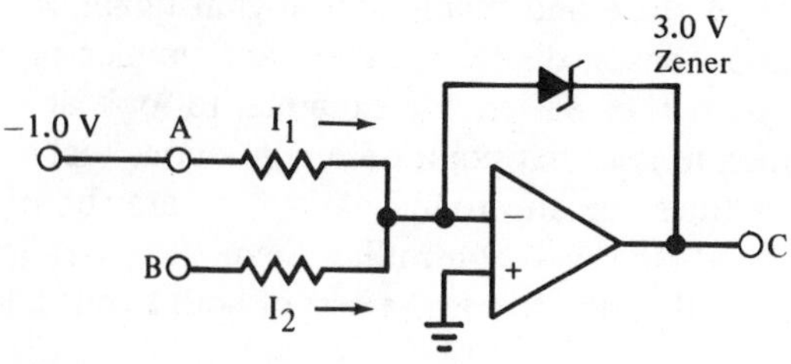

Figure 12.7 Practical Op-Amp Comparator Circuit

The circuit shown in figure 12.7 is a practical op-amp comparator circuit for use with TTL logic gates. Notice that in figure 12.7 we have taken the positive terminal of the op-amp to ground. Also, the negative terminal has the two inputs to be compared. Terminal A is tied to a −1 volt. This means we will be comparing the input at B against the −1 at A. When the magnitude of B is greater than −1.0 volt the zener current (sum of the two currents I_1 and I_2) will be positive, and the zener will be forward-biased. The zener will have about a 0.6-volt drop across it in the forward-bias state. Therefore, the output voltage at point C will be about 0.6 volts. When B is less than −1.0 volt, the zener current will be negative, creating a reverse-bias condition. The reverse drop of the zener will be +3.0 volts (the zener voltage). The output at point C will then be +3.0 volts.

It should be noted that in the discussion above, absolute values of signal at point B were used. This means that a −4.0-volt signal is greater than a +3.0-volt signal. Also, note that the two levels of output created at point C (3.0 volts and 0.6 volts) are compatible with levels required to properly bias TTL logic gates. This is shown in table 12.2.

TABLE 12.2 Output Voltage at Point C vs. Input at Point B

Point B	*Point C*
<Point A	+3.0 volts
>Point A	+0.6 volts

12.6 A/D CONVERSION (UJT OSCILLATOR)

The circuit of figure 12.8 shows another fairly simple circuit that can be used as an analog-to-digital converter. In this circuit, two functions are combined and handled simultaneously. The circuit is a simple unijunction transistor (UJT) oscillator.

The UJT will conduct when the charge on the capacitor reaches the firing potential of the UJT. The capacitor in turn charges as a function of the RC time constant of the series R and the capacitor. Once the UJT conducts, the capacitor discharges and a voltage spike will appear across the resistor R_L. The frequency of these spikes will be related to the RC time constant. If in the circuit of figure 12.8, the resistor, is a photoresistance device (a transducer), we can assume that its resistance will be a function of the light energy seen by the device. This means that the resistance of the photoresistive device will vary as the energy of light shining upon it. This change in resistance will change the RC time constant, which will change the frequency of the output pulses. So, a change in light energy (analog signal) will cause a change in the frequency of the output voltage spikes (digital signal). Here we see the UJT acting as a simple A-D converter.

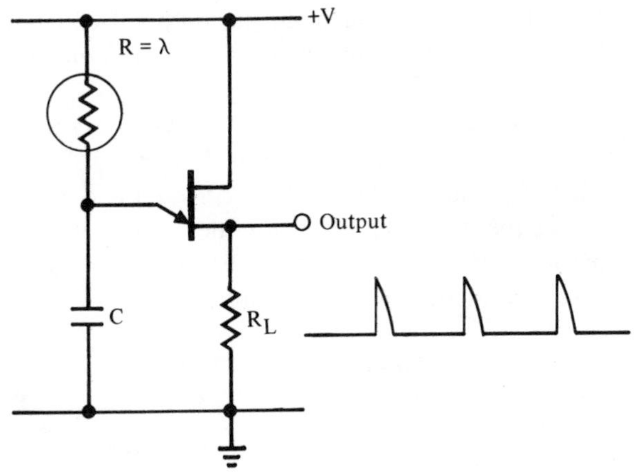

Figure 12.8 UJT Oscillator Circuit

Example Calculate the output frequency for the circuit of figure 12.9.

Solution From the specifications we see that the UJT will fire at a voltage across the capacitor of 1.89 volts. The time constant of the RC circuit is $(R \times C)$, 1.0×10^{-3} seconds. Since we see that 1.89 is 63 percent of 3.0, and we know that the capacitor will charge to 63 percent of E in one time constant, we see that the capacitor will reach the proper firing potential in 1.10^{-3} seconds. Therefore, the output frequency is 1.0 kilohertz.

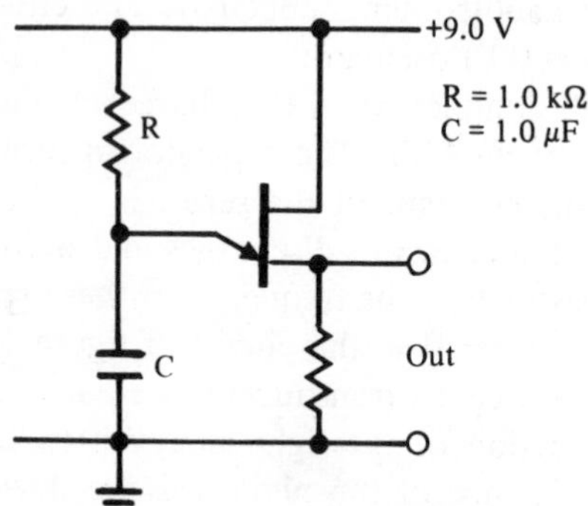

Figure 12.9 UJT A-to-D Circuit

Example Replace the R of figure 12.9 with a photoresistive device whose resistance is 1.0 kΩ at 3 footcandles and varies 250 Ω per footcandle of illumination. Now calculate the output frequency at 1 footcandle, at 3 footcandles, at 4 footcandles, and at 10 footcandles.

Solution We know that $e = E\,(1 - \epsilon^{-T/\tau})$ where e is the charge on C required to fire the transistor. Also, E is E applied, 3.0 volts in this case. The value of τ (tau) is the RC time constant and T is 1/F. Therefore, if we solve the equation for T for the various values of RC, we will have the output frequency of the oscillator for the various values of footcandle illumination on the photoresistor.

Footcandles	*Values of R**	*Value of RC***	*T****	*F*
1.0	500 Ω	0.5×10^{-3}		
3.0	1.0 kΩ	1.0×10^{-3}		
4.0	1.25 kΩ	1.25×10^{-3}		
10.0	2.75 kΩ	2.75×10^{-3}		

* R = 1.0 kΩ at 3 footcandles and varies 250 ohms/footcandle.
R at 1 footcandle = 1.0 kilohm − (500) = 500 ohms

** RC is simply R × C

*** To solve for T, solve
$e = E(1 - \epsilon^{-x})$ where $x = T/RC$ and $T = (X)\ (RC)$
The e required to fire the UJT is 1.89 volts. The E applied is 3.0 volts.
For 1 footcandle we have
$e = E(1 - \epsilon^{-x}) \rightarrow 1.89 = 3.0\,(1 - \epsilon^{-x})$

Step 1: Solve for x.

$$\frac{1.89}{3.0} = 1 - \epsilon^{-x} \rightarrow \left(\frac{1.89}{3.0}\right) - (1.0) = -\epsilon^{-x}$$
$$-0.37 = -\epsilon^{-x}$$
$$0.37 = \epsilon^{-x}$$
$$-0.99 = -X$$
$$0.99 = X$$

Step 2: Solve for T.

$$\begin{aligned} T &= (X)\ (RC) \\ &= (+0.99)\ (0.5 \times 10^{-3}) \\ &= 495 \times 10^{-6} \end{aligned}$$

Step 3: Solve for frequency.

$$F = \frac{1}{T} = 2.02 \text{ kilohertz}$$

Complete the table of the example. This should give you an idea of how to construct a chart showing the relationship of the device output (frequency) to the device input (footcandles). By connecting the output to a frequency counter, you can easily measure the output frequency of the UJT oscillator. A scaling procedure or a conversion chart will allow you to convert this frequency directly to the footcandle input. By using a conversion directly on the output dial of a test instrument, you could read directly in footcandles. This is similar to the method used for reading ohms on an ohmmeter. The meter is really responding to changes in current. The face of the meter, however, is calibrated in ohms (so many ohms restrict the current flow and create a reading), which is read directly from the instrument.

The UJT oscillator is a very simple form of analog-to-digital converter. Its simplicity makes it very useful for many types of applications. A simple change of the resistive element in the circuits creates an oscillator that responds to many types of analog input. For example, if the resistive element were a strain gauge, we would have an A/D converter that would give a change in output frequency that was proportional to pressure or tension.

12.7 RAMP TYPE A/D CONVERTER

A slightly more sophisticated and more accurate type of A/D converter is the ramp type shown in figure 12.10. The operation of the ramp type

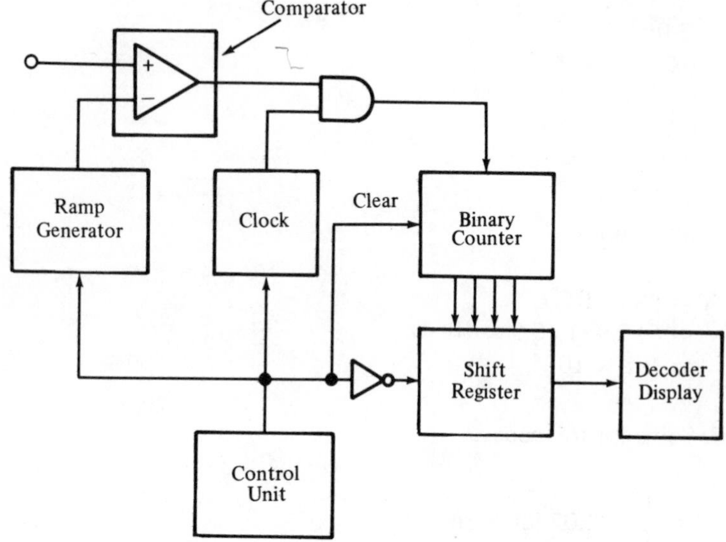

Figure 12.10 Ramp Type A/D Converter

A/D is fairly simple. It does require a control circuit and calibration. The ramp generator and the clock are controlled via a control unit. The control unit is also connected to the binary counter. The control unit may be a manually operated device such as a switch, or a relay, or it may be automatic. Let us suppose the control unit is an astable multivibrator. Next we will assume that on the negative-going transistor of the astable, the following happens:

1. The ramp generator starts to rise (from 0).
2. The binary counter is cleared to 0.
3. The clock pulses are synchronized to the ramp start.

Figure 12.11 gives a timing diagram of the function of the control unit and its relation to the other signals.

Theory of Operation

The operation of the ramp type A/D converter is as follows. An input signal (unknown voltage) is applied to the positive comparator inputs. The control unit starts the ramp generator, clears the binary counter, and synchronizes the clock. At this time, the comparator has a voltage (input signal) at its positive input and 0 volts (the ramp has just started

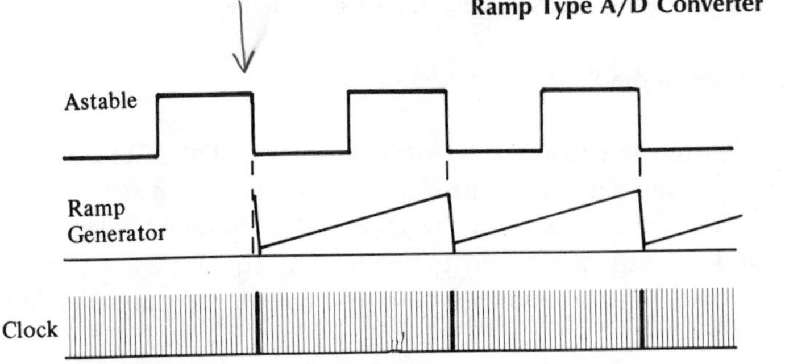

Figure 12.11 Timing Diagram for Figure 12.10

at its negative input). Under these conditions, the output of the comparator is positive. See figure 12.12. This positive is applied to one of the AND gate inputs while the clock pulses are applied to the other input of the gate. The clock pulses then are gated through to the binary counter. When the ramp generator level reaches the level of the unknown input signal, the output of the comparator drops to 0. This inhibits the AND gate and the counter will stop counting. The count stored in the binary counter is an indication of the unknown input voltage. On the positive-going pulse from the astable, the output of the counter is sent to a shift register from where it can be displayed until the next positive astable pulse. The shift register keeps the display from trying to follow the charges of the counter. It also allows the display to be seen for a longer period of time. All that is seen on the display are the final counts, one count each full period of the astable.

Notice in figure 12.12 that the output of the comparator actually goes negative after the ramp voltage and V_{in} cross. This condition does not matter since the gate will still be inhibited.

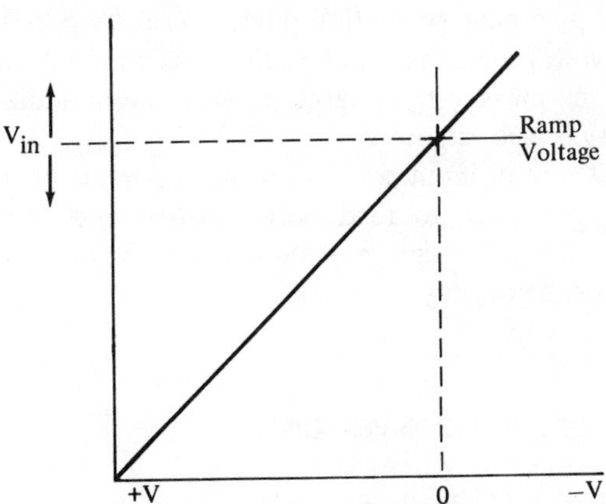

Figure 12.12 Relative Comparator Output

Calibration of the Ramp Type A/D Converter

To calibrate the converter is not too difficult. To calibrate for a 1-volt input, for example, adjust the slope of the ramp so it takes 1 second for it to reach 1 volt. Next, set the clock at a 100-hertz rate. Now, at the end of 1 second if V_{in} and the ramp are equal ($V_{in} = 1.0$ volt), the counter will have a count of 100. We can decode and display this as 1.00 volt.

If the ramp is fairly linear, it will take it 2 seconds to reach 2 volts, in which case the counter will have time to count up to 200 hertz. This will be decoded and displayed as 2.00 volts.

If the input voltage were, say, 1.5 volts, the linear ramp would need 1.5 seconds to reach this level. At a 100-hertz rate, the counter will receive 150 pulses in 1.5 seconds. This will be decoded and displayed as 1.50 volts.

To arrive at greater accuracy or resolution, the rate of rise of the generator, as well as the clock frequency, would need to be adjusted.

EXERCISES

1. Design a ramp type D/A converter similar to figure 12.10 that will give readouts to three decimal places for voltages from 0 to +10 volts.
2. Redesign the above for negative input voltages. The A/D converter of figure 12.10 has some limitations. The accuracy is very dependent on both the linearity of the ramp voltage being generated and the stability of the clock. The actual circuitry must also have protection devices (zeners possibly) to prevent the voltages at the input from damaging the comparator chip. For the measuring of large input voltages additional input circuitry is also required.
3. What type of input circuits are needed to enable the A/D ramp-type converter to read both positive and negative voltages? Draw a block diagram of such a unit. Remember to change the size of the display.

12.8 DUAL-SLOPE A/D CONVERTERS

Figure 12.13 is a block diagram of another type of analog-to-digital converter. This converter is called a dual-slope converter. As with the

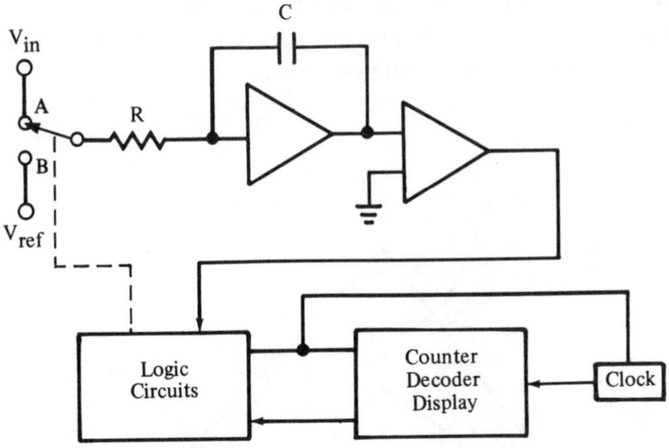

Figure 12.13 Dual-Slope A/D Converter

ramp-type converter, the main component in the dual-slope circuit is the operational amplifier. In this case, there are two op-amps used. One is used, as before, in the comparator mode. The other is used as an integrator, or ramp generator.

The operation of the dual-slope A/D converter is similar in some ways to the ramp-type converter. Each relies on a ramp generator and a clock for the conversion.

Theory of Operation of the Dual-Slope Converter

The operation of the converter starts with the switch in position A. The logic circuits have cleared the display and counters and started the clock. An input signal of unknown value (a positive dc voltage for our example) is applied via the switch to an op-amp configured as an integrator. The logic circuitry holds the switch in position A for a fixed period of time and allows the output of the op-amp to reach a level dependent on the value of the input signal. This is shown in figure 12.14.

The larger the value of input voltage, the greater the slope. Since time T_1 is a constant, the voltage output of the integrator can vary depending on the value of input voltage. At the end of time T_1, the logic circuits change the switch to position B and start the clock. The clock is started because at this time the comparator signal out is still positive. The integrator will start to discharge during time 2. The greater the starting voltage (as determined by the input voltage) the longer it will take to discharge to 0 volts, as shown in figure 12.14. When the zero level is reached, the comparator inhibits the clock (through the logic circuits) and the count is stored. The binary count, as was the case in the ramp-type converter, may now be decoded and displayed to provide a digital readout for the input analog voltage.

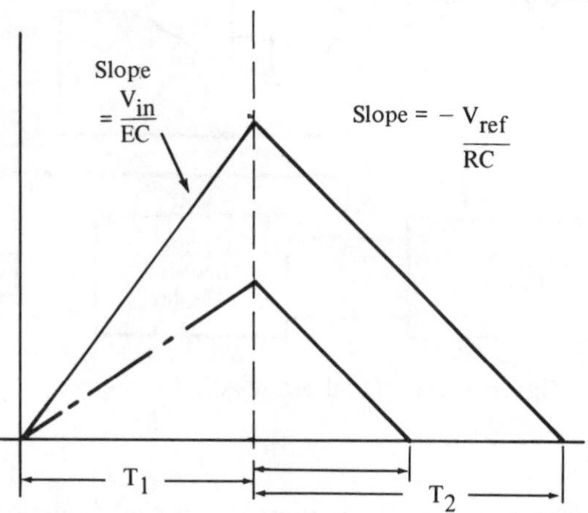

Figure 12.14 Voltages Produced by Dual-Slope Conversion

Calibration of the Dual-Slope Converter

Calibration of the dual-slope converter is very similar to the calibration of the ramp-type converter. Each one required a very linear ramp voltage to be generated, and each required a very stable clock. The dual-slope requires some extra circuitry in the logic circuits block. There is also the requirement for an accurate time base for time T_1. This is probably going to be a monostable. There are, however, some dual-slope converters that use an accurate counter for this time base and check the most-significant bit. This bit is used to switch this switch and to restart the counter. (Actually, all of the other bits in the counter have reset.)

12.9 THE SUCCESSIVE-APPROXIMATION A/D CONVERTER

Figure 12.15 shows another type of A/D converter, the successive-approximation converter. One important feature of this type of converter is that it takes a set number of clock pulses (or periods), n, to produce an n-bit output, regardless of the magnitude of the input voltage.

In order to understand the operation of the system, let's assume that the input voltage can range from 0 to 15 volts. Thus, the output of the feedback D/A converter will range from 0 to 15 volts in 1-volt steps, as shown in figure 12.15. We shall apply 11 volts to the input and go through the sequence of operations.

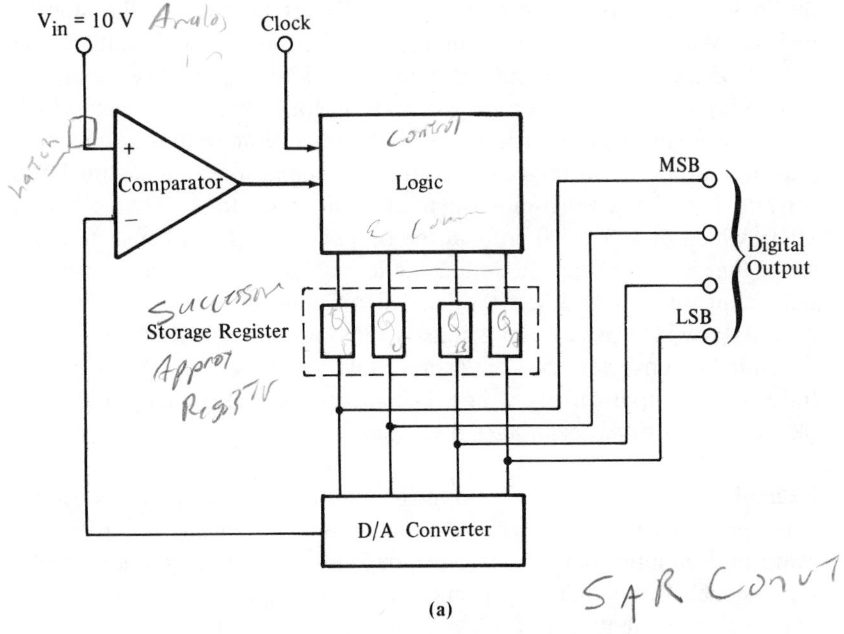

(a)

Digital Output				D/A Converter Ouput
0	0	0	0	0
0	0	0	1	1
0	0	1	0	2
0	0	1	1	3
0	1	0	0	4
0	1	0	1	5
0	1	1	0	6
0	1	1	1	7
1	0	0	0	8
1	0	0	1	9
1	0	1	0	10
1	0	1	1	11
1	1	0	0	12
1	1	0	1	13
1	1	1	0	14
1	1	1	1	15

(b)

Figure 12.15 Successive-Approximation A/D Converter

The sequence starts with the most-significant bit in the register being set to 1 by the logic block so that the digital output is 1000. From figure 12.15 we see that the output of the D/A converter is 8 volts. Since that is *less* than the analog input, the output of the comparator is high. The logic block interprets the high comparator output to mean that the digital output is too small; therefore, on the next clock pulse, the logic block sets the second-most-significant bit to 1 in the storage register. We now have the output 1100, which produces 12 volts at the output of the D/A converter. Because that 12 volts is larger than the analog input signal, the comparator output goes low. At the next clock pulse the logic block interprets the low comparator output as meaning that the output is too large; therefore, the second-most-significant bit is *reset to 0* and the third-most-significant bit is set to 1. That gives us 1010, which produces 10 volts at the output of the D/A converter. Because that is lower than the input, the logic block leaves the third-most-significant bit at 1 and sets the least-significant bit to 1. The output is now 1011, which produces 11 volts at the output of the D/A. Thus, for the four-bit converter, the conversion took only four clock periods. The following example will show how fast successive approximation is compared with the other techniques discussed.

Example An n-bit successive-approximation converter requires only n clock periods for its sequence of operations. The time required by the ramp and counter ramp converters depends on the magnitude of the input signal. The worst case occurs when the input signal is at its full-scale value. In that case, the ramp and counter ramp converters have to count from 0 to full scale. For an n-bit converter, full scale is $2^n - 1$. For a 10-bit output, compare the worst case conversion times of the successive approximation and counter ramp converters. Assume the clock frequency is 1 millihertz or 1 microsecond per pulse.

Solution The 10-bit successive approximation converter requires 10 clock periods. Because each clock period is 1 microsecond, the successive approximation converter takes 10 microseconds. The counter ramp converter must count up to $2^n - 1 = 1023$ for full scale. Hence, it takes 1023 clock periods or 1.023 millisecond. The successive approximation converter is, therefore, more than 100 times faster than the counter ramp converter for the worst-case condition.

EXERCISE

1. Construct a circuit that converts a digital signal to an analog signal.

Many devices within a digital system require an analog signal for their operation. The digital number or word must be converted to an equivalent voltage. There are several methods for converting a digital signal to an analog signal. Perhaps the simplest is the weighted resistor ladder method. The basic network shown in figure 12.16 consists of a decade counter with weighted resistors on the outputs and an operational amplifier used as a summing amp.

You will recall from our discussion of op-amps that the weight of the input voltage depends on the ratio of the input resistors (R_A, R_B, R_C, and R_D) and the feedback resistor (R_F). If the inputs are to represent the binary numbers 1, 2, 4, and 8 then the input resistors must be weighted accordingly. Since the formula for determining the output voltage of a summing amp is given by:

$$e_o = \frac{R_F}{R_{in}} e_{in}$$

The input resistor for a binary 1 (R_A) must be eight times as large as the input resistor for a binary 8 (R_D). R_A must be four times as large as R_C and twice as large as R_B. The increment voltage is the amount that the analog output voltage changes for one input pulse. For instance, if the increment voltage is designed to be 1 volt, then after three pulses the output voltage would be 3 volts, and so on. For most applications this increment must be within 10 percent or less for each step. An example is given to further illustrate this method.

Example If the voltage for a 1 condition at the output of the decade counter is 4 volts (and is the same at all outputs), and $R_A = 8$ kΩ, $R_B = 4$ kΩ, $R_C = 2$ kΩ, and $R_D = 1$ kΩ then what size should R_F be to obtain an increment voltage of 0.1 volt?

$$e_o = \frac{R_F}{R_A} e_{in} \qquad \text{volts} = \frac{\text{kilohms}}{\text{kilohms}} \times \text{volts}$$

$$0.1 = \frac{R_F}{8} 4$$

$$R_F = \frac{0.1 \times 4}{8}$$

$$= \frac{0.4}{8}$$

$$= 0.05 \text{ k}\Omega$$

$$= 50$$

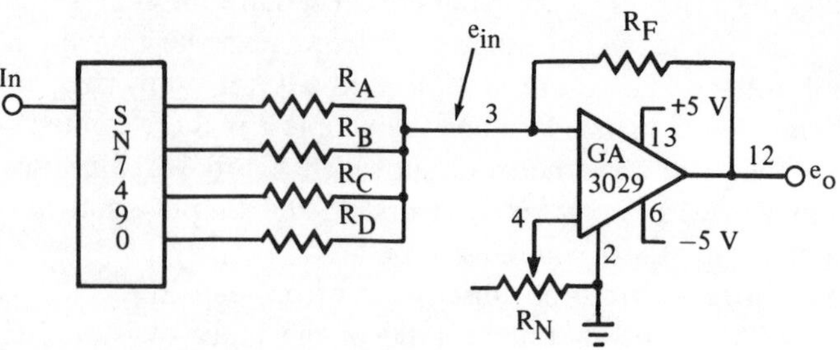

Figure 12.16 Digital-to-Analog Converter

The null resistor (R_N) is necessary to zero the output at the count of zero. It must be adjusted whenever any of the other resistor values are changed.

Using the circuit given in figure 12.16, construct a D/A converter using the values given below.

$R_A = 800\ k\Omega$
$R_B = 400\ k\Omega$
$R_C = 200\ k\Omega$
$R_D = 100\ k\Omega$
$R_N = 1\ M\Omega$

Determine e_{in} by direct measurement. What value should R_F be to obtain an increment voltage of 0.2 volt? Fill in table 12.3.

TABLE 12.3

Count	e_o	*Increment voltage**
0		
1		
2		
3		
4		
5		
6		
7		
8		
9		

* Increment voltage = $e_0 - e_1$, $e_1 - e_2$, etc.

Using the same input resistors what value must R_F have to obtain an increment voltage of 0.1 volt? Fill in table 12.4.

TABLE 12.4

Count	e_o	*Increment voltage*
0		
1		
2		
3		
4		
5		
6		
7		
8		
9		

QUESTIONS AND PROBLEMS

12.1 Define what is meant by analog and digital signals.

12.2 For figure 12.3, if we wish to have a maximum output voltage of 12.0 volts, calculate the values of R_F and 2R when R = 2.0 kiloohms.

12.3 For figure 12.3, what would be the maximum output voltage if R = 4.7 kiloohms and R_F = 10.0 kiloohms?

12.4 Using problem 12.2 and table 12.1, determine the output voltage if the binary input were a 4 (0100).

12.5 Explain the difference between the terms accuracy and resolution.

12.6 Which of these two is the most important when constructing a D/A circuit? Why?

12.7 Rework example 12.4 if the R (photoresistive device) has an initial value of 1.3 kiloohms and varies 300 Ω per footcandle.

12.8 Design a circuit that will give a digital readout of light intensity in footcandles using the circuit of problem 12.7 and appropriate counters.

12.9 Explain the operation of a ramp-type A/D converter.

12.10 How does the dual-slope converter differ from the ramp-type A/D?

12.11 Which of these two would be the more accurate? Why?

12.12 Using figure 12.13 as a starting point, design a dual-slope A/D.

The logic circuit block must be able to control the switch. Use T_1 equal to 1.0 second. Write a circuit description and operation for your design.

12.13 Using figure 12.15 as a starting point, design a successive-approximation A/D. Use any type D/A you wish. Include a seven-segment display that will not flicker.

12.14 Of the various types of A/D circuits shown, which is the fastest?

12.15 How would the op-amp comparator be classified, A/D or D/A?

12.16 Define linearity.

12.17 Using figure 12.4, explain how the human body (and mind) are similar in operation to this system.

12.18 Explain the advantages and disadvantages when comparing the circuits of figures 12.1 and 12.3.

Section IV

13

Microcomputers

OBJECTIVE: To present the organization, design considerations, and the applications of a microcomputer.

Introduction: The advances of large-scale integration technology have made it possible to place an ever-increasing amount of circuitry on a single integrator circuit chip. One of these circuits, the complete central processing unit (CPU) of a digital computer has opened up the new world of microprocessors and microcomputers. The microprocessor is a part of the complete system of the microcomputer. A microcomputer then is a system, a small (micro) computing system. It contains all of the characteristics of a computing system. There are the input-output (I/O) devices, memory devices, control devices and the ALU (arithmetic-logic-unit). Figure 13.1 shows a block diagram of the typical digital computer. Figure 13.1 is shown for reference and comparison of the advances made with the advent of the new large-scale integration technology. Figure 13.2 shows in block diagram form the basic structure of a microprocessor. This is a single chip as opposed to the system of figure 13.1 where several chips make up the system. Here on the single chip we see the timing and control sections, the arithmetic unit, and some memory storage (the accumulator). This microprocessor chip has also the capability to decode instructions fed to it from an external source. It is this feature that distinguishes one microprocessor chip from another. Although each has the ability to decode instructions, each has a different set of instructions from which it operates. This brings us to another distinguishing feature of the microcomputer. It is usually a special-purpose system. The microcomputer is generally not thought of

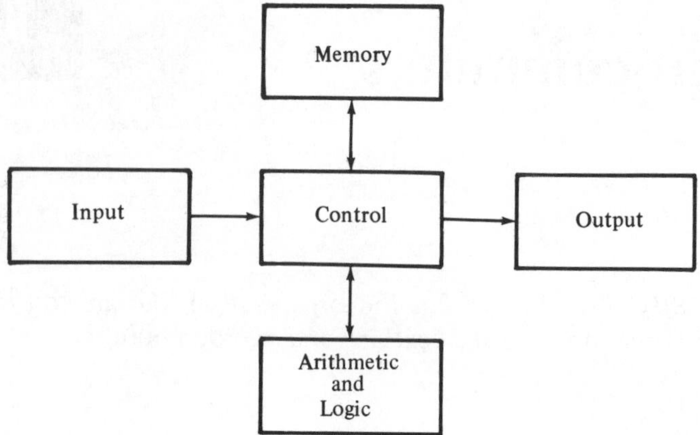

Figure 13.1 Typical Digital Computer

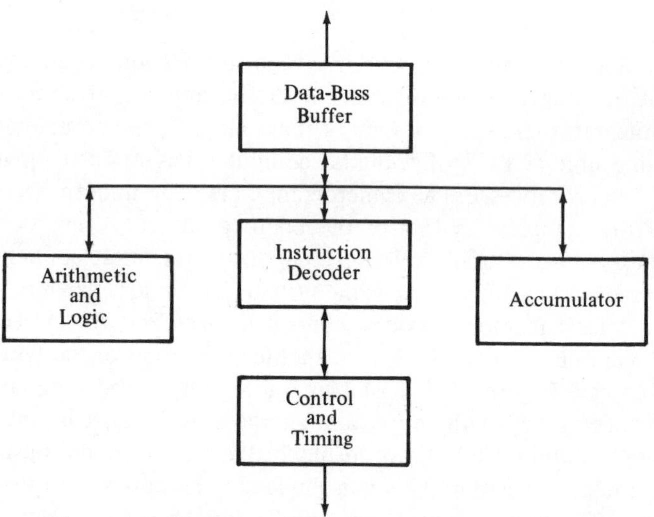

Figure 13.2 Basic Microprocessor

as a problem-solving system as is its big brother the general-purpose computer. For example, a microcomputer may function as a process controller for temperature control in a home or office or it may find its way under the hood of your automobile as a part of the ignition and carburetor system. This is where the instruction set becomes important. A designer can develop a microprocessor with a very specialized instruction set since it will only need to perform very specialized functions.

Microprocessors (and microcomputers) are now very much a part of our life. An understanding of their functions, capabilities, and limitations is a must for the electronic technician. The remainder of this chapter will present an overview of these devices and systems.

13.1 MICROCOMPUTER ARCHITECTURE

Architecture is the term used to signify the manner in which the microprocessor (or CPU) chip is configured internally and also how this chip communicates (is connected) to the external chips that make up the microcomputer system. The first chip we will look at is the CPU chip itself. Figure 13.3 shows a simplified block diagram of a popular CPU (microprocessor) chip. Notice that this chip has 40 external pins or connections. These will be discussed later. For now, we will concentrate on the internal blocks. The CPU shown contains two accumulators (some CPU chips have only one accumulator). These accumulators are merely temporary storage registers. The accumulators are labeled A_1 and A_2. The accumulators are used mainly to hold data going to and from the internal arithmetic logic unit (ALU). The index register (IR) is used mainly in a limited capacity, to store a particular memory address for use in program control of the microcomputer. The program counter (PC) is another register. Its function is to hold the address of the next byte of data to be taken from memory. (For our system, a byte is eight bits of data.) The stack pointer (SP) is a two-byte register that is used to store the first address (in memory) of the seven memory locations (addresses) into which the contents of the other five CPU registers would be stored if the CPU registers were needed to perform some function other than the present program on which it is operating. Figure 13.4 shows the sequence in which these registers would be moved to the "stack" and back again. Remember the stack is merely a set of memory locations used to temporarily hold the contents of the CPU registers. These locations are found in the external memory devices of the microcomputer system.

An example of how this might work is given here. Suppose the microcomputer is doing an arithmetic problem and has data stored in accumulator A and B. The program counter is set to get the next set of data from a location in memory. The microcomputer receives a signal telling it that it has data waiting that must be processed immediately.

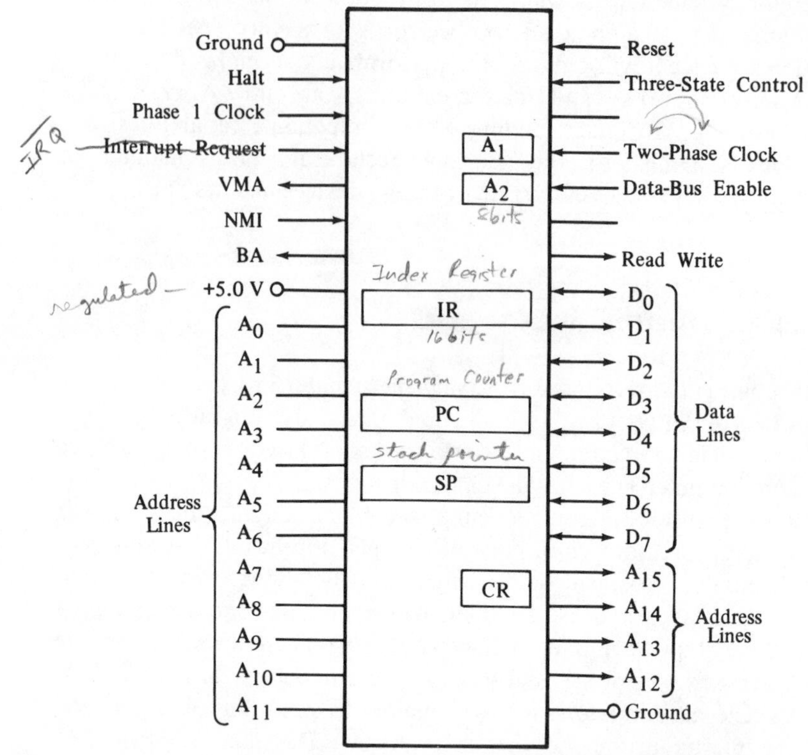

Figure 13.3 CPU Simplified Block Diagram

Operating Sequence of the Stack Poiner

Into Stack	Stack Pointer Address	MPU Register Data Saved In Stack	From Stack
1	Starting Address	Lower byte of Program Counter	7
2	Address – 1	Higher byte of Program Counter	6
3	Address – 2	Lower byte of Index Register	5
4	Address – 3	Higher byte of Idex Register	4
5	Address – 4	Contents of Accumulator A	3
6	Address – 5	Contents of Accumulator B	2
7	Address – 6	Contents of Condition Code Register	1

Figure 13.4 Operating Sequence of the Stack Pointer

At this point, the microcomputer "dumps" its registers to the stack. After that it processes the external data it needs to finish its arithmetic problem. So it returns to the stack and reloads its registers and continues the problem. As you can see, the use of the stack is very convenient. It enables the microcomputer to interrupt slow jobs while only having one ALU on board. The last register we will look at is the CR or condition code register. After the execution of an instruction, the microcomputer checks this register. The eight bits contained here are codes that act as condition instructions to the microcomputer. More on this register later. Figure 13.5 shows the data contained by this register and a description of what each bit of data means to the microcomputer.

It is the job of the CPU chip to perform the arithmetic of the microcomputer and to handle all of the logic functions required by the microcomputer. The CPU chip communicates with the other chips of the microcomputer system primarily by way of the sixteen address lines (A0–A15) and the eight data lines (D0–D7) shown in figure 13.3. Notice that the data lines are shown as two-way lines, whereas the address lines are one-way lines. This tells us that the CPU chip can accept data or transmit data that it has temporarily stored. The address lines are the means by which the CPU chip locates or addresses external memory locations.

External Memory

Figure 13.6 shows a typical bus interface between external memory chips and the CPU. Figure 13.6 shows three blocks of memory, each consisting of 256 words. By our definition, a word for our microcomputer is one byte made up of eight binary bits. Therefore, each of the memory blocks is actually a 2048-bit memory device. The arrangement of the memory device is of course 256 eight-bit words. A review of the section on memory might be helpful at this point if you have forgotten the basics of memory devices. These devices as used with the microcomputer system are solid-state and not magnetic memory. Each of the three memory blocks is tied to the CPU via three bus lines, the data line, the address line, and the control line. Each of the words of memory is given a unique address (0000–FFFF in hexademical code). For our system, block 1 could contain 0000–00FF, or 256 locations. Block 2 would then contain 0100–01FF, still 256 locations, and block 3, 0200–02FF. Each of these blocks is sometimes called a *page of memory*. Our system has three pages of memory, page 0 (0000–00FF), page 1 (0100–01FF) and page 2 (0200–02FF). By using the sixteen address lines, it is possible to address over 65,000 memory locations, each having eight bits of data. [$2^{16} = 65{,}536$]

Bit #	Bit Name	Description
0	C bit Carry/Borrow	During addition of two binary numbers this bit gets set (C = logic 1) to indicate that a carry has occurred (the binary result exceeds an 8-bit magnitude). If this bit is a logic 0, the carry bit is considered reset and no carry has occurred. Carry/Borrow Example: A = 1000 0000 B = 1000 0000 A + B = 1 0000 0000
1	V bit (Overflow)	This bit is set (V = logic 1) whenever two's complement overflow results from an arithmetic operation and is reset (V + logic 0) if two's complement overflow does not occur. (Two's complement arithmetic is used by the MPU to perform binary subtraction. Thus, if a two's complement overflow occurs, the result will be a negative number.)
2	Z bit (Zero result)	This bit is set (Z = logic 1) if the result of an arithmetic operation is zero and is anything but zero.
3	N bit (Negative)	This bit is set (N = logic 1) if bit 7 of an arithmetic operation is set (equal to 1). This indicates that the result of an arithmetic operation is negative. This bit is reset (N = logic 0) if bit 7 is equal to 0.
4	I bit (interrupt Mask)	If this bit is set (I = logic 1), the MPU cannot respond to an interrupt requested by any external device.
5	H bit (Half-Carry)	This bit is set (H = logic 1) if, during the execuition of any instruction, a carry from bit 3 to bit 4 occurs. This bit is reset (H = logic 0) if no carry occurs during execution of any instruction. Half-Carry Example: A = 0000 1000 B = 0000 1000 A + B = 0001 0000

Figure 13.5 Condition-Code Register

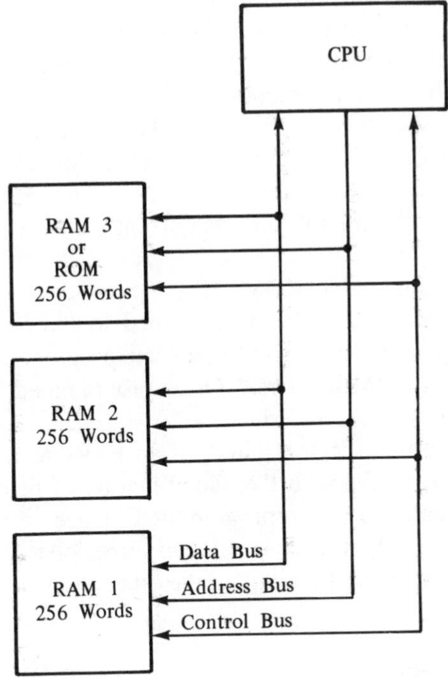

Figure 13.6 Memory Connections to CPU

The data bus is the method by which data are either sent to memory or called from memory. In this system, all data must pass through the CPU even if they are just being moved from one memory location to another. The control bus carries the signal to memory telling it whether data are being sent (written to memory) or retrieved (read from memory) at a given time. The control bus works in conjunction with the read/write line (R/W) from the CPU.

The three types of memory shown need not all be present in any particular microcomputer. They are shown here only as examples of what might be used. The types of memory shown are:

RAM, random-access memory. These devices may be written to read from in random fashion. Any of the memory locations may be addressed and their data changed. The data may be changed by the programmer or by the program itself. As an example, the programmer may store a bit of data in a memory location while writing the program. When the program begins to run, the data at that location may be changed by the microcomputer.

ROM, read-only memory. This is similar to the RAM with the exception that it is not possible for the programmer or the microcomputer to write into the memory. This chip is "programmed" by the manufacturer with either instructions or data (such as a subroutine or a math table) as specified by the user. The information stored in the ROM chip is nonvolatile, meaning it will not be lost even if there is no power applied to the system. This is not true of the RAM chip, which has a volatile memory and will lose its current status when power is removed from the device.

PROM, programmable ROM. It is almost identical to the ROM except that it is intended to be programmed in the field by the user. These chips are purchased "blank" and then programmed via PROM-programmers or "PROM-burners." Once programmed, the PROM is identical in operation to the ROM.

EPROM, erasable PROM. Think of an EPROM as having all the characteristics of a PROM with the added feature of being erasable and therefore being able to be reprogrammed. Once programmed, the EPROM can be erased by exposure to an ultraviolet (UV) light. The EPROM is more expensive than any of the other types of memory.

13.2 HOW IT WORKS

Figure 13.7 shows the CPU chip connected to memory. Notice that the data lines are two-way lines, while the address lines are one-way lines. The CPU, by way of its address lines, will address one of the memory locations of the computer. The CPU will also send a read/write (R/W) signal to the memory unit telling it whether it wishes to read from that address or write to that address. If it is a write, then data are transferred from the CPU to memory. On the other hand, a read signal will take information or data from memory and place them in the CPU. All

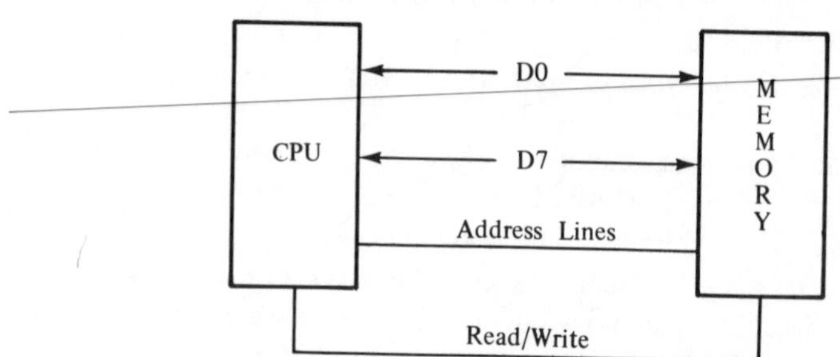

Figure 13.7 Simplified CPU-Memory Interface

of this data is transferred via the accumulator register of the CPU. That is to say, the data leaving the CPU are sent from the accumulator to a memory location and data received are sent from memory to the accumulator.

The Instruction Set

ABC	Add Accumulators
ADC	Add With Carry
ADD	Add Without Carry
AND	Logical AND
ASL	Arithmetic Shift Left
ASR	Arithmetic Shift Right
BCC	Branch if Carry Clear
BCS	Branch if Carry Set
BEQ	Branch if Equal to Zero
BGE	Branch if Greater or Equal to Zero
BGT	Branch if Greater than Zero
BHI	Branch if Higher
BIT	Bit Test
BLE	Branch if Less or Equal to Zero
BLS	Branch if Lower or Same
BLT	Branch if Less than Zero
BMI	Branch if Minus
BNE	Branch if Not Equal to Zero
BPL	Branch if Plus
BRA	Branch Always
BSR	Branch to Subroutine
BVC	Branch if overflow Clear
BVS	Branch if Overflow Set
CBA	Compare Accumulators
CLC	Clear Carry
CLI	Clear Interrupt Mask
CLR	Clear
CLV	Clear Overflow
CMP	Compare
COM	Complement
CPX	Compare Index Regiter
DAA	Decimal Adjust Accumulator A
DEC	Decrement
DES	Decrement Stack Pointer
DEX	Decrement Index Regiter
EOR	Exclusive OR
INC	Increment
INS	Increment Stack Pointer
INX	Increment Index Register
JMP J	Jump
JSR	Jump to Subroutine
LDA	Load Accumulator
LDS	Load Stack Pointer
LDX	Load Index Register
LSR	Logical Shift Right
NEG	Negate
NOP	No Operation
ORA	Inclusive OR
PSH	Push Data
PUL	Pull Data
ROL	Rotate Left
ROR	Rotate Right
RTI	Return from Interrupt
RTS	Return from Subroutine
SBA	Subtract Accumulators
SBC	Subtract With Carry
SEC	Set Carry
SEI	Set Interrupt Mask
SEV	Set Overflow
STA	Store Accumulator
STS	Store Stack Pointer
STX	Store Index Register
SUB	Subtract
SWI	Software Interrupt
TAB	Transfer Accumulators
TAP	Transfer Accumulators to Condition Code Reg
TBA	Transfer Accumulators
TPA	Transfer Condition Code Reg to Accumulator A
TST	Test
TSX	Transfer Stack Pointer to Index Regiter
TXS	Transfer Index Register to Stack Pointer
WAI	Wait for Interrupt

Figure 13.8 Executable Instructions

Each CPU or microprocessor has only one set of instructions that it understands. This means that there is one set of binary or hexadecimal codes that the microprocessor can decode and process. Figure 13.8 is a table of such an instruction set. Each of the instructions in the instruction set also has a two-bit hexadecimal code, for example:

Instruction	Description	Code
LDA	Load Accumulator	86
ADD	Add to Accumulator	8B
STA	Store Accumulator	97

This is just a partial list of the instruction set, but will be used for an illustration of how the computer operates.

Example Suppose we wish to write a program that will add two numbers and store the results in a memory location. The numbers 03 and 04 will be added. To write this program we need to do the following:

1. Select the memory locations where our program will be stored.
2. Write the program in language we can understand (English).
3. Rewrite the program using the instruction set of our microprocessor.
4. Rewrite the program in the microprocessor language (machine language).

I. We will start our program in location 0001 (hexadecimal).

II. We will write the program.

A. Load the number 03 into the accumulator.
B. Add the number 04 to the number in the accumlator.
C. Store the answer from the accumulator in a memory location.

III. This can be rewritten as:

LDAA 03	Load 03
ADD 04	Add 04 to 03
STA 00	Store answer in location 00

IV. Using the microcomputer machine language, this is rewritten as:

Memory Location	HEX Machine Language	
0001	86	Load–data that follows
0002	03	Data
0003	8B	Add data that follows
0004	04	Data
0005	97	Store accumulator in location
0006	00	Location for answer

To enter this program into the microcomputer, we would go to address 0001 and enter the hexadecimal 86 then to address 0002 and enter hexadecimal 03, and so on.

Once the program is entered, we return to address 0001 and press GO. The microprocessor will then read the contents of location 0001, the hexadecimal 86, and interpret it as a valid instruction. The 86 enters the CPU from memory via the data lines D0–D7. After it reads the 86, the CPU will wait for the next byte of memory and load it into the accumulator. The instruction 86 tells the microprocessor to load the next byte into the accumulator. After it has read the 86 from location 0001, it will address the next location (0002) via its address lines, A0–A15. Once this is done, it will address the next location (0003). Here it sees the instruction 8B. The microprocessor will continue to step from memory location to memory location in sequence performing the operations as instructed. As it steps from instruction to instruction, it will proceed to decode the instruction and perform the task assigned.

To understand how the microprocessor decodes its instructions, think back to the section on decoders. Here, a particular output was obtained from the circuit for a particular set of input data. The binary 3 (0011) was decoded by the 7445 and produced an output on the decimal 3 line. So it is with the microprocessor, each set of instructions is gated internally to produce a desired signal (output) somewhere in the systems.

13.3 INPUT/OUTPUT

To this point, we have discussed the microcomputer as a set of blocks with data coming and going with little thought as to how that data come or go. In reality, the microprocessor (or microcomputer) is of no real value unless it interfaces with the "real world." This is where the I/O (Input/Output) sections come into play.

Interfacing the I/O Devices

As shown in figure 13.9, the I/O device (or devices) are connected to the CPU in a manner similar to that of the memory devices. One other bit of information, the I/O device must also connect to the "outside world." This means that in addition to transmitting data or instructions to and from the CPU, it must shield the CPU (and the rest of the microcomputer) from the harsh noises (electrical), electrical surges, and other types of electrical interference that might show up on the lines from the outside world.

To better illustrate how one type of interface device works, the Motorola MC6820PIA is shown in figure 13.10. Figure 13.10(a) shows

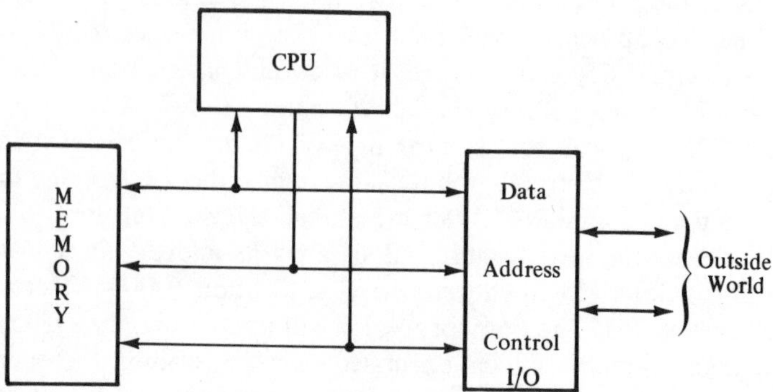

Figure 13.9 I/O Interface

a simple block diagram of the device. This device has 2 eight-bit data lines, PA0–PA7 and PB0–PB7, which connect to the real world, possibly an A/D device. One of these two sets of data lines may be selected and tied to the data lines going to the CPU. Figure 13.10(b) shows a more detailed diagram of the internal block structure of the MC6820PIA.

Pin Assignment and Functions of the MC6820PIA

Pins 2 through 17 are the connection for data transfer, either port A or port B. Pins 26 through 33 are the data transfer pins from the PIA to the CPU.

Pins 22, 23, and 24 are tied to the CPU address lines and are used to determine which (if more than one) PIA devices are being addressed by the CPU. Once a device has been selected, pin 25 is used to enable the device. Pin 34 is used to reset all PIA registers to logic 0. Pin 21 is used as the read/write line, with a logic 0 on the line enabling data to be transferred from the CPU to the PIA. A logic 1, then, will be a read signal enabling the CPU to read data from the PIA lines 26–33.

Before completing the pin assignment, we need to discuss the two control registers (CRA, CRB) housed in the PIA. Before anything else is done to the PIA, information is written to one of these registers. Suppose that we wish to use the A Port (PA0–PA7). First we need to know if we are going to gather data from the outside world or send data out of the microcomputer. To do this, we will write our instruction to the control register (CRA) and to the data direction register (DDRA). Pins 35 and 36 are used in conjunction with bit 2 of the CRA. When all of these are logic 0, the data (pins 26–33) are sent to the data direction register. This determines whether pins 2–9 are to be used as input or as

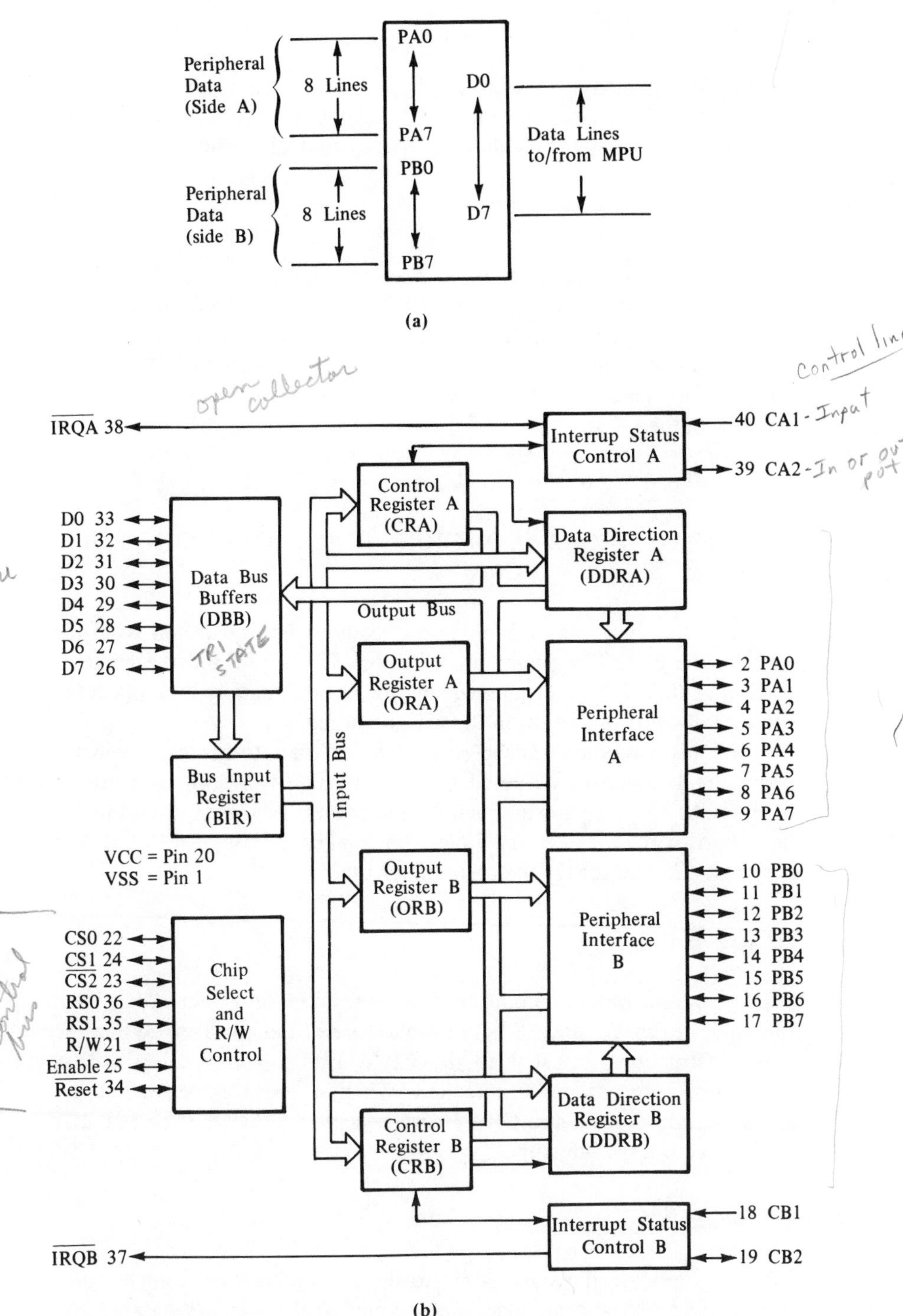

Figure 13.10 **(a)** PIA Block Diagram. **(b)** PIA Internal Structure Block Diagram

output pins. If pins 35 and 36 are logic 0 and bit 2 of the CRA is set to a logic 1, then data are sent to the peripheral data register A. If pin 35 is 0 and pin 36 is 1, then data are sent to control register A.

PIA Access

To use the PIA device properly, let's review what has taken place. First, data are set on the CPU data lines (being transferred from the accumulator). The address of the PIA device is related and the data are ready for transfer. The data are then transferred to the correct PIA, via the address lines (including CS1, CS0, and $\overline{\text{CS2}}$). These eight bits of data are now in the CRA.

7	6	5 4 3	2	1	0
1RQA1	1RQA2	CA2 CONTROL	DDRA	CA1	CONTROL

CRA

A new set of eight bits of data is now ready to be transferred from the CPU to the PIA. The combination of pins 35 and 36 and the contents of bit 2 of the CRA determine whether or not this data will be placed in the data director register as in the peripheral register.

In this way, data can be entered to the PIA to go to the outside world, or data can be accepted from the outside world all by software instructions. A software instruction, remember, is a user program and not a hard-wired set of connections. This makes the use of a PIA device very cost effective for the microcomputer user.

PIA Interrupts

Suppose we have data coming from the outside world and do not wish to accept it. Pins 37 and 38 help us take care of this problem. When the initial instructions are sent to the CRA, Bits 6 and 7 tell the PIA whether or not to accept an interrupt. When an interrupt request comes to the microprocessor the CRA, it is tested to see whether or not it is ready to accept the interrupt.

I/O Interrupts

I/O interrupts are the signals sent to the microcomputer from the outside world to let the microcomputer know that there are data at the ports that need to be processed. It is up to the programmer of the microcomputer to decide when the computer will accept this data. If

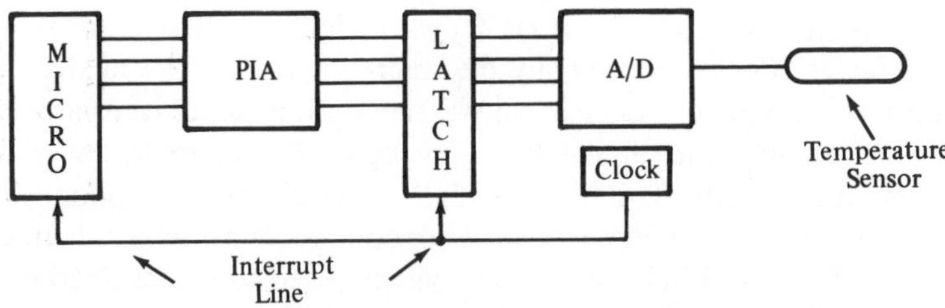

Figure 13.11 Microcomputer Interface

for example, the microcomputer is processing a temperature-control program as shown in figure 13.11, we would want to make sure that the data are ready to be processed before they are sent to the microcomputer. The simple system of figure 13.11 demonstrates how data can be processed accurately by the microprocessor. In this example, the temperature, an analog signal, is sensed and sent to the A/D converter. Here the analog signal is converted to a digital signal. The digital output of the A/D is constantly changing. The clock and the latch are used to hold the data steady at the input of the PIA. Every so many clock pulses, the data is latched. This latching pulse also is sent to the microcomputer interrupt line. This assures us that the interrupt, or request for service, will not be triggered until the data are ready to be processed.

If the microcomputer were busy doing some other task, and did not wish to be interrupted, a signal would be sent to the CRA telling it not to accept the interrupt request from the clock and not to accept the data from the A/D converter. An interrupt request is just that, a request to interrupt the normal or ongoing operation of the microcomputer. This interrupt request may or may not be acknowledged.

Another example of an interrupt request at work is seen when the microcomputer is connected to a device such as a keyboard. Here, the microcomputer must accept the data from the keyboard, enter it into the proper registers, decode it, and process it. While this is going on, it will not wish to accept another bit of data (key depression) so it uses the interrupt lines. When it is ready to accept data it signals OK on the interrupt and while it is processing the last key depression, it signals wait or do not accept data.

Serial I/O Devices

If we wish to use telephone lines (or microwave links) the data must be sent to the microcomputer in serial fashion. That is to say only one data line can be used. The same is true if we wish to store and retrieve from cassette tape. Figure 13.12 shows how serial data are handled by the microcomputer. Figure 13.12(a) shows a device called a *modem,* a serial interface device, and finally the microcomputer. The modem is a modulator-demodulator device whose function it is to translate telephone-line signals into digital logic levels, or digital logic levels into telephone-line signals. The modem must be used when telephone lines or microwave transmission are used to communicate with the microcomputer. The serial interface device will be used to convert serial signals to parallel binary logic signals, or to convert parallel binary logic signals to serial signals. More on this device later. The serial I/O device as shown in figure 13.12(b) can be used to replace the modem if telephone lines are not used. This device is either a cassette unit or a teletype unit and can connect directly to the serial interface device.

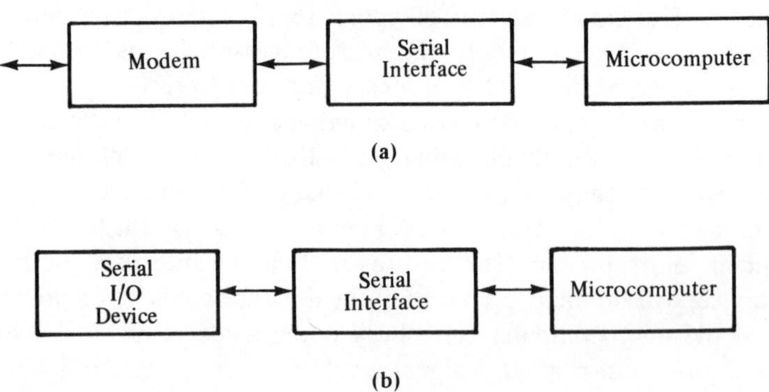

Figure 13.12 Microcomputer Serial Interface

Figure 13.13 is a pin block diagram of a typical serial interface device. Here we see eight pins used to interface with the data bus lines of the microcomputer. These are the eight data lines, D0–D7, of figure 13.13. The pin marked IOSEL is the I/O select and identifies an I/O operation of this device as being in progress. The IORW is the read/write line and its logic level tells us whether data are being read from the device (by the microcomputer) or are being written to the devices. This device will receive serial data (RD) and transfer them to parallel data ready for the microcomputer data bus. The device also uses the TD pin to transmit data it has received in parallel fashion from the microcomputer, in a serial manner.

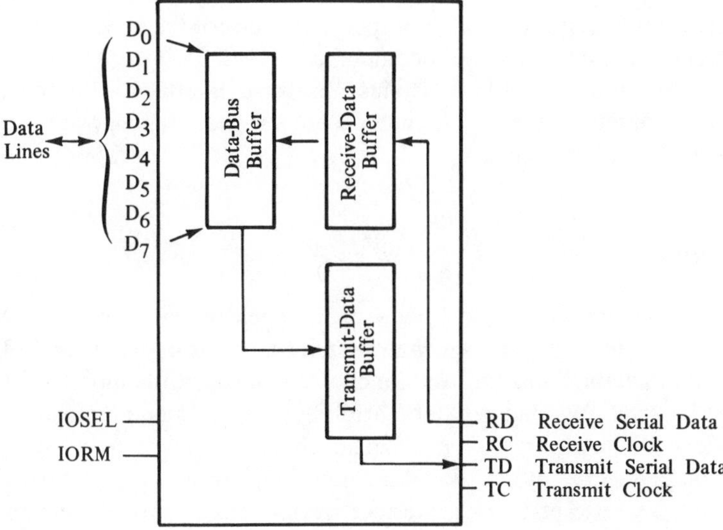

Figure 13.13 Serial I/O Interface Chip

13.4 I/O INTERFACE SUMMARY

Figure 13.14 is a block diagram of the typical I/O interface configuration of a microcomputer. Here we see the microcomputer being used as a controller element in a system. There are several analog-to-digital input sensors and several digital-to-analog outputs. The microcomputer receives data from both the analog-to-digital sensors and the operator's

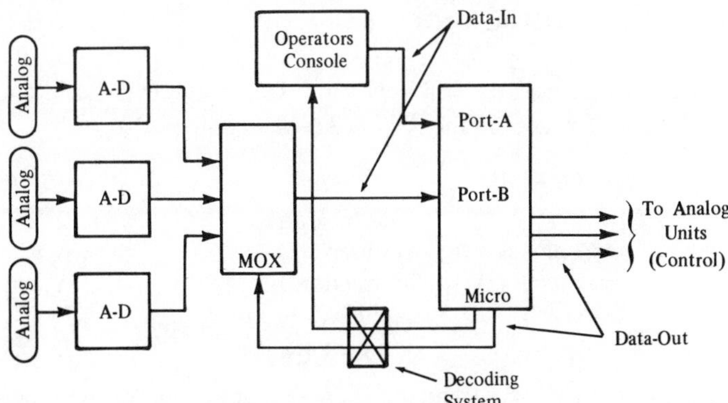

Figure 13.14 Pin-Out of SN74181 ALU

console. These data are then compared and manipulated to create the signals for the digital-to-analog outputs that control the process or function being handled by the microcomputer.

In the final analysis, it is this ability to interface with the real world that makes the microcomputer a very valuable and powerful tool in our technological society.

SUMMARY

This chapter presented an overview of the microprocessor-based microcomputer system. It was seen that the microprocessor was a single LS1 chip that contained many of the functions of a general computer but on a smaller scale. We had registers, arithmetic-logic units, and instruction decoders all incorporated on a single chip. The microprocessor chip was complemented by additional memory chips to give it a larger working area. The use of I/O or interface devices enables the system to effectively communicate with the real world. We saw how an instruction set is used by the microprocessor and how a program may be written by the user. The combination of the program and the instruction set enables the user to have the microcomputer follow a set of commands in a special sequence.

This chapter did not cover the entire topic of microprocessors and microcomputers, but rather attempted to give an overview of the topic.

EXERCISE

THE ARITHMETIC LOGIC UNIT

1. In this lab exercise, we will verify the following arithmetic and logic functions performed by the SN74181 ALU chip:

	Function-Operation	*Description*	*Logic/Arithmetic*
a.	F = A	Data transfer	L or A
b.	F = A plus B	Addition	A
c.	F = A minus B	Subtraction	A
d.	$F = A \oplus B$	Exclusive–or	L
e.	F = A	Complement	L

From the data book, it can be seen that many more arithmetic and logic functions can be performed by the 74181.

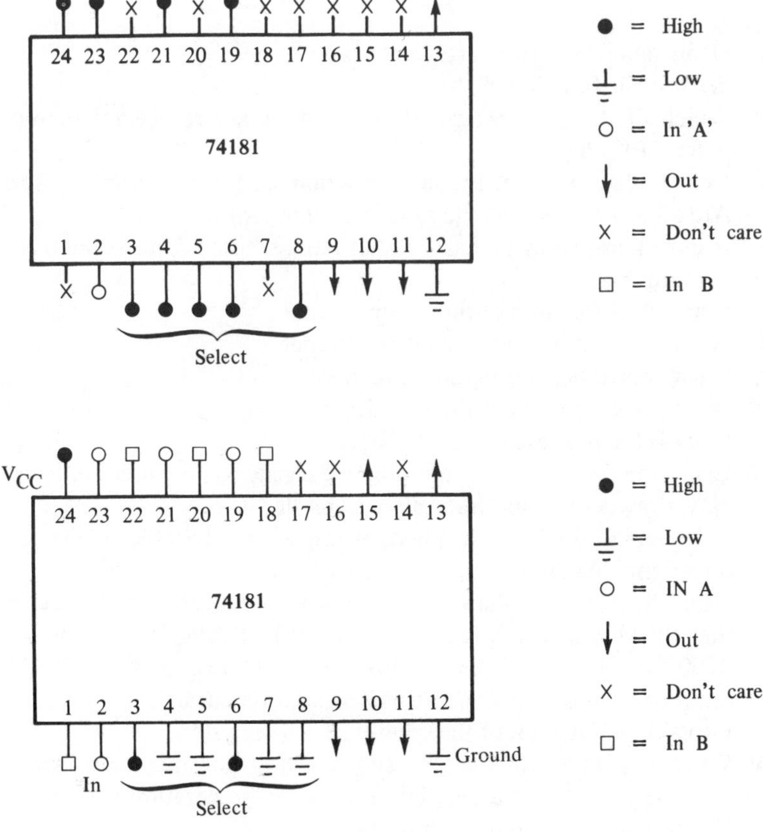

Figure 13.15 Connections of 74181 for Exercise

2. To perform the logic functions, the mode-control pin (pin 8) must be tied high (logic 1). To perform the arithmetic operation, pin 8 will be tied low (logic 0). Operation selection is done via the select lines (S0, S1, S2, S3) pins 6, 5, 4 and 3.
3. *Example 1—Data transfer*
   Figure 13.14 shows proper connection to handle the data transfer as a logic function.
   Example 2. F = A plus B (Addition)
   Figure 13.15 shows proper connection to perform the arithmetic operation of A plus B.
4. Verify each of the remaining function operations in step 1.

QUESTIONS AND PROBLEMS

13.1 Define microcomputer architecture.

13.2 What is the function of the stack?

13.3 Which digital devices are similar in function to the microcomputer CPU chip?

13.4 Define what is meant by an instruction set for a microprocessor.

13.5 Are all instruction sets the same for all microprocessors?

13.6 If our microprocessor has 77 instructions how might we increase that number?

13.7 What is the function of the accumulator?

13.8 What is meant by machine language code?

13.9 What constitutes a program sequence?

13.10 What is the function of the I/O devices?

13.11 What is the purpose of an interface?

13.12 Using the 74181 chip, and support registers, counters, and display, draw a diagram to simulate a simple CPU.

13.13 Using figure 13.6, show where it might be advisable to use bidirectional bus transceivers.

13.14 Using figure 13.6, show how you would convert the circuit so that RAM 1 is at address 0000 to 00FF, RAM 2 is at address 0100 to 01FF and RAM/ROM 3 is at address 0200 to 02FF. Show data lines from CPU as well as address lines.

13.15 Explain the function of the condition code register.

13.16 Write a program (in English) showing how you might use figure 13.11 to monitor an industrial process for temperature variations. Explain the function of each block.

13.17 Explain how the computer could be used to control the serial I/O chip of figure 13.13.

13.18 What is a modem?

13.19 What is meant by the terms hardware and software?

13.20 What is the major advantage of using a microcomputer in a control system over conventional digital control circuitry?

Appendix A
Additional Student Exercises

Section A.1 is an exercise for the student to use to put together a more complex system. The circuit has a twofold purpose; it uses 7400 logic building blocks to perform a function, and it demonstrates the concepts of the basic microprocessor chip.

Section A.2 contains another complex circuit built around a simple concept. The student will gain an understanding of control circuitry and see how to do more with available logic devices.

Note: These circuits have been added to the appendix not as an afterthought, but because their use is left to the discretion of the instructor. My thanks go to the following students whose assistance in developing these circuits is greatly appreciated: M. Broome, G. B. Wise, H. F. Robertson, C. L. Zachgo and J. M. English.

A.1 ADDITION PERFORMED BY 7400 CIRCUITS

Purpose

The purpose of this lab is to build a programmable circuit capable of adding two binary numbers. 7400 series circuits were used to perform this task.

Procedure

The circuit is built around the 7489 sixty-four-bit RAM memory. This memory stores 16 four-bit words. Both program commands and data were stored in this memory. The 7489 chip has four address pins, four data input pins, and four data output pins. The data received from the output pins have been complemented. The memory chip also has a read/write enable pin.

A 7493 four-bit counter is used to select address locations for the

RAM. The four-bit counter performs the same function as a program counter in a microprocessor chip. The counter is capable of counting to 15. This means the counter is capable of addressing all the locations in the 7489 RAM memory. The functional block diagram is shown in figure A.1. External wiring connecting output Q_a to input B needs to be performed to allow the chip to operate at the maximum count length.

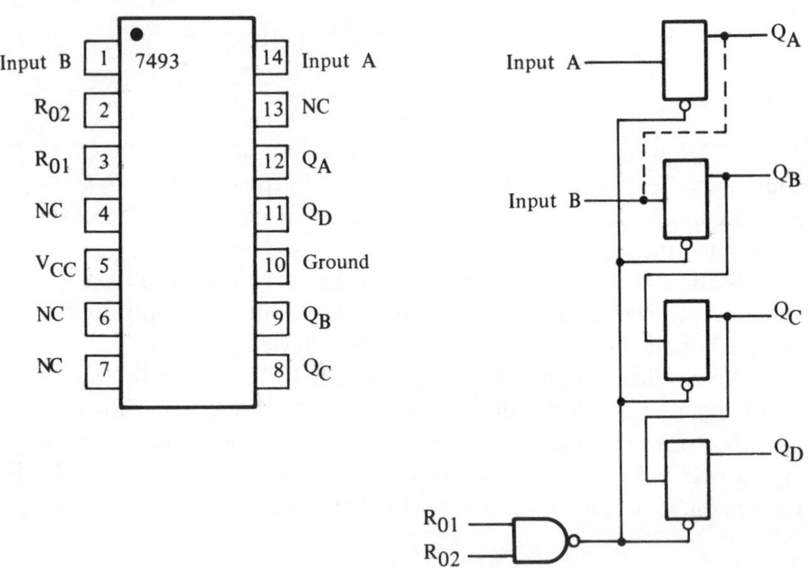

Figure A.1 Functional Block Diagram

The 7495 four-bit shift register was used as an input register in the circuit. This register performs the same function as the data bus buffer in a microprocessor chip. The shift register was operated in the parallel-in/parallel-out mode. A pin diagram for the 7495 chip is shown in figure A.2.

The last major part of the circuit is the 7483 four-bit binary full-adder. This adder is capable of adding two four-bit binary numbers. In this circuit the adder is used to add only two numbers of two-bit length. The pin diagram of the 7483 is shown in figure A.3.

In summary, this circuit performs some of the functions of a microprocessor. The 7493 counter performs the function of the program counter. Instruction decode logic and accumulator functions are performed by the 7489 RAM. Functions of the data bus buffer are performed by the 7495 shift register, and the 7483 binary adder performs as the ALU. A NAND and a NOR gate will act as control logic for the circuit.

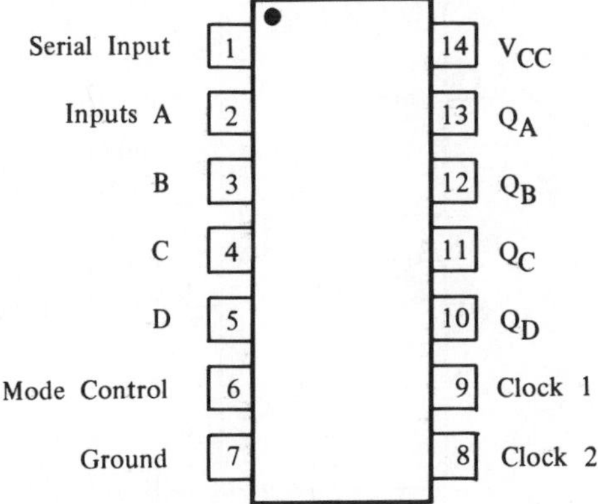

Figure A.2 Pin Diagram of the 7495

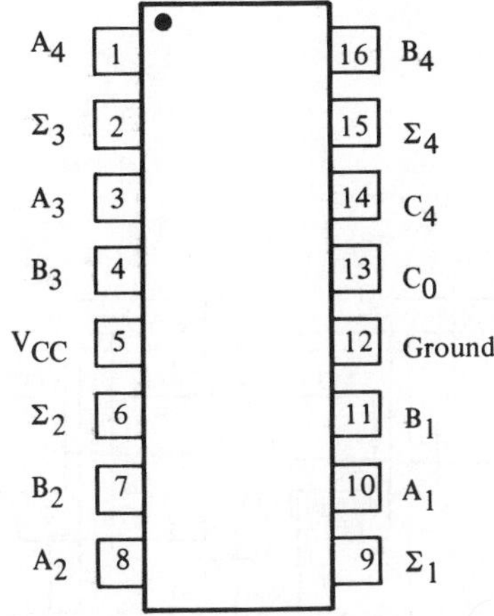

Figure A.3 Pin Diagram for the 7483

The block diagram and the pin diagram for the circuit is shown in figure A.4. These drawings are used to describe the operation of the circuit.

The RAM memory is disconnected from the other circuit components and the operating program is put into memory. This is done by grounding the $\overline{\text{WE}}$ pin of the 7489 chip and placing the complement of the program codes into memory by means of the data input pins. The reason for placing the complement of the program codes into memory

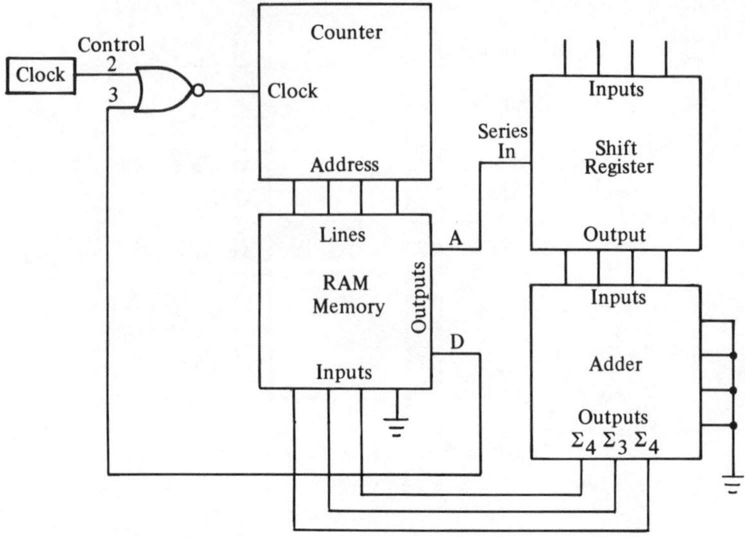

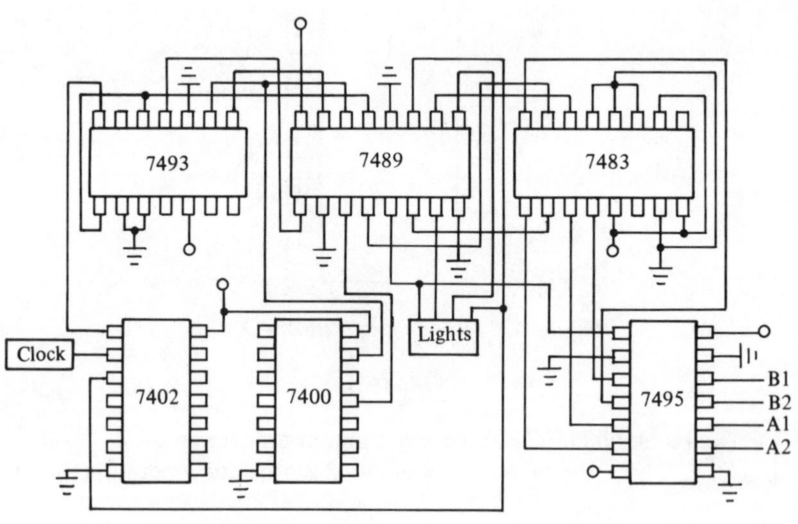

Figure A.4 Block Diagram and Pin Diagram

is because the data at the output pins are complemented. The program code is as follows:

Step	*Address*	*Output Needed*	*Input to the RAM needed to get the proper output*
1	0000	0001	1110
2	0001	0010	1101
3	0010	xxxx	xxxx
4	0011	xxxx	xxxx
5	0100	1000	0111
		MSB LSB	

The second address line connected to the RAM memory controls the read/write enable mode of the memory chip. A NAND gate acts as control logic for this function. One input to the NAND gate will be left high, while the second input will be connected to the second address line. The output of the NAND gate is connected to the $\overline{\text{WE}}$ pin of the RAM chip. A low at the $\overline{\text{WE}}$ pin will allow information to be written into the RAM memory and a high at this pin will allow information to be read from the chip. The RAM memory will normally be in the read mode. When the second address line goes high it forces the NAND gate to go low and the RAM memory switches to the write mode.

In the second step of the program, the LSB of the RAM's output goes from high to low. This gives the 7495 shift register a trailing-edge clock pulse and allows data to be shifted from the input switches to the output terminals of the chip. During the first two steps of operation the RAM memory is in the read mode.

Information from the shift register is transferred to the 7483 full-adder. The two least-significant bits of data from the shift register are added to the two most-significant bits of data from the shift register. The results of the addition and the carry can be seen at the 7483's outputs.

1010 From the shift register

10 } Are added by the 7483
10 }

100 The result at the output of the 7483

The results of the addition are transferred to the data inputs of the RAM memory.

When the third and fourth steps of the program occur the address line to the RAM forces the RAM into the write mode of operation. The result of the full-adder is then written into the memory of the RAM.

Control logic for the program counter is performed by a NOR gate. One input to the NOR gate is connected to the clock output and the other NOR input is connected to the MSB of output from the RAM.

During the first three steps of the program the MSB of output from the RAM is low and the program counter is controlled by the clock. In step five the MSB of output from the RAM is brought high. This causes the clock input to the program counter to remain low and also causes the program to stop after the fifth step.

Conclusion

In this lab exercise the student can see how different parts of microprocessor operate together. The exercise shows how each element works separately and then combine to perform special functions. The student must recognize that a processor chip is capable of more functions, but this lab demonstrates part of the microprocessor's capability.

A.2 ARITHMETIC CONTROLS

The circuits used in this experiment are composed of a ring counter, a serial adder, and the logic unit that controls the serial adder.

Purpose

The student should understand the logic functions of this control unit and be able to answer questions concerning its operation and be able to explain what functions it is capable of performing.

The student will be able to design and build a logic unit that will expand the functions presently performed by the serial adder.

Procedure

Part One

Three blocks compose the circuits used in this exercise. The control unit block is the circuit upon which most of this experiment is based.

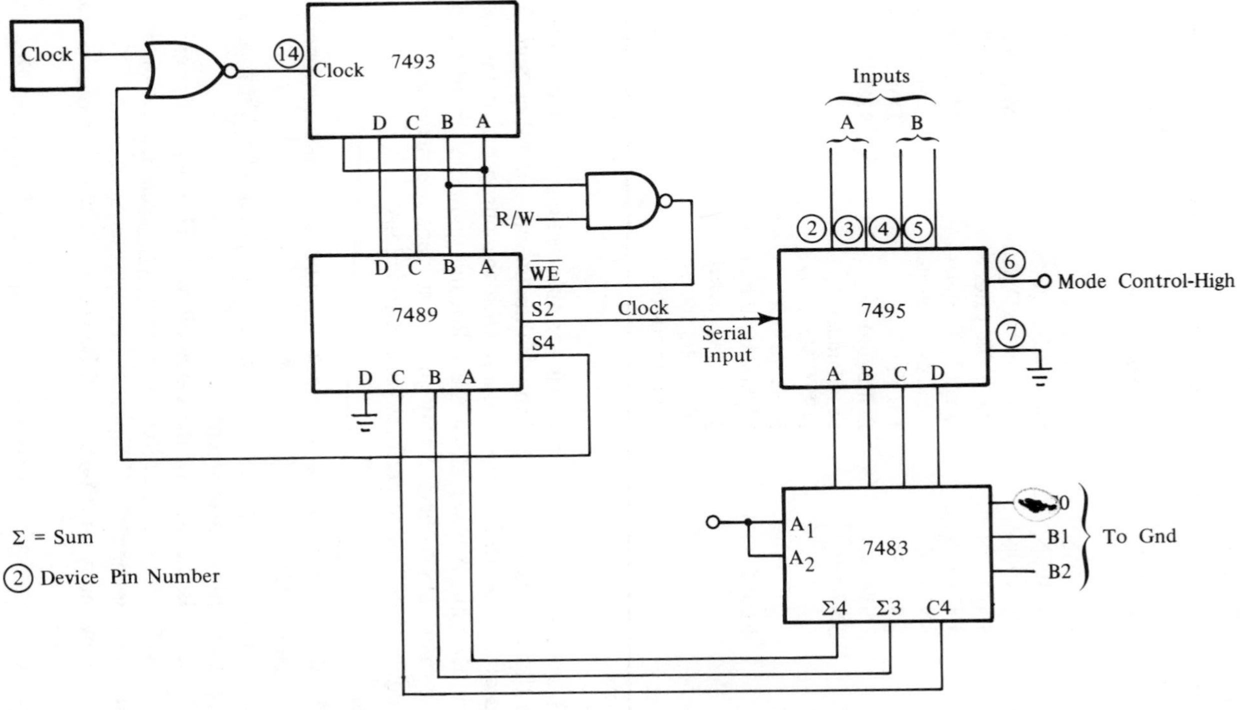

Figure A.5 Complete Circuit for Exercise A.1

TABLE A.1 Parts List

Quantity	*Part #*	*Description*	*Replacement Part #*
2	SN7404N	Hex inverter	MC3008
5	SN7408N	Quad AND gates	MC3001
3	SN7400N	Quad NAND gates	MC3000
5	SN7432N	Quad OR gates	MC3003
1	SN7402N	Quad NOR gates	MC3002
2	SN7496N	5-bit shift register	DM7496N
1	SN7473N	J-K flip-flop	DM7473N
19	SN7474N	D flip-flop	DM7474N
1	SN7480N	Serial adder	
1		DPST Mom. switch	
10		SPST switch	
10	HEP P2004	Light-emitting diodes	
40		14-pin IC socket	
2		16-pin IC socket	
3		8 x 10 perfboard	

Using the operational analysis for subtraction which is provided, follow through the operation of the arithmetic and control circuits. This should allow one to become familiar with the purpose and the use of this control unit. Answer all of the following questions about the subtraction operation to help clarify just what steps are necessary for the serial adder to be able to perform various functions.

1. What is the purpose of clock 10a and why must this operation be done?
2. Subtract (in base 2) the number 011 from 110. Show and explain all necessary steps.
3. What is the state of the carry bit at clock 9 when the circuit is performing subtraction on the numbers given in question 2?
4. What is the state of input B5 just prior to clock 10b?
5. How many shift pulses will be necessary if the answer is negative?
6. This particular control unit is capable of allowing the serial adder to perform two other functions besides subtraction. What are these two functions? Write out an operational analysis for these other two functions explaining just what operation takes place at each clock pulse.

Part Two

Design a logic control unit that will expand the number of functions performed by the serial adder. It may be a new circuit or an extension of the already existing control circuitry.

Part Three

Construct your control unit along with the serial adder and any timing that might be needed to get it to perform. If your circuit works as you had hoped, draw up the schematic and present it to the instructor with any data that you think should be included about the operation of your design. If your circuit does not work, you should begin your trouble-shooting by keeping records of the changes that you make (whether it helps the problem or not). Once you get your circuit to work submit your design to the instructor along with needed data and a copy of the changes that you needed to incorporate into your design.

Operational Analysis for the Subtraction Function

This short analysis describes the operation of the logic controls at each clock pulse. This should help the student to understand this arithmetic control unit.

Clock 1	Clears registers A and B.
Clock 2	Loads registers A and B.
Clock 3	Complements register B.
Clock 4	Shifts registers A and B.
Clock 5	" " " " "
Clock 6	" " " " "
Clock 7	" " " " "
Clock 8	" " " " "
Clock 9	(not used for subtraction)

The next steps vary depending on the state of the carry. If the carry is 1 go to the step 10a, if the carry is 0 go to step 10b.

Clock 10a	Clear register A.	Clock 10b	Complement B and light the negative indicator.
Clock 11a	Load 1 into register A.		
Clock 12a	Shift registers A and B.		
Clock 13a	" " " " "		
Clock 14a	" " " " "	Clock 11b	Stop

Clock 15a " " " " "
Clock 16a " " " " "
Clock 17 Stop

Answers to Questions

1. Clock 10a clears register A in order that the number 1 can be loaded into register A and then be added to the number that is stored in register B.
2. 110
   011
   ―――
   110
   100 Complement B
   ―――
   110
   100
   ―――
   1010 Add A to B

   Check carry: if 1, add 1 to sum to get answer
   if 0, complement sum

   In this case, the carry is 1

   010
   001
   ―――
   011 Add 1 to sum
3. The carry is a 1 at clock 9.
4. B5 is the carry, which must be 0 if the operation is to go to clock 10b.
5. Five shift pulse will be needed if the answer is negative.
6. The two additional functions performed by the serial adder are: Addition of two external numbers, addition of one external number and the number in the accumulator.

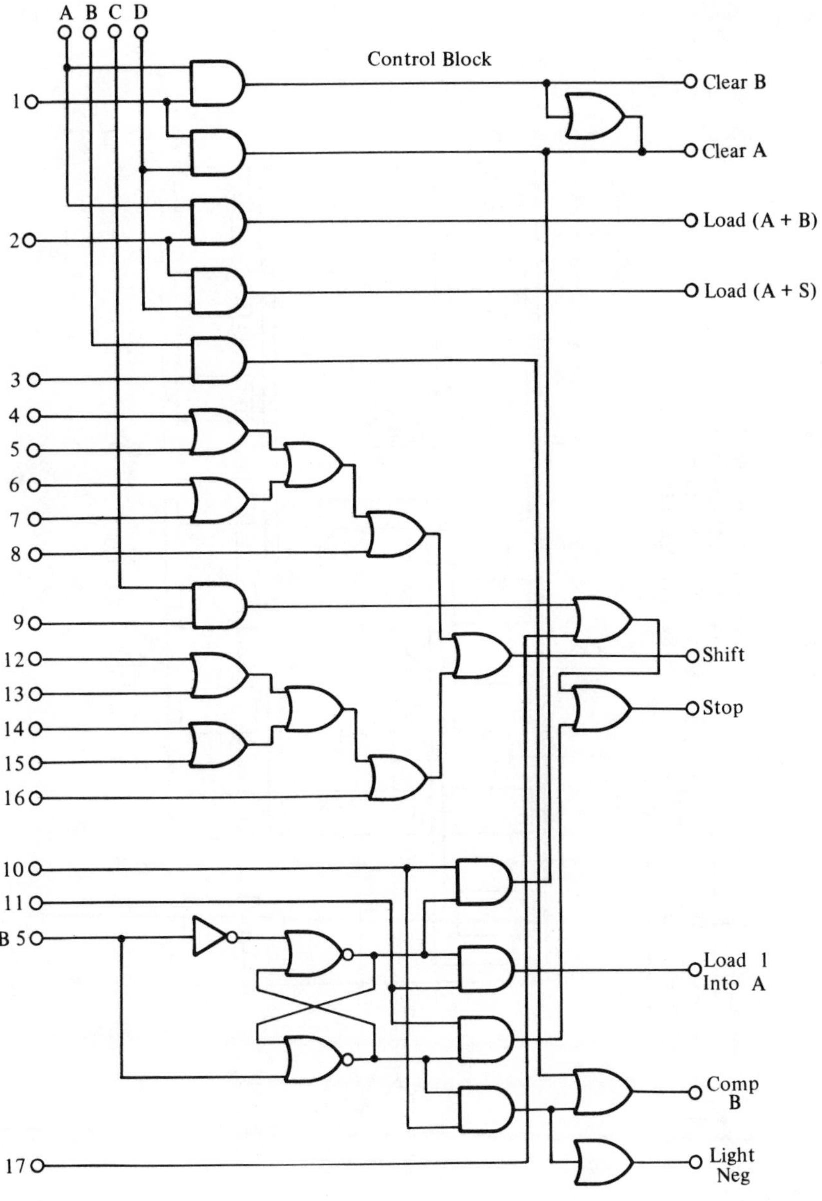
A B C D
Control Block
1
2
3
4
5
6
7
8
9
12
13
14
15
16
10
11
B 5
17
Clear B
Clear A
Load (A + B)
Load (A + S)
Shift
Stop
Load 1
Into A
Comp
B
Light
Neg

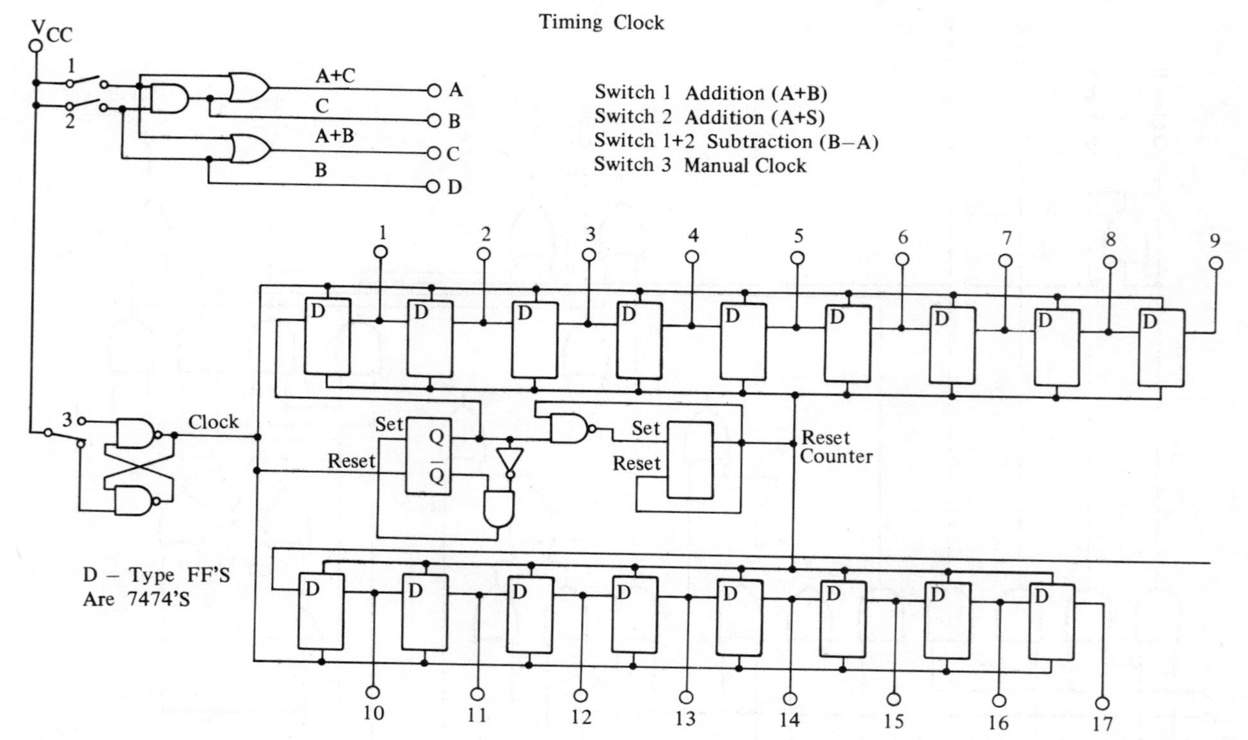
Timing Clock
V_{CC}
1
2
A+C
C
A+B
B
A
B
C
D
Switch 1 Addition (A+B)
Switch 2 Addition (A+S)
Switch 1+2 Subtraction (B–A)
Switch 3 Manual Clock
1
2
3
4
5
6
7
8
9
D
3
Clock
Set
Reset
Q
$\overline{Q}$
Set
Reset
Reset Counter
D – Type FF'S
Are 7474'S
10
11
12
13
14
15
16
17

Arithmetic Block

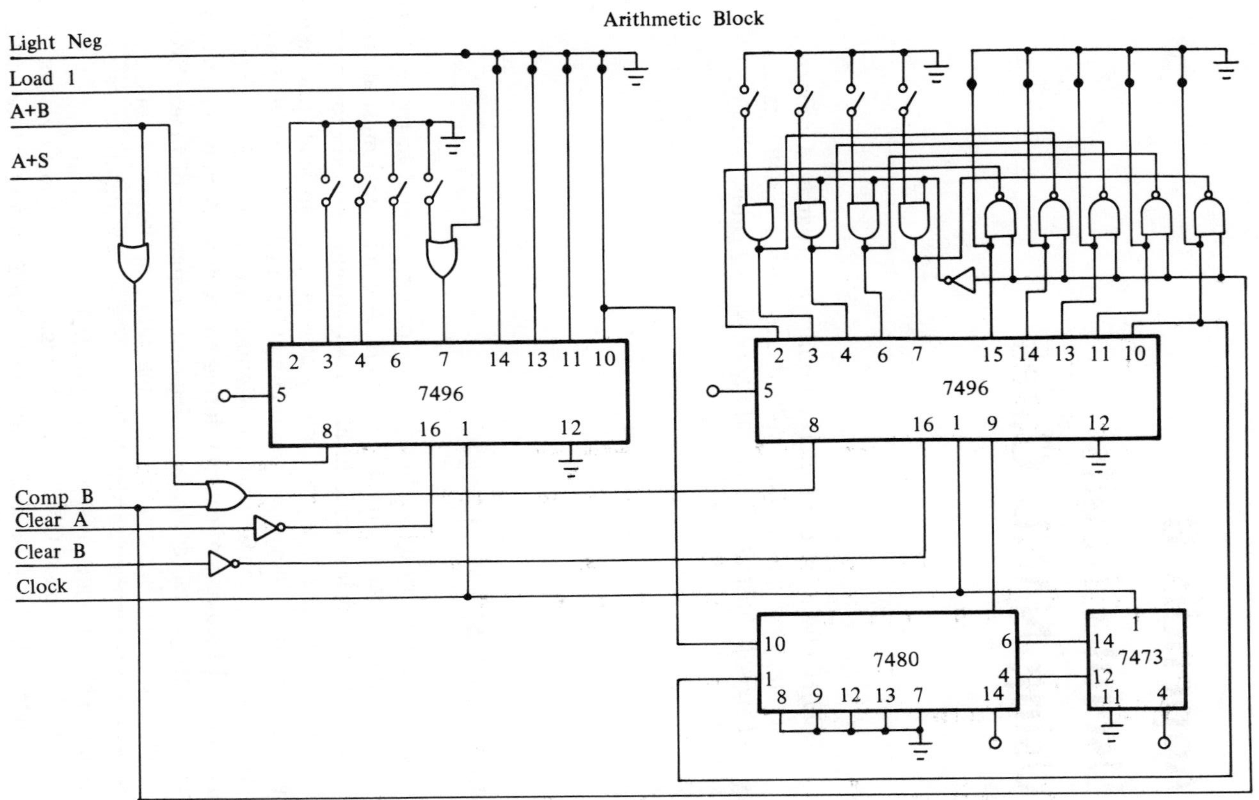

Appendix B
Useful Circuits Using NAND Gates

OBJECTIVE: To construct a bounceless switch, a monostable, and an astable multivibrator using NAND gates. This will illustrate the versatility of NAND gates.

B.1 BOUNCELESS SWITCH

Often it is necessary to produce a clock or input signal to trigger logic devices such as counters or shift registers. A clock signal must alternate between high and low logic levels. If an attempt is made to obtain a certain logic level at an input pin by attaching a clip lead from either V_{CC} or ground, a "noisy" connection will be made. The noisy connection is caused by contact bounce. Contact bounce is the uncontrolled making and breaking of contact during connection. A noisy connection will often trigger the device an indeterminate number of times. Even a standard commercial switch will often make a "bouncy" contact. To prevent contact bounce, a simple bounceless switch may be constructed from two NAND gates as shown in figure B.1.

If A is grounded Q will be high, logic state 1, and $\overline{Q}$ will be low, logic state 0. If B is grounded $\overline{Q}$ will then be high and Q will be at logic 0. The contact at A and B may be bouncy, but no bounce will occur at Q or $\overline{Q}$. Remember, an input ungrounded is a logic 1.

Construct a bounceless switch as shown in figure B.1 using NAND gates (Texas Instruments, SN7400) and fill in the Table.

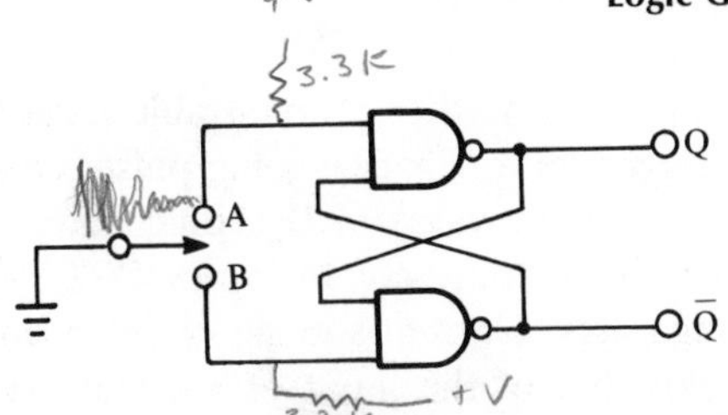

Figure B.1 Bounceless Switch Using NAND's

A	B	Q	$\bar{Q}$	
0	0	****	****	← Not allowed
0	1			
1	0			
1	1			← A was 0 last
1	1			← B was 0 last

B.2 LOGIC GATE MONOSTABLE

Often a delay is needed in such operations as clearing, triggering, or clocking. A simple delay monostable may be constructed from two NAND gates, a resistor, and a capacitor. The pulse duration of the monostable is approximately equal to the product of R and C. The circuit used is shown in figure B.2.

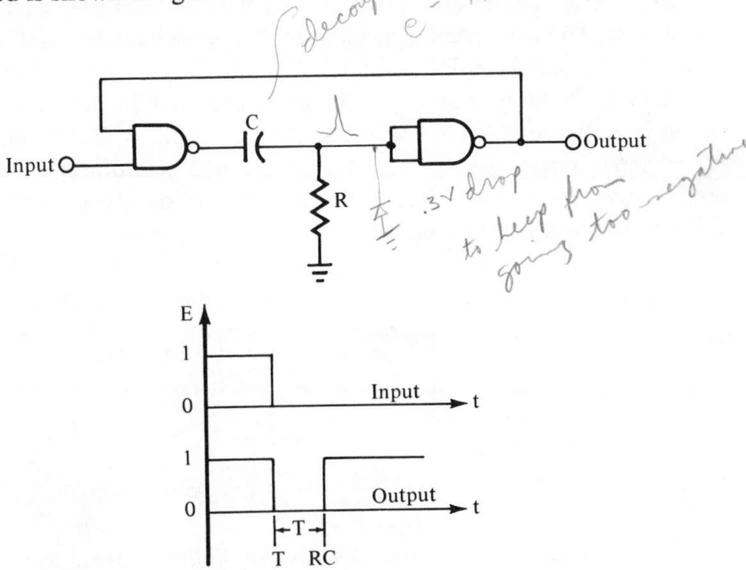

Figure B.2 Gated Stored-Charge Monostable and Waveforms

Usually the output of the delay monostable must be inverted, as most logic devices trigger on a negative-going pulse. One of the limitations of this circuit is that the value of R should not exceed 220 ohms when using TTL (transistor-transistor logic) or DTL (diode-transistor logic) current-sinking gates. Current-sinking or input-coupled logic requires that current flow out of the input of a circuit, thus the resistor must be kept small to insure that the gate's input level is not pulled down. This somewhat limits the pulse duration; however, the value of C may be changed to attain most desired pulse durations.

Construct a logic gate monostable using the circuit in figure B.2. Let R = 220 omhs and C = 100 farads. Use a bounceless switch to trigger the monostable. Draw the output waveform below.

B.3 LOGIC GATE ASTABLE MULTIVIBRATOR

An astable multivibrator may be made from logic gates as shown in figure B.3. The basic circuit is similar to the stored-charge monostable of figure B.2. The gate input resistors (R1 and R2) are chosen to put the gate input level near the logic threshold so that the charging and discharging of C1 and C2 will put the input voltages above or below the threshold level. The frequency of the multivibrator is primarily determined by the values of C1 and C2. The values of R1, R2, and R3 are chosen to optimize the oscillation, although R3 is also a fine-frequency adjustment. The frequency of oscillation is approximately ½RC, where R = R1 + R3 and R1 = R2 and C1 = C2 = C.

Construct an astable multivibrator using NAND gates (SN7400) as shown in figure B.3. Use the following values; R1 = R2 = 560 ohms, R3 = 100 ohms potentiometer, C1 = C2 = 1 microfarad. Draw the output waveform below. What is the frequency of oscillation when R3 = 0 ohms? When R3 = 100 ohms?

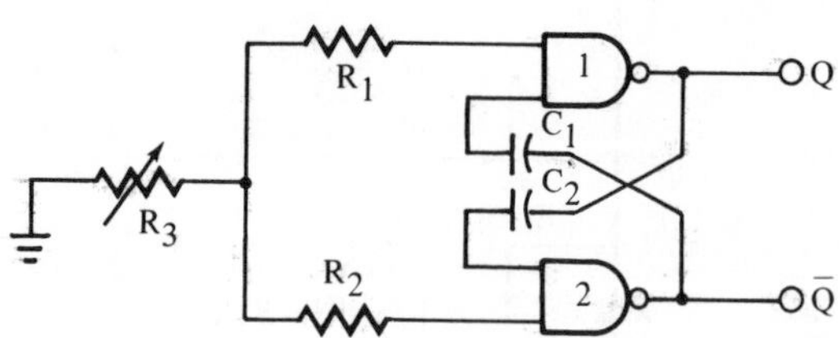

Figure B.3 Logic Gate Astable Multivibrator

Figure B.4 is an alternate method of constructing a NAND-gate astable.

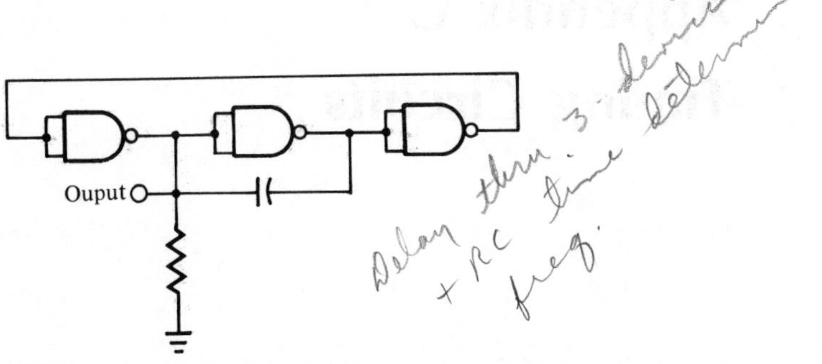

Figure B.4 Astable Multivibrator

Appendix C
Timing Circuits

Timing is one of the more important features of a digital system. Each event, being a discrete bit of data, usually has a unique time associated with it. In most systems, there is a master clock or master timing signal used to insure that each pulse or signal in the system can be referenced to a standard.

Figure C.1 shows a system clock. Here we see a clock sending out a series of square waves or pulses. The timing (frequency) of these pulses is determined by the internal structure of the clock circuit. These pulses then are sent to various places around the system. The clock may use simple RC timing (similar to the UJT shown in chapter 12) or it may be a complex crystal-controlled circuit. The clock, however, is important in our digital systems. The clock is an astable or free-running multivibrator. It produces a set of pulses whose frequency, pulse width, and duty cycle are determined by the designer. This initial clock signal may be used as is or altered by such methods as binary-divide circuits (flip-flops and counters). Figure C.2 shows an example of how this might be accomplished. Here we see the clock being fed to a divide-by-ten circuit (a circuit that will produce one output for every ten input pulses). The divide circuit might be a divide-by-two, divide-by-five, etc. These circuits are usually counters.

The clocks are usually astable multivibrators or some type of free-running oscillator. Another timing circuit is used to delay or lengthen the clock pulse. These circuits are called monostables or one-shots.

Figure C.3 shows a block diagram to illustrate the operation of a one-shot. Here we see one pulse at the input of time T, and one pulse at the output of time t. Unlike the astable, this circuit will not produce an output pulse unless there has been an input pulse. The length (t) of the output pulse is determined by external components. The one-shot then produces one pulse of a known duration for a change at its input. Monostables are usually used where accurate delays or timing pulses are needed. An example is shown in figure C.4. Here, the monostables is

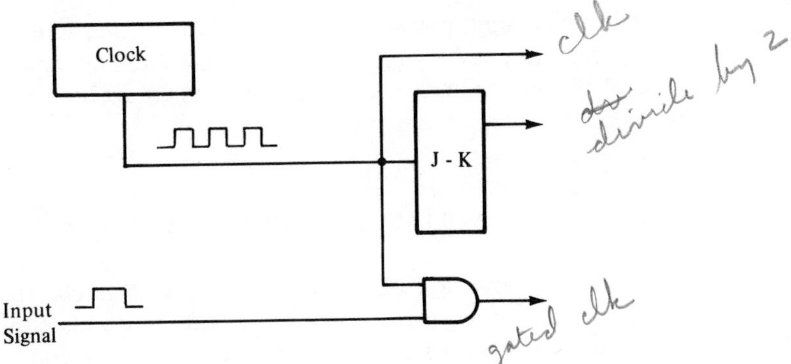

Figure C.1

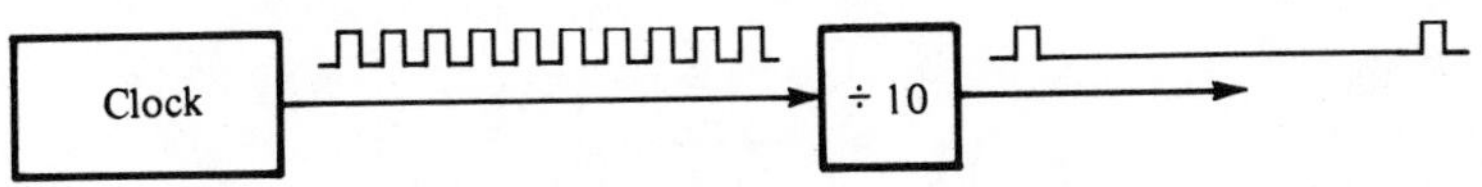

Figure C.2

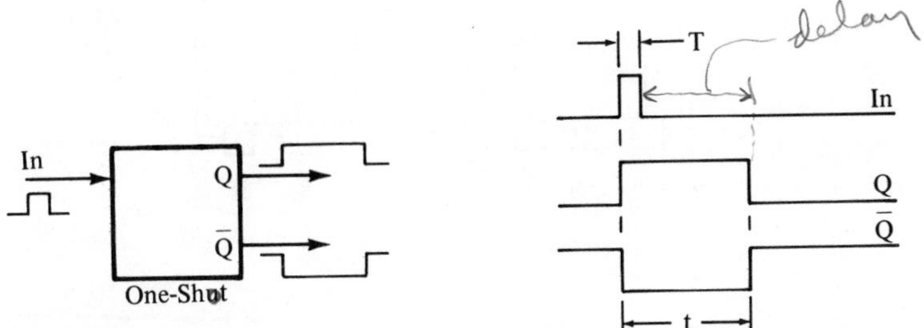

Figure C.3

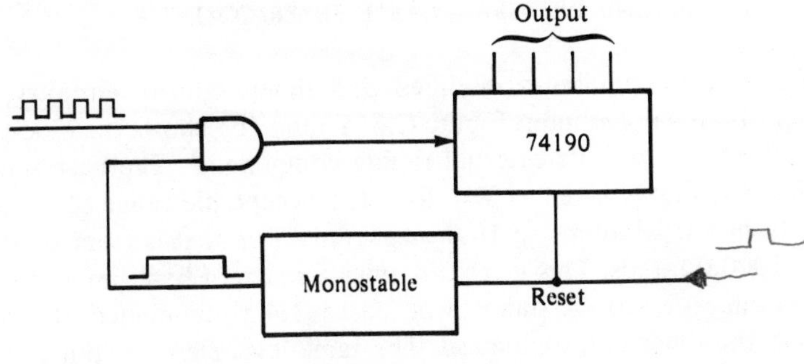

Figure C.4

used to gate input pulses to a counter. At reset, the counter goes high. It stays high for, say, 1 second. During this time, any input pulses will be gated to the counter. In this way, we will know how many pulses per second this was at the input.

C.1 THE 74124 VOLTAGE-CONTROLLED OSCILLATOR

Figure C.5 shows a TTL device that is a voltage controlled oscillator. This device can be used to develop a wide range of output frequencies for use with other TTL devices. It is quite versatile, and as you can see, a single chip contains two separate oscillators.

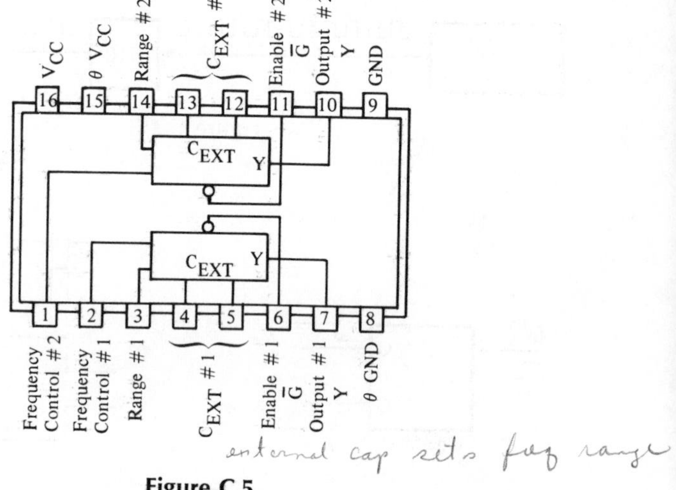

Figure C.5

C.2 THE XR 2240 PROGRAMMABLE TIMER/COUNTER

This device contains a modified 555 timer, control circuitry, and an eight-bit binary counter. Figure C.6(a)and (b) shows the block diagram of the timer with its external timing components. The base time of the circuit is simply given as T = RC. The acceptable range of values for R is from 1.0 kiloohms to 10.0 megaohms. For C it is from 0.005 farads to 1000.0 farads. This gives the timer a range of base times from microseconds to hours. A pulse of at least +1.4 volts applied at pin 11 will start the timer into oscillation. The same level signal at pin 10 will stop the oscillator. Figure D.6(c) shows the output pulses available from the

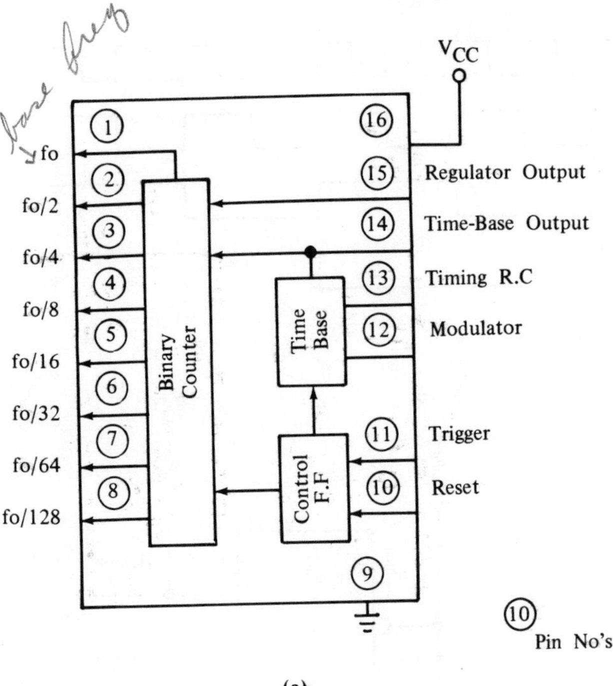

base freq
V_{CC}
fo
fo/2
fo/4
fo/8
fo/16
fo/32
fo/64
fo/128
1
2
3
4
5
6
7
8
9
10
11
12
13
14
15
16
Binary Counter
Time Base
Control F.F
Regulator Output
Time-Base Output
Timing R.C
Modulator
Trigger
Reset
Pin No's

(a)

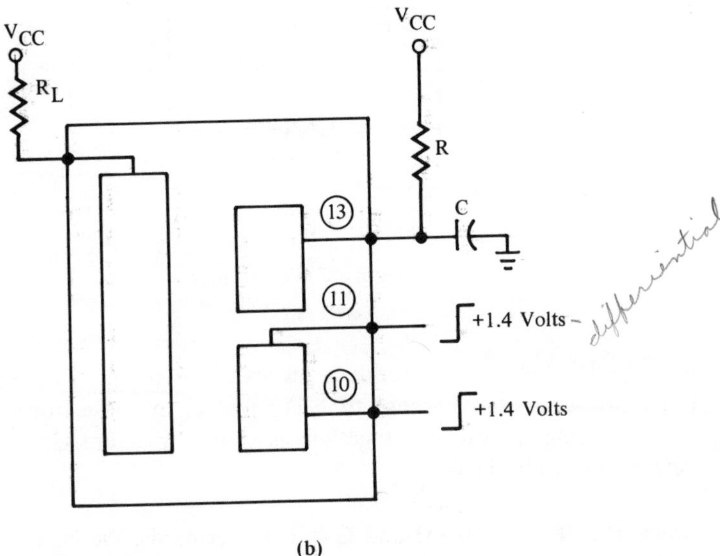

V_{CC}
R_L
V_{CC}
R
C
13
11
10
+1.4 Volts
differential
+1.4 Volts

(b)

Figure C.6

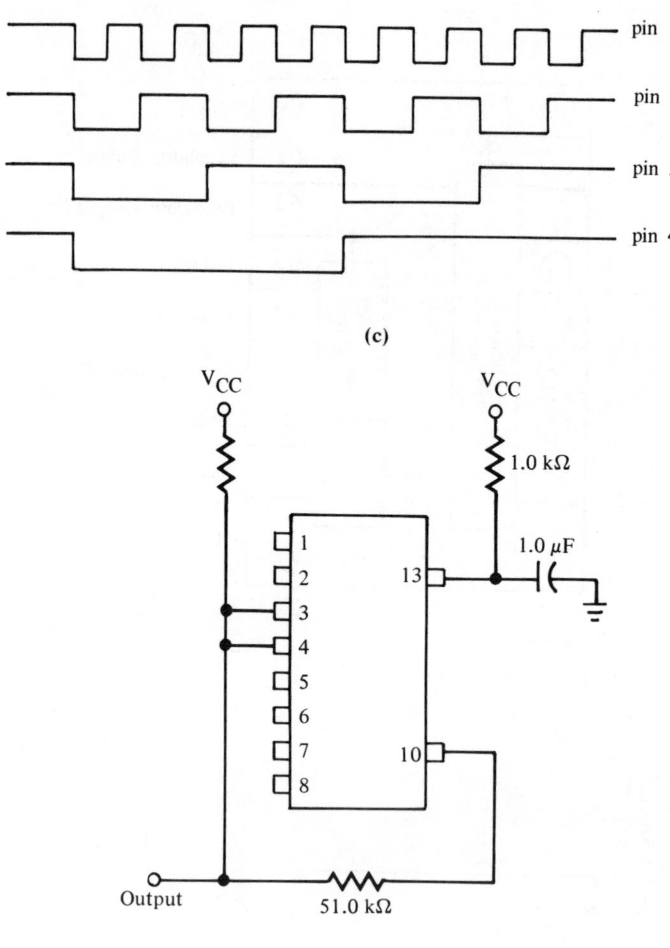

Figure C.6 *continued*

unit. Notice that pin 1 produces pulses equal to the base time, while pin 2 produces pulses whose time is equal to the base times 2, and so on. By wiring the outputs together as shown in figure C.6(d) various delays can be obtained.

Example If R = 1.0 kΩ and C = 1.0 microfaids, the base time would be 1.0 second.

By tying outputs 3 and 4 together we would establish an output pulse equal to:

Base time × Binary sum of switches

or 1 second × SW 3 + SW4

1 second × 4 + 8

1 second × 12

12 seconds

C.3 THE 74121 MONOSTABLE MULTIVIBRATOR

The 74121 shown in figure C.7 is a TTL monostable multivibrator. Its output may be controlled by only two external components, a resistor and a capacitor. By use of these components, a wide range of output times may be achieved.

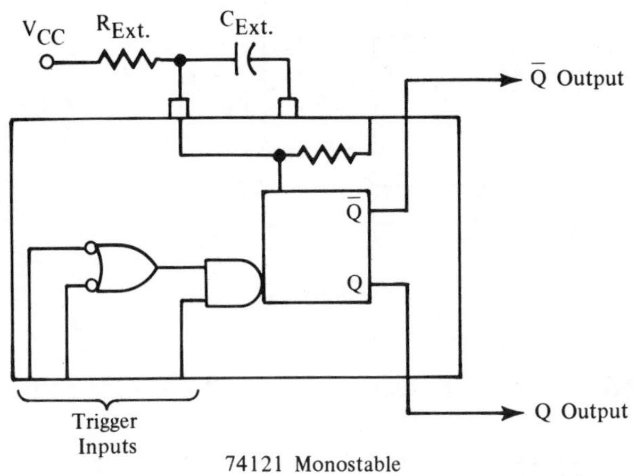

Figure C.7

C.4 THE 555 TIMER

The 555 timer is a monolithic timing circuit that can be used to produce oscillations or time delays that are both stable and accurate. The package contains trigger and reset terminals to increase the capabilities of the circuit. One external resistor and one external capacitor are the only additional components needed for operation of the circuit.

Figure C.8 shows the external pin out for a common dual-in line-package (DIP) of the 555 timer. One important feature of the 555 timer is its ability to drive TTL.

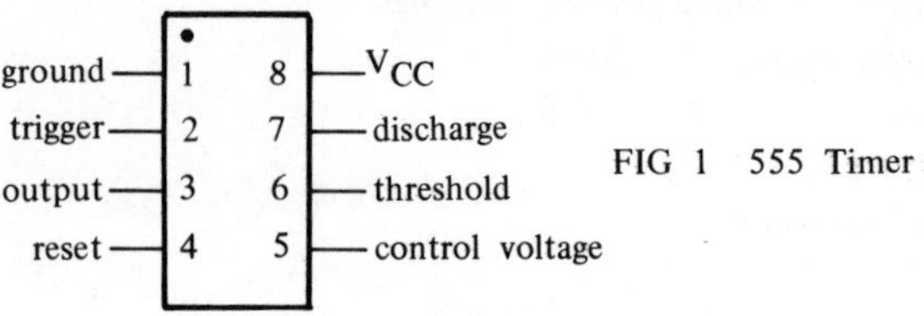

Figure C.8

Figure C.9 shows typical connections when using the 555 timer as an astable MV.

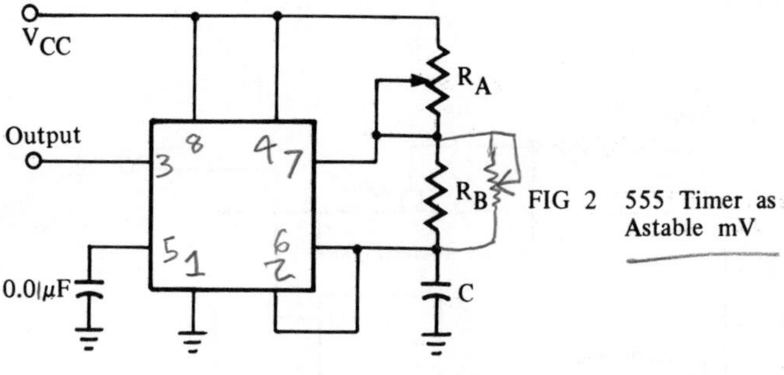

Figure C.9

The frequency of the astable of figure C.9 is given by the equation:

$$f = \frac{1.44}{(R_A + 2R_B)C}$$

and the duty cycle of the astable is given as:

$$\text{Duty cycle} = \frac{R_B}{R_A + 2R_B}$$

Figure C.10 is a monograph to be used with the astable MV. Figure C.11 shows typical connections when using the 555 timer as a monostable MV, a one-shot MV.

Figure C.12 is a monograph to be used with the monostable MV configuration of the 555 timer. The delay time of the monostable is given by the equation:

$$t = 1.1\, R_A C$$

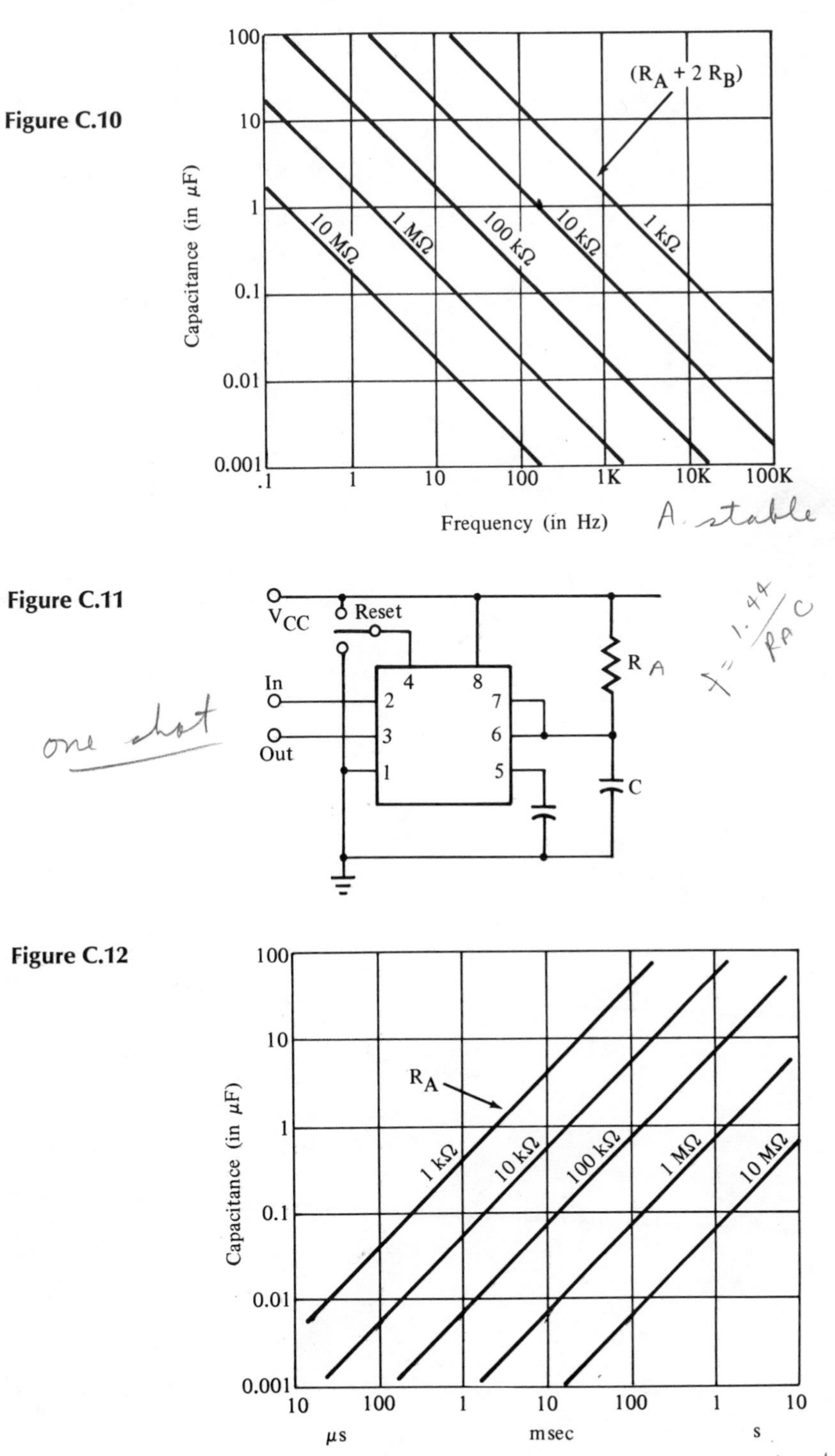

Figure C.10

Figure C.11

Figure C.12

Index

Accuracy, 263
Addend, 222
Adder Chips, 217
Addition of Binary Numbers, 8
Addressable Latch, 138
Alphanumeric Codes, 20
ALU, 285
Analog Signals, 233, 256
Analog-to-Digital Conversion, 233, 256, 261
AND gate, 24
- circuit symbol, 25
- gating action, 27
- three-input gate, 26
- truth table, 24
- positive AND gate, 26

Arithmetic unit, 205
ASCII codes, 20
Asynchronous counters, 91
Augend, 222

Base, 4
BCD–Decimal decoder driver, 123
BCD–Octal decoder, 122
Binary addition, 8, 206
Binary counters, 88
Binary numbers, 5
Binary subtraction, 9, 211
Bi-quinary counters, 107
Bistable, 60
Black-box, 101
Boolean decoding, 153
Boolean functions, 153
Borrowed, 11
Bus, 52

Cold-cathode displays, 161
Cold-cathode tubes, 123
Common-mode gain, 241
Common-mode inputs, 241
Common-mode rejection ratio (CMRR), 241
Comparators, 247
Compliment, 11
Computer arithmetic, 205
Correction factors, 19
Cost factors, 33
Cost-per-chip, 33
Counters
- asynchronous counters, 91
- bi-quinary counters, 107
- binary counters, 88
- counting registers, 88
- feedback counters, 91
- gray-code ring counters, 96
- IC counters, 101
- Johnson counters, 96
- modulus counters, 88
- non-sequentral counters, 91
- presettable decade counters, 115
- ring counters, 94
- ripple counters, 91
- self-stopping asynchronous counters, 96
- sequentral counters, 91
- switch-tail counters, 94

Counters (*Cont.*)
synchronous counters, 91
up-down counters, 112
Counting systems, 159
CPU, 287
Current sinking, 30

Data bus, 51
Data conversion, 261, 263
Data selection, 142
Decimal numbers, 4
Decimal point, 6
Decoding, 120
Demultiplexing, 147
D-flip-flops, 62
Difference inputs, 238
Digital communication, 150
Direction line, 54
Drive, 30
Dual-slope-convertor, 272
Dynamic memory, 199

Electrical characteristics, 34
Enable pulse, 78
Encoding, 120
EPROM, 198, 292
Exclusive–OR gate, 46
External memory, 289

Family, 29
Fan-in/fan-out, 30, 31
Feedback, 91, 234, 248, 257
Ferrite core memory, 176
Field programmable memory, 194, 292
Flip-flops
D flip-flops, 62
JK FLIP FLOPS, 66
master-slave flip-flops, 66
reset input, 61
RS flip-flops, 64
set input, 61
T flip-flops, 63
trigger input, 61
Full adder, 213
Fuse link, 196

Gated full adder, 219
Gating action, 27
Gray code, 138
Gray code conversion, 140
Gray code conversion circuits, 141
Ground, 236

Half adder, 45, 212
Hexadecimal numbers, 15, 295
Hex-bus driver, 53
Hex conversion, 16
HI-Z, 51
Hybrid signals, 233

IC counters, 101
IC shift registers, 79
Index register, 287
I/O devices, 299
Input coding, 109
Input/output, 295
Instruction set, 293
Interrupts, 299
Inverting input, 236
Invertors, 42

J-K flip flops, 66
Johnson counters, 96

Latch, 62
Least-significant digit, 12
LED (light emitting diode), 43
Linearity, 264
Line wise, 152
Logic family, 31
Loop-gain, 234

Magnetic disks, 174
Magnetic drum, 173

Magnetude comparators, 153
Masked ROM, 190
Master slave flip-flops, 66
Memory
  memory plane, 181
  permanent memory, 172
  temporary memory, 172
Memory cell, 185
Memory matrix, 185
Memory page, 289
Memory refresh, 200
Microcomputers, 265, 285
Microcomputer architecture, 287
Microwave links, 142
Modem, 300
Modulus counters, 88, 90
Most significant digest, 12
Multiplexing, 142

NAND gate, 28
  truth table, 28
  TTL NAND gate, 29
NDRO (nondestructive readout), 187
NIXIE tube, 161
Nor gate, 46
Not circuit, 27
Null resistor, 280
Number systems
  addition of binary numbers, 8, 206
  alphanumeric codes, 20
  ASCII codes, 20
  base, 4
  BCD, 18
  binary numbers, 5
  binary subtraction, 9, 211
  borrowed, 11
  compliment, 11
  correction factor, 19
  decimal numbers, 4
  decimal point, 6
  hexadecimal numbers, 15, 295
  hex conversion, 16
  octal numbers, 13
  quotient-remainder method, 12
Number systems (*Cont.*)
  radix, 6
  weighted value, 4

Octal numbers, 13
Odd parity, 133
Open-collector gates, 37
Operational amplifiers
  A-D conversion, 233, 256, 261
  Analogy signals, 233, 256
  common-mode gain, 241
  common-mode input, 241
  common-mode rejection ratio (CMRR), 241
  comparator, 247
  D-A conversion, 233
  data-conversion, 261, 263
  difference inputs, 238
  feedback, 234, 257
  feedback resistor, 248
  functions of external connections, 236
  ground, 236
  inverting input, 236
  loop gain, 234
  non-inverting input, 236
  null resistor, 280
  phase shift, 236
  properties, 234
  slew-rate, 246
  summing amplifier, 248
  virtual ground, 244
OR gate, 43

Parallel addition, 216
Parellel data, 152
Parallel-in/parallel-out, 80
Parallel shift register, 77
Parity, 133
Parity bit, 133
Parity generator/checkers, 133
Parity tree, 137
Permanent memory, 172
Phase shift, 236
Positional code, 138

Positive- AND logic, 26
Presettable decade counter, 115
Prime sign, 28
Priority encodes, 133
Program counter, 287
Prom-burner, 197
PROMs, 190, 192
Propagation delay, 36
Pull-up resistor, 38
Punch cards, 173

Quotient-remainder method, 12

Race, 68
Racing, 68
Radix, 6
Ramp type A-D converter, 272
Rams/Roms, 189, 291, 292
Random Access memory, 189, 291
RC-time constant, 267
Read cycle, 173
Read-only memory, 189, 292
Read/write line, 291
Read-write memory, 189, 291
Recaptured data, 147
Register clocking, 76
Remanence, 176
Reset input, 61
Resolution, 263
Retentivity, 176
Ring counters, 94
Ripple blanking, 169
Ripple counters, 91
RTL, 29

Satellite transmission, 152
Schottky barrier diode, 33
Self-stopping asynchronous counters, 96
Serial addition, 215
Serial data, 147
Serial in/serial out, 82
Serial I/O devices, 299
Serial shift register, 76
Seven-segment decoder/drivers, 123
Sequential mode counters, 91
Slew-rate, 246
Solid state memories, 185
Stack memory, 199
Stack pointer, 287
Successive approximation A-D converter, 274
Summing amplifier, 248
Switch-tail counter, 94
Synchronous counter, 91
Sync line, 152

Temporary memory, 172
Thermocouple, 265
Timing circuits, 322
Toggle flip-flops, 63
Totem-pole, 40
Transceivers, 54
Tri-state bus transceiver, 54
Tri-state devices, 50
True-false logic, 23
TTL, 29

UJT oscillator, 266
Unit load, 30
Up-down counter, 108, 112
User-programmable memory, 194, 292

Virtual ground, 244
Voltage controlled oscillator, 324

Weighted value, 4
Wire–AND, 38
Write–cycle, 173